LARGE FINITE SYSTEMS

THE JERUSALEM SYMPOSIA ON
QUANTUM CHEMISTRY AND BIOCHEMISTRY

Published by the Israel Academy of Sciences and Humanities,
distributed by Academic Press (N.Y.)

Published by the Israel Academy of Sciences and Humanities,
distributed by D. Reidel Publishing Company (Dordrecht, Boston, Lancaster, and
Tokyo)

Published and distributed by D. Reidel Publishing Company (Dordrecht, Boston,
Lancaster, and Tokyo)

VOLUME 20

LARGE
FINITE SYSTEMS

PROCEEDINGS OF THE TWENTIETH JERUSALEM SYMPOSIUM ON
QUANTUM CHEMISTRY AND BIOCHEMISTRY HELD IN
JERUSALEM, ISRAEL, MAY 11–14, 1987

Edited by

JOSHUA JORTNER

Department of Chemistry, University of Tel-Aviv, Israel

ALBERTE PULLMAN

and

BERNARD PULLMAN

*Institut de Biologie Physico-Chimique
(Fondation Edmond de Rothschild), Paris, France*

D. REIDEL PUBLISHING COMPANY

A MEMBER OF THE KLUWER   ACADEMIC PUBLISHERS GROUP

DORDRECHT / BOSTON / LANCASTER / TOKYO

Library of Congress Cataloging in Publication Data

Jerusalem Symposium on Quantum Chemistry and Biochemistry (20th: 1987)
  Large finite systems.

(The Jerusalem symposia on quantum chemistry and biochemistry; v. 20)
  Includes index.
    1. Intermolecular forces—Congresses. 2. Macro-molecules—Congresses. 3. Metal crystals—Congresses. I. Jortner, Joshua. II. Pullman, Alberte. III. Pullman, Bernard, 1919–   . IV. Title. V. Series.
  QD461.J47   1987   541.2′24   87–24345

ISBN-13: 978-94-010-8275-4   e-ISBN-13: 978-94-009-4001-7
DOI:10.1007/978-94-009-4001-7

Published by D. Reidel Publishing Company,
P.O. Box 17, 3300 AA Dordrecht, Holland.

Sold and distributed in the U.S.A. and Canada
by Kluwer Academic Publishers,
101 Philip Drive, Assinippi Park, Norwell, MA 02061, U.S.A.

In all other countries, sold and distributed
by Kluwer Academic Publishers Group,
P.O. Box 322, 3300 AH Dordrecht, Holland.

TABLE OF CONTENTS

PREFACE

The Twentieth Jerusalem Symposium reflected the high standards of these distinguished scientific meetings which convene once a year at the Israel Academy of Sciences and Humanities in Jerusalem to discuss a specific topic in the broad area of quantum chemistry and biochemistry. The Twentieth Jerusalem Symposium marked an auspicious occasion, commemorating two decades of this scientific endeavour. The topic at this year's Jerusalem Symposium was Large Finite Systems which constitutes a truly interdisciplinary subject of central interest in the broad areas of chemistry, physics, astrophysics and biophysics.

The main theme of the Symposium was built around bridging the gap between molecular, surface and condensed matter chemical physics. Emphasis was placed on the interrelationship between the properties of large molecules, van der Waals complexes and clusters, focusing on the structure, dynamics of nuclear motion, quantum and thermodynamics size effects, the nature of electronic states and excited-state energetics and dynamics of large finite systems. The interdisciplinary nature of these research areas was deliberated by intensive and extensive interactions between scientists from different disciplines and between theory and experiment. This volume provides a record of the invited lectures at the Symposium.

Held under the auspices of the Israel Academy of Sciences and Humanities and the Hebrew University of Jerusalem, the Twentieth Jerusalem Symposium was sponsored by the Institut de Biologie Physico-Chimique (Fondation Edmond de Rothschild) of Paris. We wish to express our deep thanks to Baron Edmond de Rothschild for his continuous and generous support, which makes him a true partner in this important endeavour. We would also like to express our gratitude to the Administrative Staff of the Israel Academy, and in particular to Mrs. Avigail Hyam, for the efficiency and excellency of the local arrangements.

Joshua Jortner
Alberte Pullman
Bernard Pullman

THE QUANTUM YIELD OF NON-RESONANT LIGHT SCATTERING

Pieter J. de Lange and Jan Kommandeur
Laboratory for Physical Chemistry
Nijenborgh 16
9747 AG Groningen
The Netherlands

Abstract

    After properly defining the quantum yield as the ratio between the short time emission and the photon loss we show, that conventional Bixon-Jortner model of radiationless transitions does not predict the experimentally expected rise of the quantum yield to one upon non-resonant excitation. Postulation of an intermediate state in the model yields a freqeuncy dependence of the non-radiative transition rate and thus remedies this shortcoming.
    Some speculations as to the applicability of this model and to its relation to the arbitrary variations of quantum yields of organic molecules have been added.

1. Introduction

    Radiationless transitions in large molecules are by now very well understood. Most of the predictions of the earliest theory by Bixon and Jortner (1) have been borne out and a multitude of experiments have since been performed which corroborate many of the explicit and implicit predictions of that work.
    In this paper we want to focus on the quantum yield. In the Bixon-Jortner framework a low quantum yield is due to the rapid development of the amplitude of the "prepared" state $|s\rangle$ into a set of "dark" background states $\{|l\rangle\}$. If the set l is sufficiently dense this development looks irreversible on the time scale available for an experiment and the energy absorbed from the photon field can largely stay "inside" the molecule in the form of highly excited vibrations of the ground state or of the triplet state.
    For a resonant excitation of $|s\rangle$ the Bixon-Jortner theory adequately describes the phenomena observed. Little attention has so far been paid to non-resonant excitation of a "light" state, coupled to a "dark" manifold. Usually, the frequency distribution of the exciting source is not explicitly taken into account, which then amounts to excitation with a "white" source and one then implicitly

1

J. Jortner et al. (eds.), Large Finite Systems, 1–9.
© 1987 *by D. Reidel Publishing Company.*

always treats the resonant problem. Nevertheless, the non-resonant
problem is interesting, in particular with respect to the quantum
yield. In this paper we show that the conventional theory of
radiationless transitions (1,2) leads to the result that the quantum
yield is independent of the excitation frequency. This would mean that
molecules not showing substantial fluorescence would also not show
much Rayleigh and Raman scattering. The reverse is true, Raman
spectroscopists look for non-fluorescent molecules, so as not to have
their spectra obscured by the fluorescence emission!

We will first show that the erroneous prediction comes about in
the conventional model and then make a suggestion how the model can be
amended to give the answers required. First, however, we shall have to
pay some attention to the proper definition of the quantum yield.

2. The quantum yield

Usually the quantum yield is defined as the ratio of the amount
of light emitted to the amount of light absorbed.

By absorption one then usually means the number of photons
removed from a light beam after passing through a sample. That
quantity is, however, due to absorption and scattering. A more useful
definition of the quantum yield therefore seems: the number of photons
scattered out of the beam in a finite time divided by the number of
photons removed from it (the "photon loss"). Photons remaining
"inside" the molecule for longer times contribute to a low quantum
yield, and for non-resonant light scattering (NRLS) there should be
very few of these photons and the quantum yield will go to one.

A proper model of the quantum yield should therefore give a low
value at resonance, rising to one upon detuning. In the next section
we will show that the conventional theory of radiationless transitions
falls short in this respect.

3. Quantum Yield in the conventional theory

For a "light" singlet state $|s\rangle$ interacting with a "dark"
manifold $\{|1\rangle\}$ one can write for the ω-dependence of the singlet
amplitude (2):

$$A_s(\omega) = \cfrac{1}{\omega - \omega_s - \sum_l \cfrac{v_{sl}^2}{\omega - \omega_l} + i\,\Gamma_r/2} \tag{1}$$

where ω_s, ω_l are the energies of $|s\rangle$ and $\{|1\rangle\}$, respectively, v_{sl} are
the coupling elements and Γ_r is the radiative rate of $|s\rangle$. $A_s(\omega)$ is
the total singlet amplitude. With an exciting source, such as a laser,
that is not "white" but confined to a certain region of ω-space we
have for the singlet amplitude really excited:

$$C_s(\omega) = \cfrac{\chi(\omega)}{\omega - \sum_l \cfrac{v_{sl}^2}{\omega - \omega_l} + i\Gamma_r/2} \tag{2}$$

where $\chi(\omega)$ denotes the laser envelope in ω-space, referred to ω_s, which has been put to zero.

The photon loss is equal to the "amount" of singlet character created by the light. Calling the photon loss S, and normalizing this quantity to 1 for a "white" excitation ($\chi(\omega)$ = 1), we have:

$$S = \frac{\Gamma_r}{2\pi} \int_{-\infty}^{+\infty} |c_s(\omega)|^2 \, d\omega \qquad (3)$$

The emission, i.e. the scattering in general is due to the property of spontaneous emission, inherent in $|c_s(\omega)|^2$. Its time behaviour is given by $|\hat{c}_s(t)|^2$, where $\hat{c}_s(t)$ is the Fourier transform of $c_s(\omega)$.

$$E(t) = \Gamma_r \, |\hat{c}_s(t)|^2 \qquad (4)$$

And the total emission is given by

$$E = \Gamma_r \int_{-\infty}^{\infty} |\hat{c}_s(t)|^2 \, dt \qquad (5)$$

But by Parseval's rule

$$\frac{1}{2\pi} \int_{-\infty}^{\infty} |c_s(\omega)|^2 \, d\omega = \int_{-\infty}^{\infty} |\hat{c}_s(t)|^2 \, dt \qquad (6)$$

Therefore, the ratio between S and E will always be 1 and so will the quantum yield for every laser envelope function $\chi(\omega)$. This is indeed so, if we could measure for infinite time, i.e. if the integration could be carried out to infinity.

As is well known, this is not the case. We have to take into account that the determination of a quantum yield is a mark of experimental limitation. After a fast "dephasing" of the states, there is still energy in the molecule, but it comes out as U.V. photons so slowly that any apparatus will fail to measure it. (We explicitly exclude emissions in other parts of the spectrum here). Therefore, to find the experimental quantum yield we have to integrate time space only up to a finite time $T \approx 3 \times (2\pi\langle v^2\rangle\rho)^{-1}$, after which the experimenter sees no further light coming out. It is then possible to find a quantum yield lower than one.

We therefore define the "short-time" emission E^S as

$$E^S = \Gamma_r \int_{-\infty}^{T} |\hat{c}_s(t)|^2 \, dt, \text{ with } \infty > T \gg (2\pi \langle v^2 \rangle \rho)^{-1}$$

We will now treat the resonant and non-resonant cases.

i) First we consider very broad band excitation, i.e. a δ-pulse in time and take for simplicity $\chi(\omega)$ = 1. Then we have for the photon loss:

$$S = \frac{\Gamma_r}{2\pi} \int_{-\infty}^{\infty} |C_s(\omega)|^2 \, d\omega$$

$$= \frac{\Gamma_r}{2\pi} \cdot \frac{2\pi}{\Gamma_r} = 1$$

For the short time emission we get

$$E^S = \Gamma_r \int_{-\infty}^{T} |FT \{ \frac{1}{\omega - \sum_l \frac{v_{sl}^2}{\omega - \omega_1} + i\Gamma_r/2} \}|^2 \, dt \qquad (7)$$

We can evaluate this integral by replacing the sum over $\{|l\rangle\}$ by an integral and going to an imaginary representation we have, with $\Gamma_{nr} = 2\pi \langle v^2 \rangle \rho$

$$E^S = \Gamma_r \int_{-\infty}^{T} | FT \{ \frac{1}{\omega + i(\Gamma_{nr} + \Gamma_r)/2} \}|^2 \, dt =$$

$$= \Gamma_r \int_{0}^{T} | \exp [- \frac{(\Gamma_{nr} + \Gamma_r)}{2} t] |^2 \, dt =$$

$$\approx \Gamma_r \int_{0}^{\infty} \exp [-(\Gamma_{nr} + \Gamma_r)t] \, dt$$

$$\approx \Gamma_r / (\Gamma_{nr} + \Gamma_r) \qquad (8)$$

The quantum yield $Q = E^S/S = \Gamma_r(\Gamma_r + \Gamma_{nr})^{-1}$, which is the result obtained many years ago by Bixon and Jortner (1).
ii) We now consider a non resonant block laser centered at Δ and of width W in ω-space. Then we obtain from eqns. 3 and 2 for the photon loss:

$$S = \frac{\Gamma_r}{2\pi} \int_{\Delta - \frac{W}{2}}^{\Delta + \frac{W}{2}} | \frac{1}{\omega - \sum_l \frac{v_{sl}^2}{\omega - \omega_1} + i \Gamma_r/2} |^2 \, d\omega \qquad (9)$$

We should now <u>not</u> replace the sum by an integral, since all the seperate eigenstates of $|s\rangle$ and $\{|l\rangle\}$ contribute ever so little to the total photon loss. If we would measure the photon-loss of a comb, we should not simply replace it by a rectangular object. Similarly, replacing the multitude of molecular eigenstates (ME's) by a broad Lorentzian leads to erroneous results. Or putting it yet another way, using the expression with Γ_{nr} and Parseval's theorem we would neglect the very, very weak but long time emission, which does, however, play a role in the absorption.

Evaluation of S should therefore be done retaining the summation sign. A computer can do it, but not very accurately since ω-space has to be sampled very finely, since particularly for large detunings the ME's become extremely narrow.

If, however, we take the levels $|1\rangle$ to be equally spaced and when we use the r.m.s. value of v_{s1}^2: $\langle v^2 \rangle$ we can evaluate S analytically for large detuning.

We have for the discrete sum:

$$\sum_1 \frac{v^2}{\omega - \omega_1} = \pi v^2 \rho \ ctg(\pi \rho \omega) = \frac{\Gamma_{nr}}{2} ctg(\pi \rho \omega) \qquad (10)$$

Eqn. 9 then becomes

$$S = \frac{\Gamma_r}{2\pi} \int_{\Delta - \frac{W}{2}}^{\Delta + \frac{W}{2}} \left| \frac{1}{\omega - \frac{\Gamma_{nr}}{2} ctg(\pi \rho \omega) + i \ \Gamma_r/2} \right|^2 d\omega \qquad (11)$$

Because of the periodicity of $ctg(\pi \rho \omega)$ we may also write

$$S = \frac{\Gamma_r}{2\pi} \int_{-\frac{W}{2}}^{\frac{W}{2}} \left| \frac{1}{\Delta + \omega - \frac{\Gamma_{nr}}{2} ctg(\pi \rho \omega) + i \ \Gamma_r/2} \right|^2 d\omega \qquad (12)$$

In the limit where the detuning of the laser is much larger than its width ($\Delta \gg W$) we have

$$S \approx \frac{\Gamma_r}{2\pi} \int_{-\frac{W}{2}}^{\frac{W}{2}} \left| \frac{1}{\Delta - \frac{\Gamma_{nr}}{2} ctg(\pi \rho \omega) + i \ \Gamma_r/2} \right|^2 d\omega \qquad (13)$$

This can after rearrangement be rewritten to

$$S \approx C \int_{-\frac{W}{2}}^{\frac{W}{2}} \frac{\sin^2(\pi\rho\omega) \ d(\pi\rho\omega)}{(k_1^2 + k_2^2) \sin^2(\pi\rho\omega) + 2k_1 \cos(\pi\rho\omega) \sin(\pi\rho\omega) + \cos^2(\pi\rho\omega)} \qquad (14)$$

with $C = \frac{2\Gamma_r}{\pi^2 \Gamma_{nr}^2 \rho}$, $k_1 = - \frac{2\Delta}{\Gamma_{nr}}$ and $k_2 = \frac{\Gamma_r}{\Gamma_{nr}}$

The integration can be performed analytically and one can show that in the limit for large detuning ($\Delta \gg \Gamma_{nr}, \Gamma_r$) we get

$$S \approx \frac{W}{2\pi\Delta^2} (\Gamma_{nr} + \Gamma_r) \qquad (15)$$

For the short time emission we again replace the sum by an integral and we get from eqn. 7:

$$E^S = \Gamma_r \int_{-\infty}^{T} |FT\{ \frac{\chi(\omega)}{\omega + i(\Gamma_{nr} + \Gamma_r)/2}\}|^2 \, dt \qquad (16)$$

Because of the fact that by this replacement all long-time emission is really thrown away, we may also write

$$E^S = \Gamma_r \int_{-\infty}^{\infty} |FT\{ \frac{\chi(\omega)}{\omega + i(\Gamma_{nr} + \Gamma_r)/2}\}|^2 \, dt \qquad (17)$$

which equals (Parseval)

$$E^S = \frac{\Gamma_r}{2\pi} \int_{-\infty}^{\infty} |\frac{\chi(\omega)}{\omega + i(\Gamma_{nr} + \Gamma_r)/2}|^2 \, d\omega$$

$$= \frac{\Gamma_r}{2\pi} \int_{-\frac{W}{2}}^{\frac{W}{2}} |\frac{1}{\Delta + \omega + i(\Gamma_{nr} + \Gamma_r)/2}|^2 \, d\omega \qquad (18)$$

In the limit for large detuning ($\Delta \gg W$, Γ_{nr}, Γ_r) we can approximate this by

$$E^S \approx \frac{\Gamma_r}{2\pi} \frac{W}{\Delta^2} \qquad (19)$$

And the experimental quantum yield can be calculated:

$$Q = \frac{E^S}{S} = \frac{\Gamma_r}{\Gamma_{nr} + \Gamma_r} \qquad (20)$$

Both Δ and W have been chosen arbitrarily (as long as Δ is larger then $\Gamma_{nr} + \Gamma_r$) and we can conclude that the quantum yield is independent of the frequency of excitation, a maybe unexpected result, but certainly in contradiction with experiment.
In the next section we will see what we can do to remedy the situation.

4. Consecutive Decay

Returning to equation (2) for $C_s(\omega)$ and replacing the sum by the imaginary term we have:

$$C_s(\omega) = \frac{\chi(\omega)}{\omega + i(\Gamma_{nr}/2 + \Gamma_r/2)} \qquad (21)$$

In retrospect it may not be too surprising that the quantum yield of a system leading to this equation is independent of frequency. $C_s(\omega)$ simply consists of a laser envelope multiplied by a Lorentzian of width ($\Gamma_{nr} + \Gamma_r$), where Γ_{nr} represents the loss of photons to the inner machinery of the molecule. Since it is not dependent on frequency this loss occurs equally well at all frequencies, and the quantum yield is invariant.

It might be better if a Γ^*_{nr} could be constructed that was dependent on frequency. Such an illustrious goal is fairly easily attained by letting $|s\rangle$ decay through a state $|1\rangle$ into a manifold $\{|k\rangle\}$, i.e., consecutive decay through an intermediate state $|1\rangle$.

By arguments essentially similar to those leading to eqn. 2 we then have

$$C_s(\omega) = \cfrac{\chi(\omega)}{\omega - \cfrac{v_{s1}^2}{\omega - \omega_1 - \sum_k \cfrac{v_{1k}^2}{\omega - \omega_k}} + i\,\Gamma_r/2} \qquad (22)$$

where ω_1 is the strongly coupled state and $\{\omega_k\}$ are the manifold of weakly $|1\rangle$-coupled background states. Obviously $v_{s1} \gg v_{1k}$

Let us replace the sum over $\{|k\rangle\}$ by an integral and go to the imaginary representation. We then have:

$$C_s(\omega) = \cfrac{\chi(\omega)}{\omega - \cfrac{v_{s1}^2}{\omega - \omega_1 + i\,\Gamma_1/2} + i\,\Gamma_r/2} \qquad (23)$$

wich after rearrangement leads to

$$C_s(\omega) = \cfrac{\chi(\omega)}{\omega - \cfrac{v_{s1}^2\,(\omega-\omega_1)}{(\omega-\omega_1)^2 + \Gamma_1^2/4} + i\left[\cfrac{v_{s1}^2\,\Gamma_1/2}{(\omega-\omega_1)^2+\Gamma_1^2/4} + \Gamma_r/2\right]} \qquad (24)$$

with $\Gamma_1 = 2\pi \langle v_{1k}^2\rangle\,\rho_k$.

Comparing this to eqn. 21 we now have a radiationless transition rate Γ^*_{nr} which is a function of frequency:

$$\Gamma^*_{nr} = \cfrac{v_{s1}^2\quad\Gamma_1}{(\omega-\omega_1)^2 + \Gamma_1^2/4} \qquad (25)$$

It has a peak at $\omega = \omega_1$ but rapidly diminishes upon detuning from that frequency. Of course, its effect is strongest when $\omega_1 \approx \omega_s = 0$, i.e. when the intermediate state lies close to the doorway state.

In that case the quantum yield is low around resonance, but away from resonance the value of Γ^*_{nr} becomes negligible and the quantum yield goes to one!

For an equally spaced manifold as treated before (see section 3) and with $\omega_s = \omega_1$ we have also proven this analytically (3), but for our purpose here the present arguments probably suffice.

It is a simplification to postulate only one intermediate state, postulating more leads to a similar result. Then Γ^*_{nr} becomes:

$$\Gamma^*_{nr} = \sum_l \frac{v^2_{sl} \ \Gamma_l}{(\omega - \omega_l)^2 + \Gamma^2_l/4}$$

(26)

Now the quantum yield will show dips whenever $\omega = \omega_l$, but if they are far away from ω_s the effect would be hard to find, since the absorption is very low then.

Nevertheless, with present day lasers it may well be possible to find dips in the quantum yield at non-resonant frequencies, if photon loss and emission can be measured properly.

It may even be, that the effect of the increased quantum yield at non-resonance has already been demonstrated. Some years ago Jonkman, Drabe and Kommandeur (4,5) showed that the "instantaneous" light scattering in pyrazine was of about the same magnitude when the excitation was resonant with a rotational line as when it was non-resonant. At first glance one would imagine the non-resonant scattering to be of much lower intensity than the resonant scattering. But at resonance, pyrazine has a quantum yield of 1 or 2%. And if the quantum yield in non-resonant excitation is about one, the relatively strong scattering (compared to the fluorescence) would be readily explained, since it would be enhanced by a factor of one hundred!

Conclusion

It is clear that the conventional model for radiationless transitions can not explain the behaviour of the quantum yield in that it does not predict the rise from a low value at resonance to a very high value at non-resonance.

Postulating an intermediate state (or a discrete manifold of such states) which is strongly coupled to the absorbing state and through which the energy leaks to the background states provides a way out of the dilemma. The effect seems to be quite general. It is therefore tempting to postulate that a low quantum yield is always due to an intermediate state close to or essentially at resonance with the absorbing state.

This might explain the seemingly arbitrary variation of quantum yields in organic molecules. If, for instance, we consider a vibrationless singlet excited state, couplings such as Herzberg-Teller, would be predominant to a state in the ground or triplet vibronic manifold, which has only one quantum of a particular H-T-active vibration. In its turn this state through non-harmonic interaction would be coupled to the remainder of the vibrational manifold. Such a state may or may not be very close to the singlet excited state and thus lead to a low or high quantum yield.

Careful narrow laser determinations of quantum yields may reveal such intermediate states.

Acknowledgements

The authors are very indebted to professor Mukamel for suggesting the presence of an intermediate state. This work was supported by the Netherlands Foundation for Chemical Research with financial aid from the Dutch Organization for the Advancement of Pure Research (Z.W.O.)

References

1) M. Bixon and J. Jortner, J. Chem. Phys. 48, 715 (1968)

2) A. Tramer and R. Voltz in "Excited States" (E.C. Lim., ed.) 4, 281, Academic Press 1978

3) P.J. de Lange, K.E. Drabe and J. Kommandeur, J. Chem. Phys. (submitted)

4) H.Th. Jonkman, K.E. Drabe and J. Kommandeur, Chem. Phys. Lett. 116, 357 (1985)

5) J. Kommandeur, B.J. van der Meer and H.T. Jonkman in Intramolecular Dynamics, Eds. J. Jortner and B. Pullman, Reidel (1982)

Excited States of Molecular Dimers

Edward C. Lim

Wayne State University, Department of Chemistry, Detroit, MI 48202, USA

Abstract

We present here a progress report on the study of structure and spectra of molecular dimers of aromatic hydrocarbons.

INTRODUCTION

Clusters consisting of a small number of molecules are important intermediates in many chemical and physical processes including those in the atmosphere and in living tissues. The observation that such clusters appear frequently in seeded beams has greatly stimulated research in this area.[1] Most of these studies concern clusters in their electronic ground state. The simplest excited states of clusters are those in which one molecule in the cluster is excited. Since the presence of such an excitation will affect the intermolecular forces, it will tend to change the structure of the cluster. Different classes of excited clusters will appear if more than one molecule in the cluster is excited. The study of such systems has barely started.

We are presently investigating how electronic excitation affects the structure and binding of molecular clusters. The simplest cluster is a dimer: hence it seems natural to start with a study of dimers. Specifically, we choose dimers of aromatic hydrocarbons. Following standard nomenclature we refer to the ground state cluster as the van der Waals dimer and the dimer with one molecule excited as the excimer. For the dimer with two molecules excited, we have recently coined the name bicimer[2] (short for bi-excited dimer). Singlet excimers of aromatic hydrocarbons have been known for a long time,[3] but the formation of triplet excimers[4] and bicimers[2] has been reported only very recently, and the van der Waals dimer is known only for anthracene.[5] To investigate these systems, we are using both experimental and theoretical methods.

This paper is a progress report on our efforts to identify and characterize the three major classes of molecular dimers of aromatic hydrocarbons.

J. Jortner et al. (eds.), Large Finite Systems, 11–18.
© *1987 by D. Reidel Publishing Company.*

BACKGROUND INFORMATION AND THE RESULTS TO DATE

Our interest in the van der Waals dimers, excimers and bicimers of aromatic hydrocarbons began with a theoretical study of molecular excimers[6] which indicated that the geometry of triplet excimers may be very different from that of the corresponding singlet excimers. Thus, while the preferred conformation of singlet excimers is a symmetrical sandwich structure (or one very close to it), the theoretical analysis showed that the triplet excimers with symmetrical sandwich (or cofacial) configurations are either unstable or have extremely small binding energies. This prediction was dramatically confirmed by Chandross and Dempster[7] who showed that a sandwich dimer of naphthalene in rigid glass at 77K exhibits structured phosphorescence resembling that of monomeric naphthalene, despite the fact that its fluorescence is distinctly excimer-like in character. Although these results demonstrate the instability of triplet excimer of parallel sandwich-pair conformation, they do not rule out the formation of triplet excimers whose conformation is significantly different from the sandwich structure. We have therefore set out to look for phosphorescence and transient absorption from fluid solutions of naphthalene and 1,n-di-α-naphthylalkanes (n = 1-4) that could be attributed to triplet excimers. This study[4,8,9] has shown that concentrated solutions of naphthalene and dilute solutions of the dinaphthylalkanes exhibit phosphorescence (and transient absorption) which are identical to each other but different from the monomeric phosphorescence (and monomeric triplet-triplet absorption) exhibited by dilute solutions of naphthalene (Figures 1 and 2).

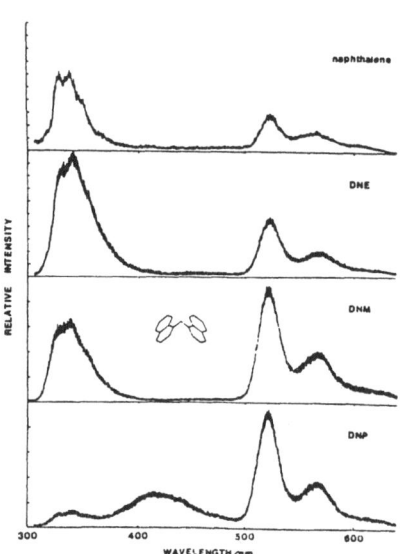

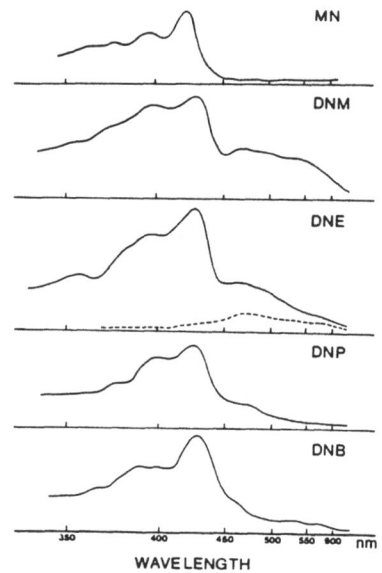

Figure 1 Excimer phosphorescence of naphthalene, 1,1-di-α-naphthylmethane (DNM), 1,2-α-naphthylethane (DNE), and 1,3-di-α-naphthylpropane (DNP) in isooctane at room temperature. Concentration is ~10⁻5 M except for naphthalene, which is ~10⁻4 M.

Figure 2 Laser-induced transient absorption spectra of dilute solutions (~10⁻5 M) of 1,n-di-α-naphthylalkanes and α-methylnaphthalene in isooctane at room temperature. The time delay between the excitation and probe pulses was 5 μs, except for the dashed curve in DNE, which was 150 μs.

The temporal characteristics of the emission as well as those of the transient absorption are consistent with the assumption that these new spectral features arise from the triplet excimers which are formed by the association of naphthalene in its lowest triplet state with a molecule in its ground state[4]. The major conclusion that can be reached from these observations is that the intermolecular triplet excimer of naphthalene and the intramolecular triplet excimers of the dinaphthylalkanes have identical, or very nearly identical, conformation that can be attained by 1,1-di-α-naphthylmethane. Thus, a likely geometry of the triplet excimer is the symmetric L-shaped conformation of two naphthalene molecules in which the long in-plane axes lie parallel and the short in-plane axes make an angle of ~109.5° with one another[10] (Figure 3).

symmetric conformation of   endoxy dimer
1,1-di-α-naphthylmethane

Figure 3

An important consequence of this geometry was predicted[11] to be a breakdown of the σ-π separation in the triplet excimer, which leads to a large enhancement of the radiative and nonradiative decay rates relative to those in the corresponding triplet monomers. These conclusions have been directly supported by the observation[12] of a short-lived phosphorescence from dilute rigid glass solutions of 7,14-endoxy-7H,14H-cycloocta [1,2,3-de: 5,6,7-d'e'] dinaphthalene 7-ol,[13] (here-after referred to as the endoxy dimer - see Figure 3 for structure), which closely mimics the excimer phosphorescence from fluid solutions of naphthalene and 1,n-di-α-naphthylalkanes.[12] A quantitative comparison of the quantum yields and lifetimes of the low-temperature luminescence (fluorescence and phosphorescence) of the rigid endoxy dimer with those of naphthalene indicates that the radiative and nonradiative decay rates of the triplet excimer are at least two orders of magnitude greater than those of the corresponding triplet monomer.[14]

Intuitively, the L-shaped structure of the naphthalene triplet excimer, vis-a-vis the sandwich pair structure of the corresponding singlet excimer, can be rationalized by considering the sources of their respective binding energy. The stability of the sandwich-pair geometry of the singlet excimers is primarily due to exciton resonance (M*M↔MM*),[6,15-18] which is the largest for the face-to-face arrangement of the two aromatic molecules. For the triplet excimers, however, exciton resonance is not expected to be the major source of their binding energy since the transition moments associated with singlet-triplet transitions are extremely small (exciton resonance interaction is proportional to the square of the transition moment[19]). The two likely contributions to the attractive forces between a molecule in its triplet state and a molecule in its singlet state are electrostatic and dispersion forces.[20] In aromatic hydrocarbons, the leading contribution to the classical electrostatic interactions between the molecular charge distributions comes from quadrupole-quadrupole interactions. Recent calculations of the quadrupole moments of naphthalene, using a double-zeta basis set, indicate that the normal (i.e., perpendicular to the molecular plane) component of the quadrupole tensor is negative, while the in-plane long-axis component is positive, for both the ground and triplet states (the ground state in addition, has a positive in-plane short-axis component).[21] Thus, the electrostatic interactions would preclude sandwich structure, and favor a T-shaped structure for triplet excimer. For the T-shaped excimer, however, the dispersion force is not expected to be nearly as large as in a cofacial (or a coplanar) structure. The preference for an L-shaped structure of the naphthalene triplet excimer might therefore represent a compromise between the dispersive force which favors a cofacial (or a coplanar) structure and the electrostatic force which favors a T-shaped structure.[10] The past theoretical attempts to predict the geometry of the naphthalene triplet excimer have not been successful.[22]

Unlike the interaction between an electronically excited molecule and a ground state molecule, leading to the formation of excimers, very little is known about the interactions between pairs of electronically excited molecules. With the powerful short light pulses available from some picosecond and nanosecond lasers, it is now possible to electronically excite pairs of molecules in close proximity and study interactions between them by time-resolved spectroscopy. A class of molecules that is particularly well suited for such a study is two aromatic molecules covalently linked to each other by methylene and other flexible chains. Using this approach, we have very recently observed the first example of a triplet bicimer in 1,1-di-α-naphthylmethane (DNM), through luminescence generated by intense laser excitation[2] (Figures 4 and 5).

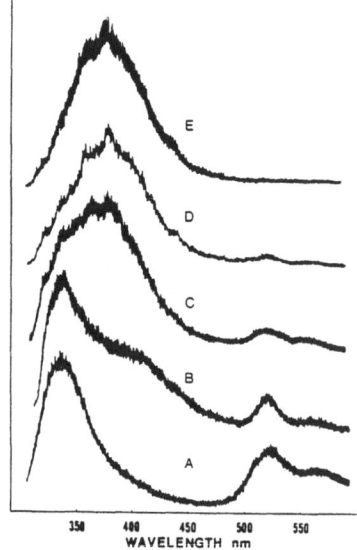

Figure 4 Excitation intensity dependence of the time-resolved emission spectra of dinaphthylmethane (DNM) in isooctane at room temperature. Approximate laser power: (A) < 0.01 mJ, (B) 0.01 mJ, (C) 0.03 mJ, (D) 0.1 mJ, (E) 0.3 mJ. Time delay between excitation and probe was ≈ 140 μs.

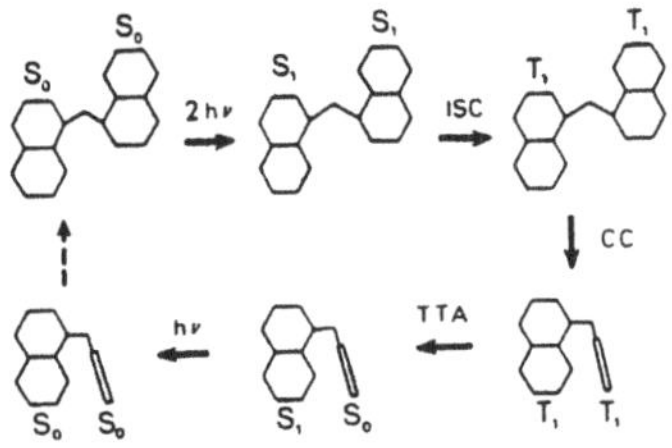

Figure 5 Schematic illustration of the sequence of events leading to the delayed emission from bicimer.

Although the exact geometry of the triplet bicimer could not be deduced from this experiment, the limited range of the conformation accessible to DNM and the fact that emission does not appear in the symmetric rigid model compound (e.g., the endoxy dimer) make it possible to conclude that both the long in-plane and the short in-plane axes of two naphthalene molecules are non-parallel in the bicimer.[2] The geometry of the triplet bicimer is therefore significantly different from that of the triplet excimer or that of the singlet excimer.

It is perhaps surprising that the least known molecular association in polycyclic aromatic hydrocarbons is that between the two ground state molecules leading to the formation of van der Waals dimer. In fact, the only known example of the ground state dimer is that of anthracene.[5] Intuitively, one might expect the structure of the dimer to resemble that of the corresponding triplet excimer since the attractive forces leading to their formation are thought to be qualitatively similar. It is therefore interesting to note that the spectroscopic studies of the ground-state anthracene dimer indicate that the long in-plane axes of the molecules are parallel and the short in-plane axes of the molecules make an angle of 60±15° in the dimer.[5] Unfortunately, the triplet excimer of anthracene (and the ground state dimer of naphthalene) has not been identified and characterized, so that the validity of this intuitive feeling remains untested. Preliminary experiments[24] on jet-cooled 1,2-di-α-naphthylethane (DNE) indicate that this species, unlike DNM, shows spectral features that can be attributed to the interaction between two naphthalene moieties (Figure 6).

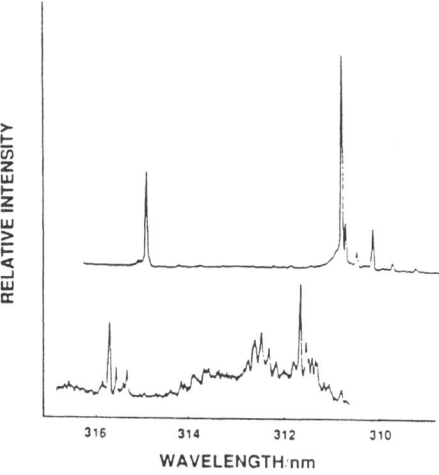

Figure 6 A portion of fluorescence excitation spectrum of jet-cooled α-ethylnaphthalene (top) and that of 1,2-di-α-naphthylethane (bottom).

Since the angle between the two short in-plane axes is much smaller in the symmetric conformation of DNE (~ 45°) than in that of DNM (~ 110°), the result can be interpreted as implying that the dimeric form of DNE has a structure similar to the ground-state anthracene dimer.

Future Work

From what has been described, it should be clear that our understanding of molecular associations in aromatic hydrocarbons is very fragmented even for the most extensively studied case of naphthalene. Much more data on the structure and spectra of dimers (ground and excited states), excimers, and bicimers, for a variety of molecules, are needed to develop theoretical understanding of the intermolecular forces responsible for the three classes of the molecular associations. In as much as the forces between molecules control the progress of molecular collisions, determine the bulk properties of matter, play an important role in the conformation and functions of a number of biologically important materials, and influence photodimerization and soot formation in hydrocarbon combustion, etc., the understanding of molecular associations in the ground and excited states is a central problem in chemistry, physics, and biology, and even more so at the interface of these scientific disciplines.[20]

We have therefore initiated an experimental and theoretical study of the structure and binding of dimers, excimers, and bicimers of aromatic hydrocarbons. The specific questions we are addressing are: 1) Is the triplet excimer formation common to aromatic hydrocarbons and are the triplet excimers generally L-shaped in linear polyacenes? 2) What is the general structural feature of the ground-state dimers of linear polyacenes and how is it related to that of the corresponding triplet excimer? 3) Is the bicimer formation general and what are the geometries of the naphthalene bicimers (singlet and triplet)? 4) Can theory account for the observed structures of the three classes of dimers in naphthalene?

Experimentally, answers to these questions will be sought through a systematic investigation of the time and wavelength resolved emmission and absorption spectroscopy on various rigid and nonrigid double molecules (diaryalkanes and their rigid analogs) of a number of aromatic hydrocarbons. Experiments in both condensed phase and in supersonic jet are planned. For practical reasons, a detailed experimental study is possible only for a small number of systems. To gain the necessary generalization, a theoretical framework is needed. We are initiating such a study in collaboration with Willem Siebrand and his co-workers at NRC, Ottawa. In principle, judiciously chosen quantum-chemical methods will allow us to construct potential-energy surfaces of two interacting molecules as a function of their relative orientation. The complexity of the problem excludes *ab initio* methods from serious consideration: they would either be prohibitively expensive or severely limit the region of parameter space that can be sampled. Of the available semi-empirical methods, we choose the PPP/WK method, i.e., the Pariser-Parr-Pople method as modified by Warshel and Karplus.[25] Although basically a π-electron method, it takes account of the core of σ-electrons through molecular mechanis modeling and includes nonbonding interactions. Its careful para-metrization makes it suitable for optimization of geometry. For SCF wave-functions with adequate CI it has been shown to reproduce experimental geometries quite accurately and to yield satisfactory harmonic force fields. It is hoped that the combination of these experimental and theoretical studies will result in detailed characterization of molecular dimers of aromatic hydrocarbons.

References

1. See, for example, proceedings of this symposium.

2. R.J. Locke and E.C. Lim, Chem. Phys. Lett. 134, 107 (1987).

3. See, for review, J.B. Birks, Rep. Prog. Phys. 38, 903 (1975); B. Stevens, Adv. Photochem. 8, 161 (1971).

4. See, for review, E.C. Lim, Acc. Chem. Res. 20, 8 (1987).

5. J. Ferguson, A.W.H. Mau and J. Morris, Aust. J. Chem. 26, 103 (1973); E.A. Chandross, J. Ferguson and E.G. McRae, J. Chem. Phys. 45, 3546 (1966).

6. A.K. Chandra and E.C. Lim, J. Chem. Phys. 49, 5066 (1968).

7. E.A. Chandross and C.J. Dempster, J. Am. Chem. Soc. 92, 704 (1970).

8. P.C. Subudhi and E.C. Lim, J. Chem. Phys. 63, 5491 (1975); P.C. Subudhi and E.C. Lim, Chem. Phys. Lett. 44, 479 (1976); S. Okajima, P.C. Subudhi and E.C. Lim, J. Chem. Phys. 67, 4611 (1977).

9. P.C. Subudhi and E.C. Lim, Chem. Phys. Lett. 56, 59 (1978).

10. B.T. Lim and E.C. Lim, J. Chem. Phys. 78, 5262 (1983).

11. A.K. Chandra and E.C. Lim, Chem. Phys. Lett. 45, 79 (1977).

12. E.C. Lim, R.J. Locke, B.T. Lim, T. Fujioka and H. Iwamura, J. Phys. Chem. 91, 1298 (1987).

13. A.C. Agosta, J. Am. Chem. Soc. 89, 3505 (1967).

14. B.T. Lim, R.J. Locke and E.C. Lim, unpublished results.

15. E. Konijnenberg, Doctoral Thesis, Free University of Amsterdam, Holland, 1963.

16. J.N. Murrell and J. Tanaka, Mol. Phys. 45, 363 (1964).

17. T. Azumi, A.T. Armstrong and S.P. McGlynn, J. Chem. Phys. 41, 3839 (1964).

18. M.T. Vala, I.H. Hiller, S.A. Rice and J. Jortner, J. Chem. Phys. 44, 23 (1966).

19. See, for example, G.J. Hoytink, Z. Elektrochem. 64, 156 (1960).

20. See, for example, G.C. Maitland, M. Rigby, E.B. Smith and Wakeham, "Intermolecular Forces", Oxford University Press, New York, 1981.

21. A. Hinchliffe, J. Chem. Soc. Faraday Trans. 2, 73, 1627 (1977); A. Chablo, D.W.J. Cruickshank, A. Hincliffe and R.W. Munn, Chem. Phys. Lett. 78, 424 (1981).

22. A.K. Chandra and B.S. Subhindra, Mol. Phys. 28, 695 (1974); B.S. Subhindra and A.K. Chandra, ibid. 30, 319 (1975); E.J. Padma Malar and A.K. Chandra, Theoret. Chim. Acta 55, 153 (1980).

23. See, for example, R.L. Fulton and M. Gouterman, J. Chem. Phys. 35, 1059 (1961); 41, 2280 (1964).

24. R.J. Locke and E.C. Lim, unpublished results.

25. A. Warshel and M. Karplus, J. Am. Chem. Soc. 34, 5612 (1972); A. Warshel and M. Levitt, QCPE 247, Indiana University, 1974; A.C. Lasaga, R.J. Aerin and M. Karplus, J. Chem. Phys. 73, 924 (1983).

ROTATIONAL EFFECTS ON INTRAMOLECULAR RADIATIONLESS TRANSITIONS -
"THE STORY OF PYRAZINE"

Aviv Amirav
School of Chemistry - Sackler Faculty of Exact Sciences
Tel Aviv University, Ramat Aviv 69978 Tel Aviv, ISRAEL

ABSTRACT. The experimental rotational effects on the S_1 ($^1B_{3u}$) excited state dynamics of pyrazine are briefly summarized. These experiments impose severe constraints on the validity of the general radiationless transitions theory in the limit of intermediate level structure. A model presented accounted for all the problems confronted by the theory in terms of rotationally induced intra-triplet vibrational energy redistribution. This new type of intramolecular radiationless transition called vibrational crossing is an interstate process with intrastate vibronic coupling features. From this model two major predictions emerge, which are shown to be fulfilled. We demonstrate that coriolis coupling is an important and dominant vibronic coupling leading into intramolecular vibrational energy redistribution. The absorption spectrum, fluorescence excitation spectrum and the resulting emission quantum yield of pyrazine were studied using dye laser with 2 GHz resolution as a light source. The emission quantum yield decreased with the rotational state. The absorption contour of a single rotational state was broader than the excitation contour and a low quantum yield absorption is shown between adjacent rotational transitions. These experimental results combined with the quantitative fit of the inverse emission quantum yield and A+/A- were predicted by the vibrational crossing model and thus strongly support its validity.

I. INTRODUCTION

The S_1 ($^1B_{3u}$) state of the pyrazine molecule serves as a touchstone for critical scrutiny of the theory of radiationless transitions [1-5]. Following the pioneering work of Tramer and co-workers [6,7], conventional wisdom has attributed intramolecular dynamics in this system to the intermediated level structure (ILS) limit in the theory [1-7]. As detailed experimental results have become available, this picture became less clear and understandable as the conventional theory in the ILS limit was confronted with a large volume of conflicting pieces of experimental results. These puzzling self-inconsistencies are described in details elsewhere [8].

19

J. Jortner et al. (eds.), Large Finite Systems, 19–30.
© 1987 *by D. Reidel Publishing Company.*

The two major problems are the following [8]:

a)   The problem of the missing states arising from the confrontation
of the time resolved and emission quantum yield data with the ultrahigh resolution spectral data. The number of spectroscopically observed Molecular Eigenstates (MEs) is small (~12) and rotation independent [9-13], whereas the time resolved [14,15] and quantum yield data
[6,16] require a much higher number of MEs and a rotational dependence
of this number.

b.   The origin of short time resolved emission component. The number
of spectroscopically observed MEs is too small to enable a true dephasing in the ILS limit. This problem is further exacerbated by the
existence of this short time component in the bi-exponential time
resolved emission of pyrazine, following an incoherent nanosecond
laser excitation [6,14]. In addition to this inconsistency, it was
shown by J. Kommandeur and co-workers [9-12] that the amplitude ratio
of this short component to the long time emission (A+/A-) is sharply
maximized in the valleys between adjacent rotational transitions where
no spectral lines were observed.

According to the conventional ILS theory A+/A- should be proportional to the number of coherently excited MEs. The number of
spectrally observed MEs is too small to enable any short component
under Nsec incoherent excitation and obviously it should not maximize
in the "no man's land" where no MEs are detected.

The existence of near resonance Raman light scattering
(NRLS) was therefore invoked in order to account for the existence of
this short component [9-12], but later on the short component was
shown to be real with a lifetime of 100-120 Psec [17,18]. In order to
explain the wealth of experimental results and to reconcile between
the experimental results and the theory of radiationless transitions,
a new theoretical model was advanced by A. Amirav [8].

II. VIBRATIONAL CROSSING - A NEW TYPE OF INTRAMOLECULAR RADIATIONLESS
TRANSITION

Following Ref. [8], the relevant energy level diagram in
pyrazine is shown in Fig.1, S_0 is radiatively coupled to S_1 with J=0+1
standard selection rules. S_1 is strongly coupled to the sparse triplet
T manifold, via spin-orbit coupling. The spin-orbit selection rules
are   N=0 and no K selection rules. T is nonradiatively coupled
through an _intrastate_ K" selective Z type Coriolis coupling to T' (X,Y
Coriolis coupling can also be considered).   T' contains energy levels
of T, that due to selection rules, are not allowed in spin-orbit
coupling. T' may decay nonradiatively to S_0 in   > 350 Nsec.

From Fig.1, it is implied that a rotation induced vibrational
energy redistribution within the triplet state leads to the observed
dynamics in pyrazine. This state of affairs can be written as

$$(\alpha \langle S| + \Sigma \beta \langle T|) |V_{COR}| \Sigma \gamma |T'\rangle \tag{1}$$

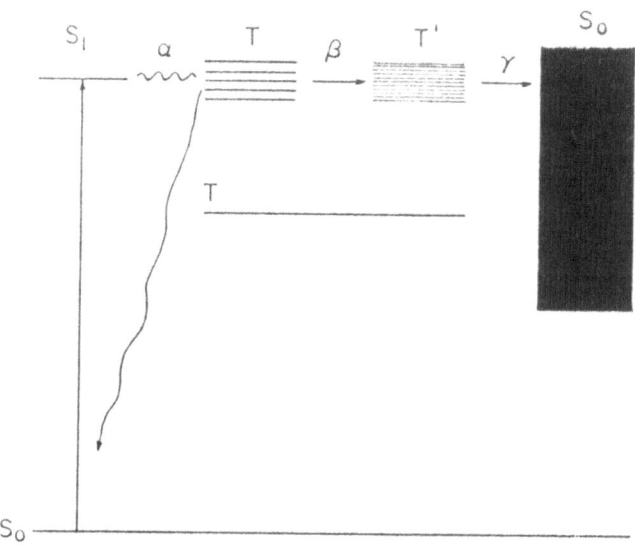

Figure 1. Energy level diagram of pyrazine. T and T' vibrational rotational states belong to the same triplet electronic state. T' states are vibronically coupled via Coriolis coupling to T and are uncoupled to S_1 via spin orbit coupling due to selection rules.

     These matrix elements represent a new type of radiationless transition, named vibrational crossing (VC). An intrastate vibronic coupling which we assume to be Coriolis coupling is also associated with increased dilution and hence interstate electronic relaxation of the singlet component of the MEs. (intersystem crossing in this case). Thus, vibrational crossing is an interstate radiationless transition with intrastate vibronic coupling features.
     This model gives a full qualitative and quantitative interpretation of all the debated aspects of pyrazine dynamics [8]. Basically MEs originating from the interstate coupling of $S_1(J',K')$ with the sparse T manifold exhibit a dual behavior with regard to their Coriolis coupling with T' states as Z type parallel Coriolis coupling

was found to be $V_{Rz} \propto K$. (K is the precessional-rotational quantum number). The MEs with $S_1 - T(J",K"=0)$ correspond to the small molecule limit (vanishing Coriolis coupling) having a long lifetime (350 Nsec) and Doppler limited spectral width. The MEs containing an $S_1 - T(J",K" \neq 0)$ admixture correspond to the large molecule statistical limit being characterized by a short (~120 Psec) lifetime, low emission quantum yield ($Y=6.10^{-4}$) and a broad homogeneous width of 1.5 GHz. The experimental bi-exponential decay observed is inhomogeneous in nature and the rotational dependence of A+/A- and the emission quantum yield is due to the 1/2J+1 statistical abundance of MEs with $S_1 - T(J",K"=0)$ character.

III. MODEL PREDICTIONS

        Two major predictions clearly emerge from this model:
1. Coriolis coupling is an important and dominant vibronic coupling in inducing intrastate vibrational energy redistribution.
2. The "missing states" are "grass", namely a continuous, broad, short-lived and low quantum yield background; hence, they are very weak in the excitation spectrum, but should be amenable for observation in direct absorption spectroscopy. Each rotational line envelope is broader in absorption than in excitation and the absorption "background" extends to the valley between adjacent rotational transitions.
        In this paper, I shall describe new experimental results demonstrating the verification of these predictions and thus giving additional credence to the validity of the vibrational crossing (VC) model.

IV. EXPERIMENTAL

        Our experimental apparatus and techniques for the measurement of absorption spectra and absolute fluroescence quantum yield [16,21,22] were extended to allow the use of our new eximer pumped dye laser as a light source. (Lambda Physik EMG53MSC+FL2002E).
        Absorption spectra and fluorescence excitation spectra of pyrazine cooled in planar supersonic expansions were simultaneously determined using the dye laser equipped with an etalon and frequency doubling crystal, having 2 GHz spectral resolution using pressure tuning. Pulsed planar jets were generated by expansion of seeded argon through a nozzle slit. Two nozzle slits were used having dimensions of 0.22x33 and 0.27x90mm. The repetition rate of both nozzles was 6-9Hz and the gas pulse duration was 200-300 μsec. The nozzle was at about room temperature (in the range of $15^{\circ}C - 40^{\circ}C$) and pyrazine was mixed with argon at the stagnation pressure up to 70 torr.
        Laser light after attenuation to a 1 μ Joule/pulse crossed the beam parallel to the slit at a distance of X=16mm from it (10mm and 30mm were also used). The light beam was split by a sapphire window.

Before heating the silicon photo diode detectors (UDT PIN 10D/SB 12mm diameter followed by a current to voltage convertor), the laser beam was fully absorbed in a red dye cell and was turned into a diffuse lower intensity red fluorescence. This was done in order to increase the detector linearity. The attenuation ΔI of the light beam due to absorption was determined from the difference in the light intensity before and after crossing the planar jet. The fluorescence intensity I_F was monitored by a photomultiplier. The number of laser pulses was doubled compared with the nozzle pulse rate, and a second differentiation was performed between the results obtained with the laser pulse that crossed the pulsed jet and the second untimed laser pulse. This procedure eliminated any stray absorption and resulted in a zero, flat, wavelength independent baseline.

The role of collisions was studied by varying both the stagnation pressure of the argon gas and the pyrazine concentration by changing the nozzle temperature. Probing the long time emission, we have controlled the collision effect which was small and rotation independent in our experiments. Absolute quantum yield was measured on the peak of R_4 transition, comparing it with that of 4-chloro-trans-stilbene in an expansion containing both molecules [21]. Time resolved decay curves were processed in a LeCroy 9400 125MHz signal averager. Spectral tuning was achieved using a pressure transducer in a homemade pressure tuning unit, using dry nitrogen gas. The resolution was 2 GHz and the long term stability was ~1 GHz/5 minutes. The spectral region near the 0-0 transition was also studied and found to contain negligible contributions from dimers, impurities, hotbands and other spectral interferences. Laser stray light was carefully filtered out and had no effect on the experimental results herewith presented.

V. CORIOLIS COUPLING AND INTRASTATE INTRAMOLECULAR VIBRATIONAL ENERGY REDISTRIBUTION (IVR)

Recently there has been considerable experimental and theoretical work on the effect of molecular rotations on electronic intrastate as well as interstate radiationless processes [8]. Most noticeable is the celebrated problem of the onset of "channel three" in benzene. Riedle et al, [23] have demonstrated the important role of rotational (J') and precessional (K') states in the radiationless transitions of benzene in the region of the onset of channel three that initiated a drastic reduction in the emission quantum yield and lifetime as the vibrational energy was increased. The K'=0 precessional states in each rotational state J'=0-14 were much narrower than the other K'≠0 states and appeared as sharp lines in the ultrahigh resolution spectra. The other J'(K≠0) states were much broader and had an order of magnitude smaller emission quantum yield resulting in a broad and unresolved background "noise" in the spectra. This important effect was attributed [19,23] to Z type parallel Coriolis coupling causing a rotational induced intrastate vibrational energy redistribution

(RIVR). On the other hand, Felker and Zewail have beautifully demonstrated quantum beats in the energy resolved emission spectra of
anthracene [24]. The existence of these quantum beats as well as the
weak rotational effect on their dephasing was invoked as a strong
argument for the dominating role of anharmonic coupling on IVR in
anthracene [25]. Further work was therefore required to clarify the
generality of Coriolis coupling on intrastate IVR. Recently Amirav et
al. have demonstrated the important role of Coriolis coupling both in
9-cyanoanthracene [26] and anthracene [27].

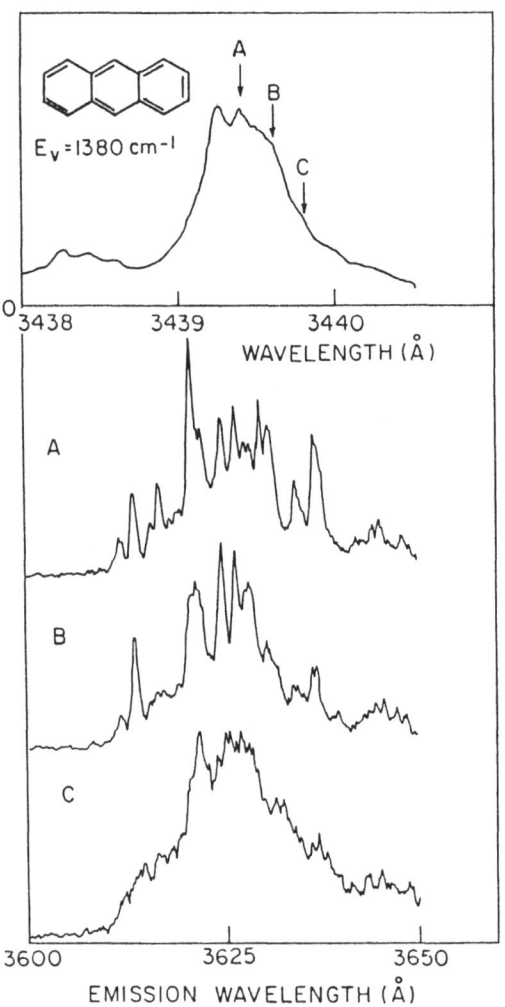

Figure 2. Excitation (upper curve) and energy resolved emission of
anthracene on the spectral range 3600-3650Å. The excitation wavelengths for traces A,B and C are marked in arrows on the upper excitation spectrum. Note the dramatic rotational dependence of the energy
resolved emission.

In Fig. 2, we show the excitation spectrum of anthracene near its $1380cm^{-1}$ vibration (upper trace) together with its energy resolved emission spectra when it was excited on different positions and rotational states of its rotational contour. For clarity, only the portion of the emission to states near $1380cm^{-1}$ ground state energy is shown. A remarkable difference is observed upon changing the excitation energy by 1.6 and $3.2cm^{-1}$. An increased rotational energy completely changes the emission trace A into the highly congested trace C, demonstrating the largely increased number of coupled background states. The emerging conclusion from these and other time resolved experiments [26] is that in addition to anharmonic coupling, Coriolis coupling is very important and it dominates intrastate intra-molecular vibrational energy redistribution.

If we accept this situation which was found in benzene [19,23], 9-cyanoanthracene [26], anthracene [27] and pyrimidine [28], then we should naturally suspect that indeed this is the case for intra-triplet IVR in pyrazine. The triplet vibrational energy in pyrazine at the S_1 origin is $3950cm^{-1}$ [29], high enough to enable RIVR. The major difference, however, is that intra-triplet RIVR in pyrazine is inherently associated with interstate electronic relaxation manifesting the essence of the new type of vibrational crossing radiationless transition.

VI. THE "GRASS" IS REAL!

Fig.3 shows the direct absorption spectrum (lower trace), fluorescence excitation spectrum (middle trace) and the resulting ratio of absorption over fluorescence traces in the first low nine R branch rotational transitions of pyrazine S_1 origin. Several important observations are made clear from Fig. 4:
a) The emission quantum yield strongly depends on the rotational quantum number, and monotonically reduces with J'.
b) The absorption spectrum shows a substantial background in between adjacent rotational transitions in contrast to the excitation spectrum.
c) The inverse emission quantum yield (1/Q = absorption/fluorescence ratio) strongly oscilates crossing each rotational transition. The ratio of the peaks to valleys can exceed a factor of 20!
d) The minima in the inverse quantum yield spectrum are on the peak of the rotational transitions, while the maxima are located in the valleys in between adjacent rotational transitions.

Comparing these results with the short gated and delayed excitation spectra of J. Kommandeur and co-workers [9-12], our absorption spectrum clearly resemble the short gated spectrum. Perhaps the most intriguing conclusion that can be drawn from this comparison is that the inverse quantum yield spectrum quantitatively matches the A+/A-spectrum of Kommandeur and co-workers [30]. These are seemingly entirely different experiments and yet the results of these two different experimental observables are practically identical.

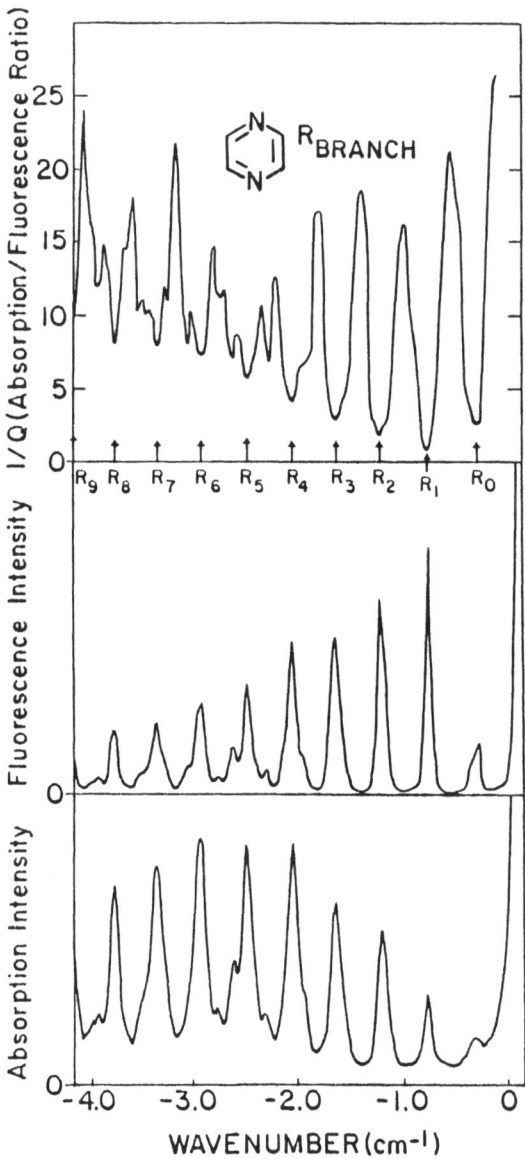

Figure 3. Absorption (lower) and fluorescence excitation (middle) spectra of the rotational R branch of pyrazine $S_1 (^1B_{3u})$ origin. The upper trace is the ratio of absorption over excitation spectra (1/Q). Laser wavelength is 3237Å and pressure tuning is used for the fine tuning. Nozzle temperature is 20°C and the argon backing pressure is 60 Torr.

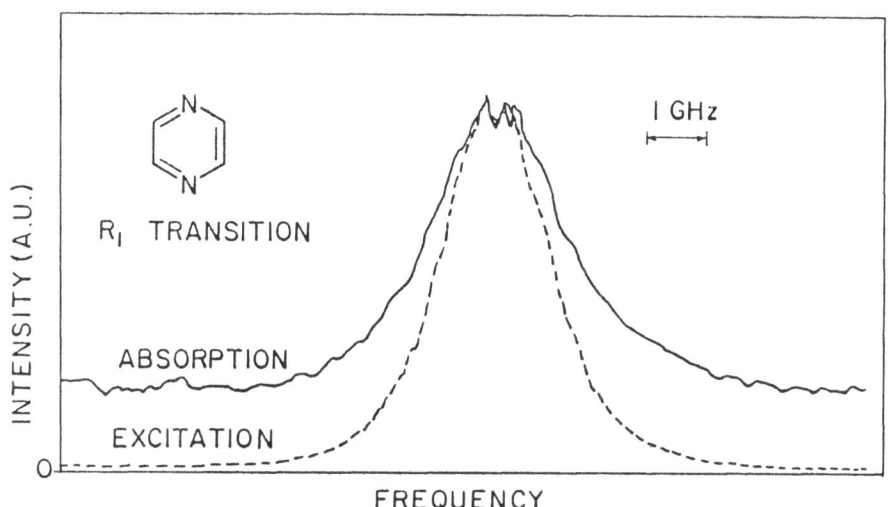

Figure 4. Absorption (full line) and fluorescence excitation (dashed line) spectra of the R_1 rotational transition of pyrazine S_1 ($^1B_{3u}$) electronic origin. Spectral resolution is 2 GHz. The rotational temperature is ~20°K. The two spectra were simultaneously measured using an etalon and pressure tuning.

In Fig. 4, the R_1 transition is magnified. This figure clearly demonstrates the fulfillment of the predictions drawn from the vibrational crossing model. The absorption of R_1 is broader than the excitation spectrum and a clear low quantum yield background absorption is observed on the two sides of the R_1 transtion. We note that the relative emission quantum yield on the two sides of R_1 transition is about 5% of its value in the center of this transition.

Two remarks are in order concerning Fig. 3 and Fig. 4:
a) The largely reduced emission quantum yield in the valleys between rotational transitions is not only in full agreement with the vibrational crossing model but also contradicts models based on near Raman light scattering (NRLS) [9-12]. Any photon which is "Raman scattered" is scattered and, hence, NRLS has an emission quantum yield of 100%, which is two orders of magnitude higher than some average value of pyrazene emission quantum yield [16]. The existence of NRLS and "no man's land" between rotational transitions implies a much higher quantum yield in the valleys. The experimental results are completely the opposite of this implication.
b) The quantitative fit between the A+/A- spectrum and 1/Q spectrum is explained as follows:

$$1/Q = A/F = (A_L + A_S)/(A_L Q_L + A_S Q_S) \tag{2}$$

where A_L and A_S are the absorption responsible for the short and long time emission component respectively. Q_L and Q_S are their emission quantum yields respectively. If we ignore the short component contribution to the fluorescence quantum yield $(A_L Q_L \gg A_S Q_S)$, we obtain

$$1/Q = 1/Q_L \ (1 + A_S/A_L) \tag{3}$$

According to the vibrational crossing model, A_S/A_L are proportional to $A+/A-$ so we can write

$$1/Q = 1/Q_L \ (1 + CA^+/A^-) \tag{4}$$

where C is a constant, depending on the detection resolution and lifetimes.

Since Q_L and $A+/A-$ are high, $1/Q_L$ can be neglected and we obtain

$$1/Q = CA^+/A^- \tag{5}$$

as is observed.

VII. CONCLUSIONS

The excited state dynamics of pyrazine shows several experimental unique features, such as a short time emission component and background low quantum yield absorption. These features cannot be accounted for by the general radiationless transitions theory in the ILS limit.

A new type of intramolecular radiationless transition, namely vibrational crossing (VC) which was invoked, could account for all the experimental results. This VC radiationless transition is based upon intra-triplet RIVR leading into intersystem crossing of the singlet character of the MEs. A unique feature of VC is that it is an interstate process with intrastate features which, in the case of pyrazine, is determined by the properties of intrastate Coriolis coupling.

In this paper, the predictions based upon the existence of VC were demonstrated experimentally. I showed that the "grass" is real and Coriolis coupling is an important and dominant coupling leading into intrastate IVR. New Manifestations of this Coriolis coupling as well as a full account on the presented new results will be described elsewhere [31].

ACKNOWLEDGEMENT

My very deep appreciation to Professor Joshua Jortner for many long, fruitful and stimulating discussions. I also thank Professor E.C. Lim for a fruitful collaboration on the subject of rotational effect on radiationless transitions. Finally, my gratitude is given to Mr. Chanan Horwitz for maintaining our new laser. The research was supported by the Fund for Basic Research of the Israel Academy of Sciences.

REFERENCES

1. M. Bixon and J. Jortner. J. Chem. Phys.,48, 715 (1968).
2. M. Bixon, J. Jortner and Y. Dothan. Mol. Phys. 17, 109 (1969).
3. J. Jortner and S. Mukamel in Molecular Energy Transfer, eds. R.D. Levine and J. Jortner (Wiley, New York, 1979) p. 178.
4. J. Jortner and R.D. Levine. Advan. Chem. Phys. 47, 1 (1982).
5. A. Nitzan, J. Jortner and P.M. Rentzepis. Proc. R. Soc. Lond. A, 327, 367 (1972).
6. A. Frad, F. Lahmani, A. Tramer and C. Tric. J. Chem. Phys. 60, 4419 (1974).
7. F. Lahmani, A. Tramer and C. Tric. J. Chem. Phys. 60, 4431 (1974).
8. A. Amirav. Chem. Phys. 108, 403 (1986).
9. K.E. Drabe and J. Kommandeur in Excited State, eds. E.C. Lim and K.K. Innes (Academic Press, New York, 1987) Vol. 7.
10. J. Kommandeur, B.J. van der Meer and H.Th. Jonkman in Intra-molecular Dynamics, eds. J. Jortner and B. Pullman (Reidel, Dordrecht, (1982), p. 259.
11. B.J. van der Meer, H.Th. Jonkman and J. Kommandeur. Laser Chem. 2, 77 (1983).
12. H.Th. Jonkman, K.E. Drabe and J. Kommandeur. Chem. Phys. Lett. 116, 357 (1985).
13. B.J. van der Meer, H.Th. Jonkman, J. Kommandeur, W.L. Meerts and W.A. Majewski. Chem. Phys. Lett., 92, 565 (1982).
14. Y. Matsumoto, L.H. Spangler and D.W. Pratt. Laser Chem. 2, 91 (1983) and Chem. Phys. Lett. 98, 333 (1983).
15. H. Saigusa and E.C. Lim. J. Chem. Phys. 78, 91 (1983) and Chem. Phys. Lett., 88, 455 (1982).
16. A. Amirav and J. Jortner. J. Chem. Phys. 84, 1500 (1986).
17. A. Lorincz, D.D. Smith, F. Novak, R. Kosloff, D.J. Tannor and S.A. Rice. J. Chem. Phys. 82, 1067 (1985).
18. J.F. Knee, F.E. Donay and A.H. Zewail. J. Chem. Phys. 82, 1042 (1985).
19. E. Riedle, H.J. Neusser, E.W. Schlag and S.H. Lin. J. Phys. Chem. 88, 198 (1984).
20. I.M. Mills. Pure Appl. Chem., 11, 325 (1965).
21. M. Sonnenschein, A. Amirav and J. Jortner. J. Phys. Chem. 88, 4214 (1984).

22. A. Amirav, C. Horwitz and J. Jortner. Submited to J. Chem. Phys.
23. E. Riedle and H.J. Neusser. J. Chem. Phys. $\underline{80}$, 4686 (1984) and
     E. Riedle, H.J. Neusser and E.W. Schlag. J. Phys. Chem. $\underline{86}$,
     4847 (1982).
24. P.M. Felker and A.H. Zewail. J. Chem. Phys. $\underline{82}$, 2961 (1985)
     and ibid, $\underline{82}$, 2975 (1985).
25. P.M. Felker and A.H. Zewail. J. Chem. Phys. $\underline{82}$, 2994 (1985).
26. A. Amirav, J. Jortner, M. Terazima and E.C. Lim. Chem. Phys.
     Lett. $\underline{133}$, 179 (1987).
27. A. Amirav. J. Chem. Phys. $\underline{86}$, 4706 (1987).
28. B.F. Forch and E.C. Lim. Chem. Phys. Lett., $\underline{110}$, 593 (1984).
29. E. Villa, M. Terazima and E.C. Lim. Chem. Phys. Lett. $\underline{129}$, 336
     (1986).
30. Ref. 10, p. 266, Fig. 10; and Ref. 12, p. 360, Fig. 3.
31. A. Amirav. In preparation.
32. M. Terazima and E.C. Lim. Chem. Phys. Lett., $\underline{127}$, 330 (1986).

THE DYNAMICS OF POLYATOMIC MOLECULES DURING AND AFTER COHERENT EXCITATION

Roberto Marquardt and Martin Quack
Laboratorium für Physikalische Chemie
ETH – Zürich (Zentrum)
CH-8092 Zürich – Switzerland

ABSTRACT. The quantum dynamics of the IR excitation of the alkyl CH-chromophore in polyatomic molecules is studied in terms of the motion of the wavepacket in a reduced coordinate space. Interesting conclusions regarding a quasi-classical behaviour on one hand and a non-classical behaviour on the other hand are drawn. The results are discussed in view of the fast vibrational redistribution existing in these systems.

1. INTRODUCTION

The vibrational dynamics of highly excited polyatomic molecules is of great importance for the understanding of vibrational energy redistribution (1,2), intramolecular statistical mechanics (3,4) and chemical reaction dynamics (5 – 9). These primary processes of chemical kinetics can be of a great complexity even for small polyatomic molecules consiting of three or more atoms, which may thus be considered as large finite systems.

The alkyl CH-chromophore is a very important group in hydrocarbons. A detailed study of its excitation by coherent, monochromatic light is of interest for several reasons:
1. A large amount of data from high resolution spectroscopy from the far IR to the visible region is available (10 – 13); these allow for the conclusion, that there is a very fast transfer of vibrational energy between the CH-stretching and bending combination states on the subpicosecond time scale, while the transfer away from the isolated vibrational modes is typically 10 to 100 times slower; thus, on the subpicosecond time scale, the CH-chromophore can be treated as isolated and the Hamiltonian describing the CH-group in the reduced, three-dimensional CH-coordinate space is taken to be the total vibrational Hamiltonian.
2. The subpicosecond redistribution of vibrational energy is a universal property of the alkyl CH-chromophore.
3. When the molecular frame CX_3 is heavy, the short time vibrational dynamics is local in the coupled CH-stretching and bending manifolds.
4. We expect the potential surface of the CH-group to be symmetric in

31

J. Jortner et al. (eds.), Large Finite Systems, 31–44.

the CH-bending coordinate, when the latter has small values (4); this
is in good agreement with the experimental observations and leads to a
further reduction of the coordinate space.

In the present paper we investigate the redistribution process and
show some computational results of the time evolution of the quantum
mechanical probability density in the CH-coordinate space of CHF_3. The
CH-chromophore is excited by a coherent, monochromatic light beam of
high intensity. We assume that the light beam is switched on sharply
and the molecular system is in its ground vibrational state at some
initial time zero. Furthermore, the electromagnetic field may be treated as a classical, coherent wave.

The simple model provides a good insight into the dynamics of such
systems and helps to clarify the discussion of quasiclassical motion
within the framework of a realistic system. This system may absorb many
IR photons through sequential excitation like a one dimensional harmonic oscillator system. The resulting motion of a harmonic oscillator
wavepacket is quasi-classical (16 - 18). It is interesting to consider
the motion of the wavepacket for a realistic, anharmonically coupled,
two dimensional system.

We also briefly review the motion of the highly excited CH-
chromophore in CHF_3 in a nonstationary state of the isolated molecule
without radiation filed.

2. THE MOLECULAR HAMILTONIAN

The molecular Hamiltonian is described by a model Hamilton operator, established to fit the data of 35 bands lying between 2'000 and
17'000 cm^{-1} of CHF_3 (see ref. 11):

$$H = \frac{1}{2} \, v_s \, (p_s^2 + q_s^2) \; + \; \frac{1}{2} \, v_b \, (p_b^2 + q_b^2) \; +$$

$$+ \; K_{sbb} \, q_s q_b^2 \; + \; K_{sss} \, q_s^3 \; + \; K_{ssbb} \, q_s^2 q_b^2 \; + \; \ldots \quad (1)$$

The dots indicate that many more terms have been included into the fit
(see ref. 19). The qs and qb are the dimensionless normal coordinates
of the stretching and the twofold degenerate bending normal modes of
CHF_3 (in eq. (1) qb^2 may stand for $qb_1^2 + qb_2^2$ and $pb^2 = pb_1^2 + pb_2^2$). Due
to the heavy frame CF_3, these are roughly the local stretching and
bending modes along the CH bond length and the CH bending angle.

The cubic coupling term Ksbb·qs·qb^2 is normally large (Ksbb $\simeq$
100 cm^{-1}) and is responsible for the subpicosecond redistribution of
vibrational energy. The other, smaller terms lead to anharmonic corrections and weaker couplings between the two manifolds. The absolute
values of the coefficients in the expansion in eq. (1) are sensitive to
the molecular frame (10 - 15). The appearance of the large cubic coupling term, however, seems to be a general property of the alkyl CH-
chromophore.

We compute the two dimensional Hamilton matrix arising from the
operator of eq. (1) in the basis set of product states of Morse oscil-

lator states (of the pure stretching manifold) and twofold degenerate
harmonic oscillator states (of the pure bending manifold). The matrix
is diagonalized in order to obtain the eigenvalues which are fitted to
the experimentally observable band centers. The eigenstates of the
molecular system are the linear combination of the the basis states.
The detailed evaluation of experimental and of ab initio potential surfaces is discussed elsewhere (19). The important point in the present
context is that we have a *realistic* molecular model Hamiltonian.

Figure 1 shows schematically the energy diagram of the lowest
eigenstates. Only the eigenstates with $l = 0$ are shown (l is the bending angular momentum, which is a good quantum number in our model, due
to the symmetry of the system). The states with $l > 0$ lie between the
multiplets shown on this diagram. They are very weak in the absorption
spectrum and do not contribute to the dynamics of excitation in our
model because of the selection rule $\Delta l = 0$. Thus we neglect them here.

The integers N (N = 0,1,2,...) on the left side of figure 1 label
multiplets of N+1 strongly interacting states. The N-th multiplet,
which we refer to as the N-th *level* of the system, has an energy corresponding to the N-th overtone of the high frequency stretching mode
($\simeq 3'000$ cm^{-1}).

The transition moments were obtained using Mecke's effective CH
dipole moment function evaluated from experimental data (20,21). In
this approach only the states of the pure stretching manifold carry

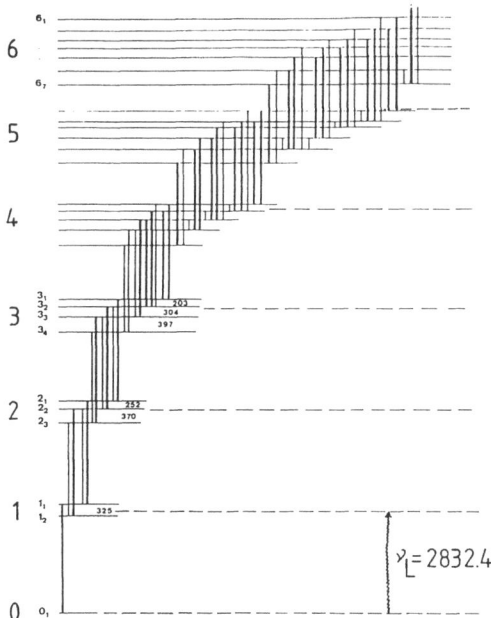

Fig. 1: Schematic energy diagram of the lowest eigenstates. The vertical lines indicate the strongest transition moments between
these states.

much oscillator strength. However, due to the strong intramolecular coupling, many eigenstates will have coupling with the light. The vertical lines in figure 1 represent the strongest transition moments between the eigenstates.

3. THE EQUATIONS OF MOTION

The main results will be observed by inspection of the time evolution of the probability density

$$|\psi(x,t)|^2 = \sum_{i,j} \chi_i^*(x) \cdot \chi_j(x) \cdot P_{ij}(t) \tag{2}.$$

x represents the two CH coordinates (i.e. x = (CH – bond length, CH – bending angle)) and the j-th eigenfunction is a linear combination of the product states basisfunctions:

$$\chi_j(x) = \sum_i c_{ij} \cdot \phi_{v_s(i)}(q_s(x)) \cdot \phi_{v_b(i)}(q_b(x)) \tag{3}.$$

The matrix $\{cij\}$ of eigenvectors is obtained from the fit-procedure (ref. 19) and the $\phi(vs)$ are Morse oscillator wavefunctions, the $\phi(vb)$ are two-fold degenerate harmonic oscillator wavefunctions. As in ref. 19, 180 product states were needed to fit the data up to $17'000$ cm^{-1}. The sums in eq.(2) and eq.(3) and the ranks of the matrices cover all these states.

The time evolution of the system is governed by the density matrix **P**, which obeys the Liouville-von-Neumann equation

$$\mathbf{P}(t) = \mathbf{U}(t) \, \mathbf{P}(t) \, \mathbf{U}^\dagger(t) \tag{4}.$$

The propagator $\mathbf{U}(t)$ is obtained from the direct integration of the time-dependent Schroedinger equation in the field-free case.

When the light is on, the Hamiltonian for the molecule interacting with the light is periodic in time:

$$i\hbar \frac{\partial \psi}{\partial t} = (\mathrm{H_{MOL}} + \mu^z \cdot \epsilon^z \cdot \cos(\omega t)) \, \psi \tag{5},$$

H(MOL) is the molecular Hamiltonian from eq. (1), $\mu(z)$ is the z-component of the electric dipole operator, $\epsilon(z)$ is the linear polarized electric field (in z-direction) and ω is the light frequency. In our model the stretching axis is parallel to the z-axis (no rotation, cf. the next section). Furthermore, possible effects due to the abrupt switching on the light are not considered in this model.

According to Floquet's theorem, one could calculate the propagator for eq. (5) at discrete multipliers of τ, the light period:

$$\mathbf{U}(n \cdot \tau) = \mathbf{U}(\tau)^n \tag{6}.$$

The necessary calculation of $U(\tau)$ is, however, very time consuming. A much faster, though not exact solution is provided by the quasiresonant approximation (9,18,22), which applies very well to the system under investigation here. Indeed, a check of our results with the method of eq. (6) yields deviations of about 5% only for the intensity and frequencies used (see below).

In the quasiresonant approximation the propagator is given by

$$U(t) = \exp(-i \cdot H_{eff} \cdot t / \hbar) \tag{7},$$

where the effective, time independent Hamiltonian $H(eff)$ describes both molecule and field (9,18,22).

4. RESULTS FOR MOLECULAR MULTIPHOTON EXCITATION AND VIBRATIONAL MOTION

The coupled vibrational modes of the CH-chromophore can easily be excited by sequential multiphoton absorption, when light of high intensity up to about 100 TWcm^{-2} is used. When one includes rotation, efficient excitation occurs at much lower intensities, but in the present model we are interested in vibrational dynamics only.

Figure 2 gives a survey of the calculated multiphoton spectrum in the region of the CH fundamental transition at around 3'000 cm^{-1}. Three different intensities are shown. When the intensity is low, well separated peaks are observed, i.e. due to a one-photon absorption at 2'712 cm^{-1} or to a sequential three-photon absorption at 2'930 cm^{-1}.

When the intensity is increased, the isolated peaks disappear. There is an overall absorption with a pronounced maximum at a frequency shifted to the red of the CH fundamental transition. At 100 TWcm^{-2} substantial populations up to the 7th multiplet are obtained, quite independently of the exact frequency of the laser (the multiphoton absorp-

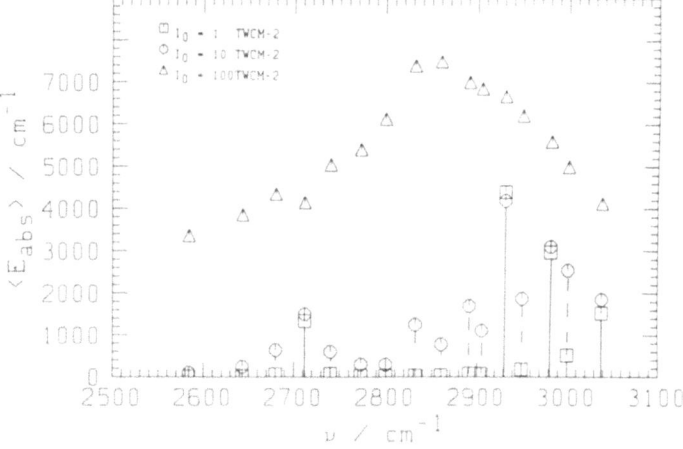

Fig. 2: Survey of the multiphoton absorption spectrum in the region of the CH fundamental transition (calculated). Three intensities are shown; vertical lines help visualizing the peaks.

tion spectrum has a pronounced maximum but is broad) in the range between 2'700 and 3'000 cm^{-1}. In the remainder of this paper the wavenumber and intensity of the laser will be fixed to 2'832.4 cm^{-1} and 100 $TWcm^{-2}$, respectively.

At time zero the CH-chromophore is in its ground vibrational state. When light is switched on, the population of the ground state (the zeroth level) decreases very rapidly. Figure 3 shows the time evolution of the populations of the lowest levels (summed over the individual states in each level).

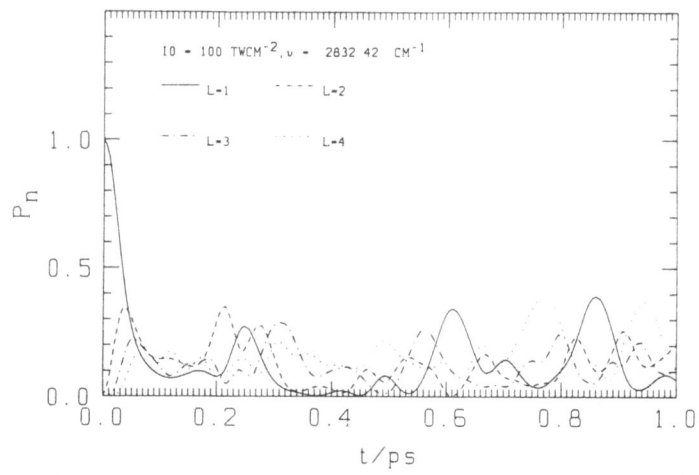

Fig. 3: Time evolution of level populations. L = 1 means the level with
        N = 0, L = 2 means N = 1 and so forth. Only the evolution of
        the four lowest levels is shown here.

The populations of each of the higher levels do not increase very much then, but the distribution becomes soon very wide.

Due to the special coupling scheme of our model, we expect the motion of the wavepacket to be confined to the stretching axis, at least during the initial part of the dynamics. It depends somewhat on the ratio of intramolecular coupling and the coupling with the light how long it will take to excite higher bending states as well. Figure 3 shows this very neatly: in the first 50 fs of irradiation, the populations of the higher levels increase roughly like a one dimensional harmonic oscillator system (see, for example ref 18, 23). Later, the evolution becomes irregular and, in great contrast to the harmonic oscillator case, there is no recurrence of the initial situation.

We now follow the motion of the wavepacket in coordinate space. Figure 4a shows a snapshot of the probability density in CH-coordinate space at time zero, which is the absolute square of the wavefunction of the CH-chromophore, the eigenfunction with lowest energy in this case. In this figure the wavepacket is represented by its contour lines (see also the description of the figure). The horizontal axis is the CH-bond length (in pm units here; this is mainly the stretching manifold), the

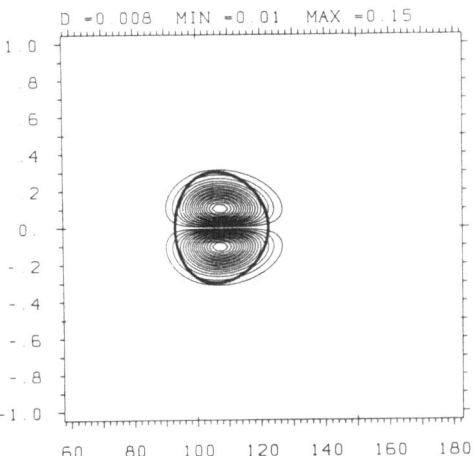

Fig. 4a: Snapshot of the probability density in CH–coordinate space of CHF$_3$ at time zero (groundstate wavepacket; contour line representation). The horizontal axis is the CH–bond length (in pm), the vertical axis is the CH–bending angle (in radians). MIN means the lowest value, MAX the highest value of contour lines drawn and D the difference between two contours. The units of the density values correspond to a total density of 1. The dark line corresponds to the potential energy contour at the expectation value of the molecular energy.

vertical axis is the CH–bending angle (in radians; this is roughly the bending manifold). The symmetry of the system is shown explicitly by allowing the CH–bending angle to take negative values as well.

   In the case of CHF$_3$, this wavepacket is essentially the probability density of finding the proton at some point in the CH–coordinate space. At time zero, it is concentrated in its equilibrium position at about 109 pm. The contour line of the potential surface at the energy of the wavepacket is drawn as well (the ground state energy in this case). This line marks the classically accessible region in CH–coordinate space. Strictly speaking the coordinates in figure 4a are best described as reduced dimensionless normal coordinates (19), but for visualization internal coordinates are more practical.

   Figure 4b shows the wavepacket at 50 fs, at a time at which some higher levels have already been excited.This has still roughly the shape of the ground state wavepacket, but the maximum of probability is shifted towards the smaller values of the stretching axis. This is nearly the coherent motion of a harmonic oscillator wavepacket in the streching direction during excitation with monochromatic, coherent light (16,18), which is known to be a coherent state (17). As already some energy has been absorbed, the classically allowed region in coordinate space is larger now compared to the ground state. Due to the wide distribution of states, the energy of the wavepacket has a standard deviation indicated by the dashed lines.

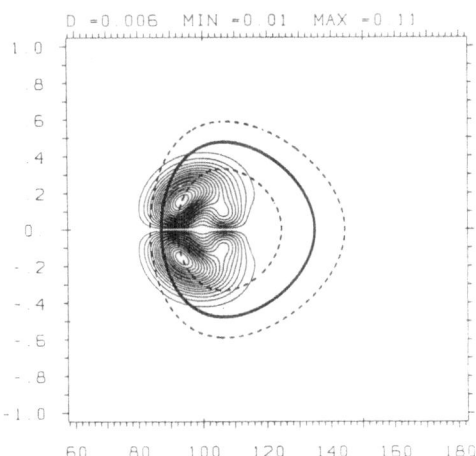

Fig. 4b: Snapshot of the wavepacket at 50 fs; see description of fig 4a

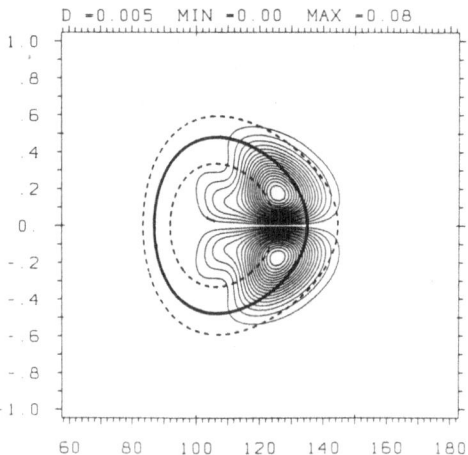

Fig. 4c: Snapshot of the wavepacket at 54 fs; see description of fig 4a

Figure 4c shows the wavepacket at 54 fs: it has now reached the opposite end of the classically allowed region along the stretching direction, without much changing its shape. This is the smooth end of the potential surface, which allows for an easier penetration by tunneling.

In the course of the next 6 fs the wavepacket will oscillate back. However, due to the anharmonicities of the system, there will be, in addition to the shift of the maximum, a gathering and concentration of probability. At 60 fs the wavepacket will reach the sharp end of the potential surface at the small values of the stretching axis; then it will dilate again to obtain its former shape and proceed with the oscillation along the stretching manifold towards the larger values (we do not show the corresponding figures here). This quasiperiodic motion has a period nearly equal to the light period (the classical, forced harmonic oscillator does also perform a periodical motion with the period of the driving field, when resonance conditions are established). The quasicoherence of the motion is a consequence of the fact, that intramolecular couplings have not yet been effective.

Now, for later times the intramolecular redistribution becomes dominant. This feature, occuring also at other frequencies of irradiation, is shown in figure 4d. This figure shows a snapshot of the wavepacket at 90 fs in a three-dimensional representation: the front axis is the CH-bending angle, the rear axis is the bond length. The wavepacket has become dilated and spread out over coordinate space.

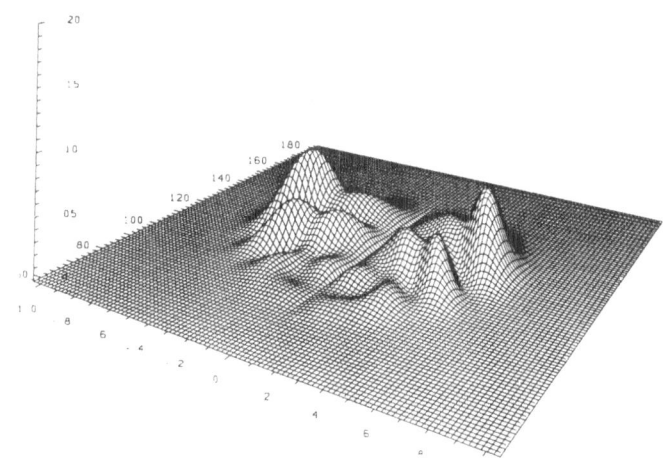

Fig. 4d: Snapshot of the wavepacket at 90 fs (three-dim. representation); the front axis is the CH-bending angle, the rear axis is CH-bond length; the third axis is the probability density; the units are as in fig 4a.

In the course of the next 4 to 6 fs we again observe a gathering of the wavepacket (see figure 4e, showing the wavepacket at 94 fs). Thereafter, it will aspread out and gather again in a quasiperiodic way. This alternating motion has now replaced the initial quasiperiodic, quasicoherent motion.

At later times, when a maximum of absorption has been reached, the wavepacket covers nearly all of the energetically allowed region in coordinate space. Figure 4f shows the wavepacket at 392 fs. The CH-chromophore is highly excited now (about 12'000 cm^{-1} mean energy, which corresponds roughly to four absorbed photons), which can be seen from the large diameter of the potential surface.

If the light is switched off now, the CH-chromophore state will be a superposition of many eigenstates and the wavepacket will still be

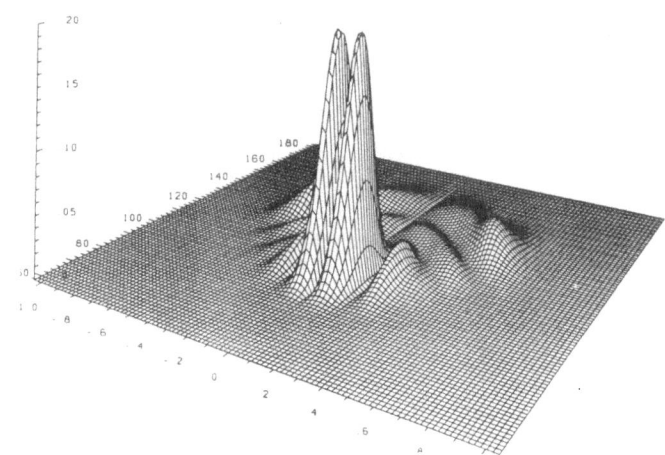

Fig. 4e: Snapshot of the wavepacket at 94 fs; see description of fig 4d

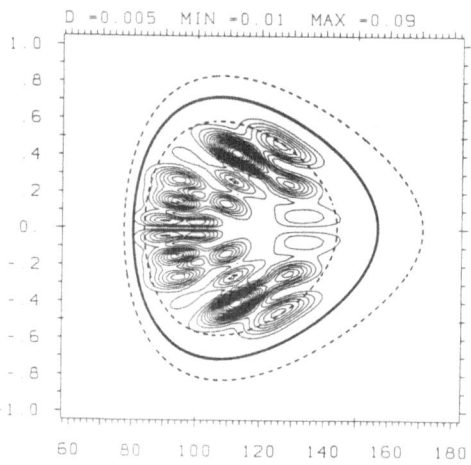

Fig. 4f: Snapshot of the wavepacket at 392 fs; see also fig 4a.

time dependent. Its evolution will exhibit an alternating gathering and spreading-out motion.

We assume now that a pure stretching state has hypothetically been excited by a strong laser pulse, for instance the $|v(s)=6,v(b)=0 >$ state, which is a linear combination of mainly the seven eigenstates in the 7th multiplet on figure 1. The laser pulse is over (the light is switched off) and our clock is reset to zero. This highly localized state is shown in figure 5a.

The wavepacket now delocalizes on a time scale shorter than 1 ps. At 216 fs the wavepacket is completely spread out (figure 5b). Apart

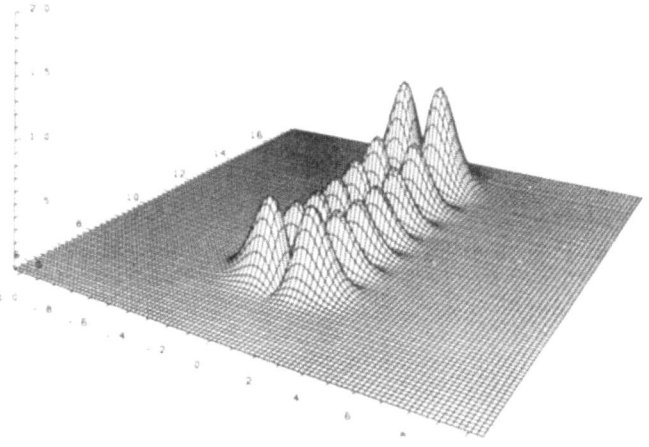

Fig. 5a: Wavepacket of the pure stretching state in the 7th multiplet.
The laser is switched off now.

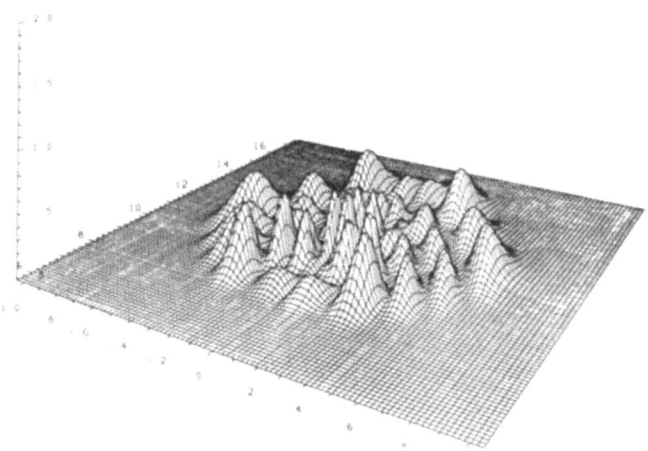

Fig. 5b: After 216 fs the wavepacket of fig 5a delocalizes; see fig 5a

from some recurrence phenomena, the wavepacket remains delocalized for
most times (4).
    On these time scales only couplings between the states of the same
multiplet are important (4). In that sense, when the light is switched
off, the states of one multiplet form a closed system of nearly uniform
energy. And, as it turns out, the delocalized probability distribution
tends to be similar to a microcanonical equilibrium destribution.
    Figure 5c shows the wavepacket of a typical member of the micro-

canonical ensemble of the 7th multiplet. This state is a linear combination of the 7 states in this multiplet with equal weights for each state but random phases. This "statistical" wavepacket covers nearly all of the energetically accessible region. It looks similar to the result of Figure 5b. Figure 5d shows the actual microcanonical distribution which is the time and phase average over densities of the type shown in Figure 5c.

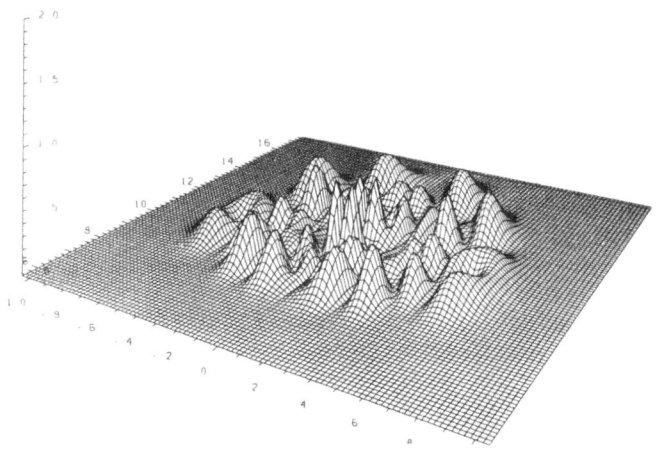

Fig. 5c: The wavepacket of a typical member of the microcanonical ensemble of the 7th multiplet; see also fig 5b

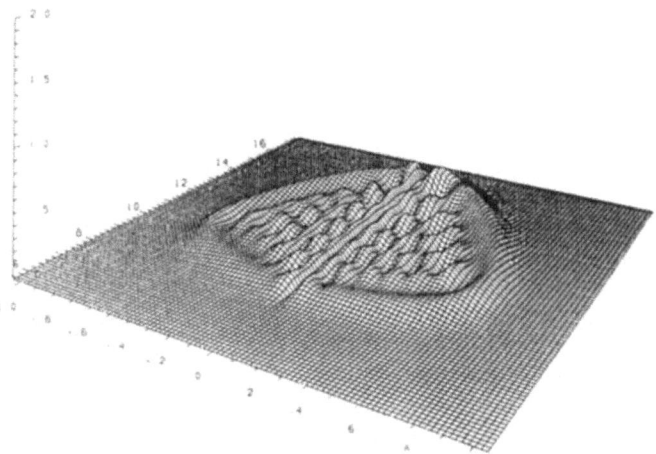

Fig. 5d: The microcanonical distribution in CH-coordinate space of the 7th multiplet; see also fig 5b

5. CONCLUSIONS

Detailed quantum dynamical calculations are feasible for large finite systems such as polyatomic molecules or subsystems in these, which are practically isolated on very short time scales.

The alkyl CH-chromophore, an important building block of many polyatomic molecules, has localized vibrational modes which can be treated separately from the molecular frame by effective Hamiltonians during time intervals of shorter than 10 ps (in general). Within this time interval a very efficient, ultrafast redistribution of energy between the localized vibrational modes occurs.

The IR excitation of the CH-chromophore at high intensities is a very good tool to understand the role of this vibrational redistribution. The sequential multiphoton absorption, induced at high intensities, initially leads to a quasi-classical behaviour of the wavepacket. This can be observed by following the motion of the probability density in CH-coordinate space. After typically 50 to 100 fs, however, the motion can no longer be described by a quasi-classical Gaussian wavepacket. This is the time scale for efficient intramolecular vibrational redistribution. Then, the wavepacket becomes spread out and covers nearly all of the classically accessible region in coordinate space, just as a microcanonical equilibrium distribution would do.

Acknowledgement: The hospitality of the Israel Academy of Sciences and Humanities the Fondation Edmond de Rothschild are greatfully acknowledged. We enjoyed helpful discussions with M. Lewerenz. Our work is supported financially by the Schweizerischer Nationalfonds and the Schweizerischer Schulrat.

REFERENCES

1) G. Herzberg, *Molecular Spectra and Molecular Structure* (van Nostrand New York, 1945, 1966), vols. II and III
2) Many inportant contributions in the present volume of the Proceedings of the 20th Jerusalem Symposium.
3) Quack, M. : *Faraday Discuss. Chem. Soc.* **71**, 359 (1981)
4) Marquardt, R., Quack, M., Stohner, J. and Sutcliffe, E. : *J. Chem. Soc. Faraday Trans. 2* **82**, 1173 (1986)
5) The survey in *Faraday Discuss. Chem. Soc.* **75** (1983)
6) S.A. Rice, in *Excited States*, edited by E.C. Lim (Academic, New York, 1975)
7) Quack, M. and Troe, J. : *Int. Rev. Phys. Chem.* **1**, 97 (1981)
8) M. Quack, in *Energy Storage and Redistribution in Molecules*, edited by J. Hinze (Plenum, New York, 1983), p. 493
9) Quack, M. : *Advan. Chem. Phys.* **50**, 395 (1982)
10) Dübal, H.R. and Quack, M. : *J. Chem. Phys.* **81**, 3779 (1984)
11) Segall, J. , Zare, R.N., Dübal, H.R., Lewerenz, M. and Quack, M. : *J. Chem. Phys.* **86**, 634 (1987)
12) Peyerimhoff, S., Lewerenz, M. and Quack, M. : *Chem. Phys. Lett.* **109**, 563 (1984)
13) Baggott, J.E., Chuang,M.C., Zare, R.N., Dübal, H.R. and Quack, M. : *J. Chem. Phys.* **82**, 1186 (1985)

14) Dübal, H.R. and Quack, M. : *Mol. Phys.* **53**, 257 (1984)
15) Amrein, A., Dübal, H.R. and Quack, M. : *Mol. Phys.* **56**, 727 (1985)
16) Kerner, E. : *Can. J. Phys.* **36**, 371 (1958)
17) Glauber, R. : *Phys. Lett.* **21**, 650 (1966)
18) Quack, M. and Suttcliffe, E. : *Infrared Phys.* **25**, 163 (1985)
19) Lewerenz, M. and Quack, M. (in preparation)
20) Mecke, R. : Z. *Elektrochemie* **54**, 38 (1950)
21) Lewerenz, M. and Quack, M. : *Chem. Phys. Lett.* **123**, 197 (1986)
22) Quack, M. and Suttcliffe, E. : *J. Chem. Phys.* **83**, 3805 (1985)
23) Marquardt, R. and Quack, M., to be published

SPECTRAL FLUCTUATIONS IN FINITE SYSTEMS

R.D. Levine
The Fritz Haber Research Center for Molecular Dynamics
The Hebrew University, Jerusalem 91904, Israel

ABSTRACT. The intensities in a discrete spectrum will often vary in an irregular manner, particularly so at higher levels of excitation. It was recently shown possible to discuss these variations as fluctuations with respect to a smooth spectral envelope. Here this conclusion is derived starting with the thermodynamic theory of fluctuations. The small number of degrees of freedom does however lead to a non-Gaussian distribution of fluctuations.

1. INTRODUCTION

Finite systems will exhibit finite fluctuations in the values of their thermodynamic variables [1,2]. Our interest here is in systems which are not in thermal equilibrium. Specifically we want to consider spectral intensities i.e., the distribution of states following light absorption. For some time we have argued [3,4] that a thermodynamic-like description is possible and useful also for finite mechanical systems. Recently this point of view has been successfully applied to spectral fluctuations [5-7] and to fluctuations in decay rates of individual quantum state [7,8]. An important parameter which emerges from such an analysis has been called [5-8] 'the number of degrees of freedom'. The origin of the name is in mathematical statistics [9]. The purpose of this discussion is to show that this parameter does have the physical significance of the number of degrees of freedom in the sense of the thermodynamic theory of fluctuations i.e., in a mechanical sense. A perhaps unexpected result is that for the strongly coupled systems of interest here this number can be (and is) smaller than the number of degrees of freedom if the Hamiltonian of the system was separable.

For large finite systems or for smaller systems at higher excitation energies the optical spectrum has a complex and seemingly erratic structure. The traditional spectroscopic procedures are then no long useful. Yet the high resolution made possible using lasers [10] provides well defined spectra, often to individual final quantum states. Considerable attention has been given to the distribution of level spacings in such spectra. (See [11-13] for recent studies). The approach discussed here is complementary in that it

45

J. Jortner et al. (eds.), Large Finite Systems, 45–52.

considers the distribution of line intensities irrespective of the line positions.

2. FLUCTUATIONS

We begin with the central result of the thermodynamic theory of fluctuations and then apply it to the problem of spectral intensities. The resulting distribution of intensities turns out to be identical to that previously derived [7]. The advantage of the present route is that the significance of the 'number of degrees of freedom' becomes very evident.

We consider a system characterized by the mean values of R variables, $<A_r>$, $r = 1,...,R$. The most probable state of the system is one of maximal entropy subject to these R mean values and to normalization $<A_0> = 1$, $A_0 = I$. The procedure of seeking the maximum of the entropy subject to $R + 1$ constraints provides $R + 1$ Lagrange multiplier λ_r, $r = 0,1,..,R$ 'conjugate' to the constraints. The values of the $R + 1$ multipliers can be determined from the $R + 1$ expectation values. In particular, imposing the normalization, $r = 0$, constraint $<A_0> = 1$ makes λ_0 a function of the other R Lagrange multipliers which are then determined as the solutions of the implicit equation [14,15]

$$<A_r> = -\partial \lambda_0(\lambda_1,..,\lambda_R)/\partial \lambda_r . \tag{2.1}$$

The central result of thermodynamic fluctuation theory is that the probability of finding the system in a microstate where the $R + 1$ constraints have the values $a_0, ..., a_R$ is *

$$P = \exp(S - \sum_{r=0}^{R} \lambda_r a_r) \tag{2.2}$$

where S is the entropy and, to conserve normalization $a_0 = 1$. The most probable value of a_r is, from (2.2), the implicit solution of

$$\lambda_r = \partial S/\partial a_r . \tag{2.3}$$

One can show [15] that the most probable value of a_r is indeed the mean. $<A_r>$. Expanding the entropy up to second order

$$S(\{a_r\}) = S(\{<A_r>\}) \tag{2.4}$$
$$+ \sum_r (a_r - <A_r>)(\partial S/\partial a_r)_{<A_r>}$$

---

*One can show [15] that everywhere

$$S - \sum_{r=0}^{R} \lambda_r a_r \leq 0$$

with the maximum being attained if and only if $a_r = <A_r>$.

$$+ \sum_{r,s}(a_r - <A_r>)(a_s - <A_s>)(\partial^2 S/\partial a_r \partial a_s)_{<A_r>,<A_s>}$$

we have, using (2.3) that

$$P \propto \exp[-\sum_{r,s,}(a_r - <A_r>)g_{rs}(a_s - <A_s>)] . \qquad (2.5)$$

which is the Einstein Gaussian approximation where g_{rs}

$$g_{rs} = -(\partial^2 S/\partial a_r \partial a_s)_{<A_r>,<A_s>} \qquad (2.6)$$

is a positive definite (metric tensor [16]), since the entropy has its maximum where $a_r = <A_r>$.

For most thermodynamic needs the Einstein approximation is perfectly adequate. The reason is that in thermodynamics proper one is seldom interested in computing moments beyond the second. The Einstein approximation gives the same results as the exact form (2.2) for moments up to third order. As we shall see, the spectral fluctuations are quite extensive. There can be large excursions about the mean hence the expansion (2.4), which is only up to second order, is not sufficient. Instead of giving up we shall use the exact result (2.2) for the probability of a microstate.

The dependence on the number of degrees of freedom is implicit in that the entropy and the mean values $<A_r>$ are, in the thermodynamic approach, extensive variables. Hence so is the exponent in (2.2), which we remind the reader, is non-positive definite. Large fluctuations are disfavored, the more so the larger the system. For our purpose however it proves useful to explicitly evaluate the fluctuations. The result is

$$(2.7)$$

$$<(a_r - <A_r>)(a_s - <A_s>)> = -\partial<A_r>/\partial\lambda_s = \partial^2\lambda_0/\partial\lambda_r \partial\lambda_s .$$

The proof is well known (see, e.g., [14,15]).

3. SPECTRUM

For a well defined initial state $|i>$, the fully resolved spectrum for the transition operator T is given by

$$S(E) = \sum_f |<f|T|i>|^2 \delta(E - E_f) . \qquad (3.1)$$

The area under the spectrum is

$$\int dE S(E) = \sum_f |<f|T|i>|^2 = <i|T^+T|i> \qquad (3.2)$$

and is unchanged even if the final energy resolution is imperfect so that the delta function in (3.1) is replaced by some window $A(E-E_f)$ of finite width (as long as the area under each window remains unity). One can therefore characterize the distribution of intensities by the probabilities

$$p_f = |<f|T|i>|^2/<i|T^{\dagger}T|i> . \tag{3.3}$$

A finite energy resolution implies a more uniform spectrum. The probabilities for the lower resolution are p_j^o

$$
\begin{aligned}
p_j^o &= \int dE \sum_f p_f A(E-E_f) \delta(E_j-E) \\
&= \sum_f p_f A(E_j-E_f) .
\end{aligned}
\tag{3.4}
$$

The smoother spectrum is thus an average over the actual spectrum. If we compute the spectrum by working in the time domain then very long integration times are necessary to obtain the well resolved spectrum $S(E)$. For finite energy resolution only shorter integration times are required. The smooth spectrum reflects the short time dynamics [17]. Hence different fully resolved spectra will share the same common smooth envelope.

Equilibrium is always defined with respect to a time scale. For molecular chemical lasers in the infrared there is, on the time scale of interest, rotational (but not vibrational) equilibrium [18]. Many mixtures are thermal (but not in chemical) equilibrium. Even at 'complete' (both thermal and chemical) equilibrium out systems are usually not in equilibrium with respect to nuclear reactions. We assume that a similar separation of time scales can be introduced for the intramolecular dynamics of the isolated molecule. We expect the 'rapid' dynamics to be over on a time scale which is of the order $h/D(E)$, where $D(E) = 1/\rho(E)$ is the mean level spacing (or $\rho(E)$ is the density of states) at the energy range of interest. We take the smooth spectrum to be the spectrum after this initial dephasing is over. The p_j^o's are then our 'equilibrium' spectral distribution.

Any particular well resolved spectrum represents a possible deviation from 'equilibrium'. The probability of the spectrum is given by (2.2) or, in the present notation

$$P(\mathbf{p}) = \exp[S(\mathbf{p}) - \sum_j \mu_j p_j] \tag{3.5}$$

where

$$<p_j> = p_j^o \tag{3.6}$$

and $\mathbf{p}$ denotes the set of normalized intensities given by (3.3). The μ_j's are the Lagrange multipliers, determined as in (2.1) by the values of the smooth intensities $\mathbf{p}^o$.

The distribution of spectral fluctuations given by (3.5) is exactly the one previously derived (equations (3.17) and (3.25) of [7]) directly using the procedure of maximal entropy. To obtain explicit results we need the dependence of the entropy $S(\mathbf{p})$ on the normalized intensities. It is here that the statistical 'number of degrees of freedom' denoted by ν comes in. Say the probabilities p_j can vary in ν independent directions[*]. Then, $\exp(S(\mathbf{p}))$ being the volume of the region sampled by the p_j's

$$S(\mathbf{p}) = \ln\Pi_j p_j^{\frac{\nu}{2}-1} = (\frac{\nu}{2}-1)\sum_j \ln p_j .$$

(3.7)

Then

$$p(\mathbf{p}) \propto \exp[\sum_j ((\frac{\nu}{2}-1)\ln p_j - \mu_j p_j)] = \Pi_j P(p_j)$$

(3.8)

with

$$P(p_j) = (\mu_j^{\frac{\nu}{2}}/(\frac{\nu}{2}-1)!)\exp[(\frac{\nu}{2}-1)\ln p_j - \mu_j p_j]$$

(3.9)

where the factor in front of the exponent is due to normalization. The value of the Lagrange multiplier μ_j is now determined by the implicit equation

$$<p_j> = p_j^o = \int p_j P(p_j) dp_j$$

(3.10)

or

$$\mu_j = \nu/2<p_j> .$$

(3.11)

Hence the final, explicit form is

$$P(p_j) = p_j^{\frac{\nu}{2}-1} \exp(-\nu p_j/2<p_j>)/ \left(\frac{2<p_j>}{\nu} \right)^{\frac{\nu}{2}} \Gamma(\frac{\nu}{2})$$

(3.12)

known in statistics as a χ-square distribution with ν degrees of freedom.

According to (3.8), the lines fluctuate independently of one another:

$$<(p_j-<p_j>)(p_k-<p_k>)> = \delta_{j,k} 2<p_j>^2/\nu .$$

(3.13)

The variance of the fluctuation in units of $<p_j>$ is $2/\nu$ just as for the thermodynamic

---

[*]Which would be the case if, e.g., the p_j's were a linear combination of the squares of ν independent random variables.

theory when ν is the number of degrees of freedom. Indeed, also from (3.9) we see that the exponent of $P(\mathbf{p})$ is proportional to ν, again as expected. The larger is ν, the more centered is the distribution about the most probable value. However, because for realistic molecules ν can be quite small, fairly large fluctuations are possible and hence the Gaussian approximation (2.5) is typically not accurate enough. In the limit when ν is large (so that $<(p_j - p_j^o)^2> \ll p_j^{o\,2}$), one readily verifies that (3.12) can be approximated as a Gaussian

$$P(p_j) = (4\pi p_j^{o\,2}/\nu)^{-1/2}\exp\left(-\frac{\nu}{4p_j^{o\,2}}(p_j - p_j^o)^2\right) \tag{3.14}$$

and the width (cf. (3.13)) $\sigma^2 = 2p_j^{o\,2}/\nu$. Figure 1 shows the distribution (3.12) *vs.* $p_j/<p_j>$ at increasing values of ν.

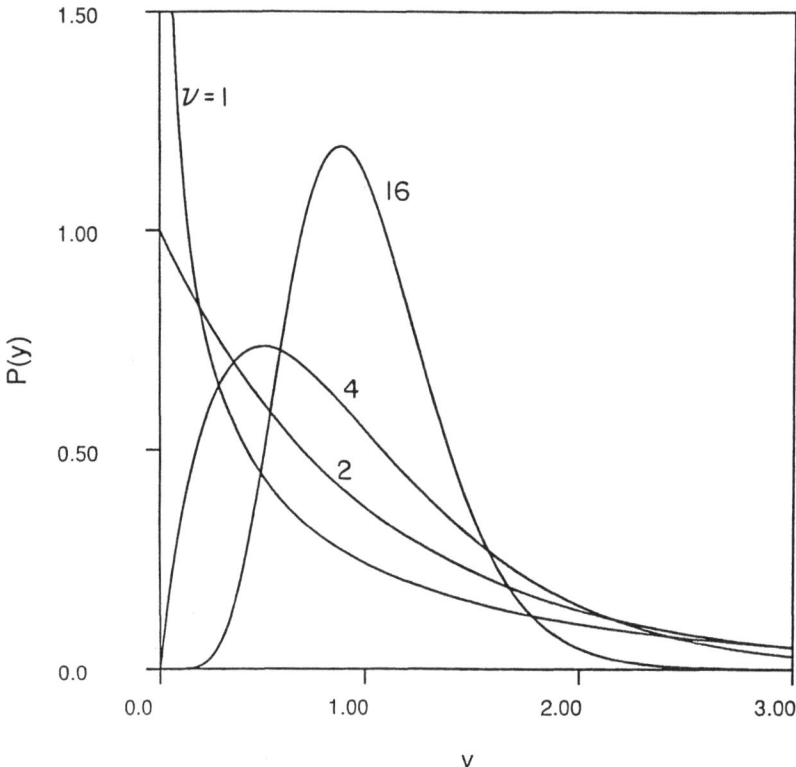

Figure 1. The distribution (3.12) plotted *vs.* $y = p_j/<p_j>$. At low ν values it is quite skewed but becomes more Gaussian-like as ν increases.

4. THE ONSET OF CHAOS

In the limit when the underlying classical dynamics is fully chaotic one expects that $\nu \to 1$ [5,7]. The distribution (3.12) is then known as the Thomas-Porter form and has been derived from random matrix theory [19,20]. That the discrete spectrum of a classically chaotic system reflects just one degree of freedom is not unreasonable. Those degrees of freedom which act independently for the low lying states are, at higher excitations, so strongly coupled that one cannot confine the motion to any subspace of the entire phasespace. At lower levels of excitation the situation is more complex and depends also on the initial state. To simplify, consider the spectrum averaged over all initial states in a narrow energy interval. [7]. Such an averaged spectrum will have a higher value of ν reflecting the effective number of independent degrees of freedom. Restricted intramolecular vibrational energy redistribution can therefore be discerned by a higher value of ν. The practical determination of ν from a spectra, the error bars on ν and the construction of the spectral envelope are discussed in [6,7].

Acknowledgment

Our work on spectral fluctuations is supported by the U.S. Air Force Office of Scientific Research under Grant AFOSR 86-0011 and by the Stiftung Volkswagenwerk. The Fritz Haber Research Center is supported by the Minerva Gesellschaft für die Forschung, mbH, München. BRD.

References

1. D. Landau and E.M. Lifshitz, *Statistical Physics*. Pergamon Press. Oxford 1980.

2. H.B. Callen, *Thermodynamics*, Wiley, New York 1960.

3. R.D. Levine. Adv. Chem. Phys. **47**, 239 (1981). R.D. Levine *in* Theory of Reactive Collisions. M. Baer. ed., CRC Press, Florida 1986.

4. The first ever review of this point of view has been published in the proceedings of the VIth Jerusalem symposium: R.D. Levine *in* Chemical and Biochemical Reactivity, E.D. Bergmann and B. Pullman, eds. Reidel, Dordrecht 1974.

5. Y. Alhassid and R.D. Levine, Phys. Rev. Lett. **57**, 2879 (1986).

6. J. Brickmann, Y.M. Engel and R.D. Levine, Chem. Phys. Lett., in press.

7. R.D. Levine, Adv. Chem. Phys. **70** (1987).

8. Y.M. Engel, R.D. Levine, J.W. Thoman, J.I. Steinfeld and R. McKay, J. Chem. Phys., in press.

9. M.G. Kendall and A. Stuart. *The Advanced Theory of Statistics*. Hafner. New York 1958.

10. See the contributions to this volume e.g., by J. Kommandeur, A. Amirav. E. Riedle and A. Trammer.

11. C.E. Hamilton, J.L. Kinsey and R.W. Field, Ann. Rev. Phys. Chem. **37**, 493 (1986).

12.   L. Leviander, M. Lombardi, R. Jost and J.P. Pique, Phys. Rev. Lett. **56**, 2449 (1986).

13.   J.P. Pique, Y. Chen, R.W. Field and J.L. Kinsey, Phys. Rev. Lett. **58**, 475 (1987).

14.   E.T. Jaynes *in* Statistical Physics (Brandeis Lectures 1962) Benjamin 1963.

15.   R.D. Levine *in* Maximum Entropy and Bayesian Methods in Applied Statistics, J.H. Justice, ed., Cambridge University Press 1986.

16.   R.D. Levine, J. Chem. Phys. **84**, 910 (1986).

17.   E.J. Heller, J. Chem. Phys. **72**, 1337 (1980).

18.   A. Ben-Shaul, Y. Haas, K.L. Kompa and R.D. Levine, *Lasers and Chemical Change*, Springer Verlag, Berlin 1981.

19.   C.E. Porter, *Statistical Theory of Spectra: Fluctuations* Academic Press, N.Y. 1965.

20.   T.A. Brody, J. Flores, J.B. French, P.A. Mello, A. Pandey and S.S.M. Wong, Rev. Mod. Phys. **53**, 385 (1981).

Time Dependent Quantum Mechanical Calculations of the Dissociation Dynamics of the Cluster He_n-I_2.

R. Kosloff and A. Hammerich

Department of Physical Chemistry and The Fritz Haber Research Center for Molecular Dynamics, The Hebrew University, Jerusalem 91904, Israel.

M. A. Ratner

Department of Chemistry and Materials Research Center, Northwestern University, Evanston, Illinois 60201 USA.

Abstract

A time dependent quantum mechanical framework is used to examine the dissociation dynamics of van der Waals clusters, in particular the He_n-I_2 system. The time dependent approach exploits the time scale separation between the He motion and the I_2 vibration. The formalism used is the Time Dependent Self Consistent Field (TDSCF). In this picture, in which the He degrees of freedom are moving in the average field of the I_2 molecule and vice versa, the equations of motion are solved by the Fourier grid method which calculates the operation of the operators constituting the Hamiltonian locally. The result is a very fast convergence with respect to grid size. The TDSCF approximation is tested for the collinear $He-I_2$ system by comparing to an exact time dependent propagation. Good results were obtained for low vibrational excitation of the I_2 bond. For higher excitations the TDSCF approximation could not account for the fast dephasing part of the autocorrelation function, nevertheless the long time behavior responsible for the dissociation was represented well. The TDSCF approach was then applied to calculate the dissociation of T-shaped and X-shaped He_n-I_2 clusters. The basis of this approximation is the weak interaction between the He atoms, and the extra averaging due to increase in the number of particles. Results show very small dependence of dissociation rate on cluster size in contrast to an RRKM picture. The symmetry of the He wavefunction to exchange is investigated. A scheme to incorporate part of the correlations responsible for collective motion which are missing in the simple TDSCF approach is presented. This scheme is based on a projection operator approach and the time dependent variational principle. On the basis of symmetry it is predicted that the dissociation rate of a cluster consisting of He_3 will be faster than a cluster of He_4.

J. Jortner et al. (eds.), Large Finite Systems, 53–66.

I. Introduction

Dissociation of van der Waals clusters has received considerable attention both experimentally and theoretically. The introduction of the seeded supersonic beam, in combination with new laser spectroscopic techniques, enabled the detailed study of the dissociation dynamics of these clusters [1]. These studies of a well-defined isolated system inspired considerable theoretical work [1][2]. The "text book" cluster of He_n-I_2, which was also one of the first to be produced, is the subject of this study. While this system has been studied both classically [2] and quantum mechanically [3], nevertheless it is still a source of fundamental questions on the nature of these van der Waals clusters.

In this work a time dependent quantum mechanical approach is used to elucidate the following problems:
1) The relation between the cluster size and dissociation rate.
2) The role of quantum effects at the low temperatures of the experiment and in particular the role played by the symmetry to particle exchange of the He_n part of the wavefunction.
3) Classical studies [4][5] have found that the energy transfer is dominated by bottlenecks. How does this behavior manifest itself in a quantum regime?

The study of van der Waals clusters through a time dependent picture is appealing because of the ability to interpret the dynamical process while following the evolution. Moreover the time dependent treatment is directly comparable to classical trajectory results which allows an easy identification of quantum effects. Numerical time dependent techniques have advanced considerably recently, enabling exact quantum mechanical detailed studies. The time dependent quantum mechanical method used in this work is the Fourier method [6][7]. This method has fast convergence properties and is very flexible. Numerical efficiency is obtained because of the Fast Fourier Transform (FFT) algorithm which is well suited for parallel and vector computers.

The time dependent approach is also the source of a dynamical mean field approximation. The Fourier method allows a simple direct implementation of a Time Dependent Self Consistent Field (TDSCF) formulation [8]. The time dependent approach gives a superior mean field approximation to the static SCF due to the ability to include as a time correlation many of the omitted static correlations. This approximation is found to be accurate in the study of van der Waals clusters because of the natural time scale separation between the molecular degrees of freedom and the surrounding rare gas atoms. It is also expected that the quality of the approximation should increase with the cluster size due to the extra averaging with the increase in the number of cluster atoms. Classical and semiclassical TDSCF studies of similar systems have been tried previously [9][10], but the validity of these approximations could not be tested directly, therefore a full quantum mechanical time dependent study of the collinear dissociation of the He-I_2 molecule has been carried out. In that study a direct comparison of the exact and TDSCF wavefunctions was done [11]. That work proves the validity of the TDSCF approximation for treating wan der Waals clusters. This study is directed at understanding the behavior of He clusters at low temperature where a quantum picture is necessary. In particular we want to shed light on the three fundamental questions posed previously.

This paper is divided as follows. Section II describes the use of the Fourier method and the development of the TDSCF approximation for collinear He -I_2. Section III describes the dissociation dynamics of large T-shaped and X-shaped clusters. Section IV develops a systematic procedure to include correlations omitted in the simple TDSCF approach. Section V summarizes the results.

II. The collinear He–I$_2$ as a test of the validity of the TDSCF approximation.

The dissociation of the collinear He–I$_2$ system is used to demonstrate the quality of the time dependent self consistent field approximation for van der Waals clusters. This is done by comparing an exact solution of the time dependent Schrödinger equation for the collinear system to the approximate TDSCF approach.

The starting point of this comparison is the two-dimensional Schrödinger equation in atomic units:

$$i\frac{\partial \psi}{\partial t} = \hat{H}\psi \tag{2.1}$$

subject to the initial condition in which the vibrational bond of I$_2$ is excited. The Hamiltonian can be written

$$\hat{H} = \frac{1}{2}\left(\hat{P}_{r_1}^2 + \hat{P}_{r_2}^2\right) + \hat{V}(r_1, r_2) \tag{2.2}$$

using a mass scaled coordinate system for which: $r_1 = aR_{\text{He}-\text{I}_2}$ and $r_2 = bR_{\text{I}-\text{I}}$, where:

$$a = \left(\frac{m_1(m_2 + m_3)}{(m_1 + m_2 + m_3)}\right)^{\frac{1}{2}} \qquad b = \left(\frac{m_3(m_1 + m_2)}{(m_1 + m_2 + m_3)}\right)^{\frac{1}{2}} \tag{2.3}$$

$m_1 = m_{\text{He}}$ and $m_2 = m_3 = m_{\text{I}}$.

Examining equation (2.3) one can identify the r_1 coordinate as predominantly a He motion and r_2 as a relative I$_2$ motion.

A direct numerical solution of the Schrödinger equation is based on a discretization scheme in which all operators constituting the Hamiltonian operators are calculated locally. This is done by representing the wavefunction on a grid. In configuration space the potential operator is local and therefore its operation is just a multiplication of the value of the wavefunction at a grid point with the value of the potential at that point. The kinetic energy operator is calculated locally in momentum space. The transformation from coordinate space to momentum space is done by a discrete Fast Fourier Transform (FFT). As a result the classical phase space is discretized by rectangular cells of area h . The time propagation operator is expanded in a Chebychev polynomial:

$$\hat{U}(t) = e^{-i\hat{H}t} = \sum_n^N a_n \phi_n(-i\hat{H}t/R) \tag{2.4}$$

where ϕ_n is the complex Chebychev polynomial which is calculated by its recursion relation, a_n are expansion coefficients, and R is the range of eigenvalues of $\hat{H}$ represented on the grid multiplied by t. The details of the method can be found elsewhere [12]. The initial wavefunction was chosen as a product form in the r_1 and r_2 coordinates where, for the He degree of freedom, a Morse initial wavefunction was chosen and for the I$_2$ coordinate the eigenfunction problem was solved on a one-dimensional grid by a relaxation method described previously [13]. This procedure overcame the numerical difficulties of calculating high order Morse wavefunctions [14][15].

The TDSCF equations for the collinear He$-$I$_2$.

The TDSCF idea in quantum mechanics manifests itself by the product form for the wavefunction:

$$\psi(r_1, r_2) = \phi(r_1)\chi(r_2)e^{i\sigma(t)} \tag{2.5}$$

and normalization:

$$<\phi|\phi> = 1 \quad , \quad <\chi|\chi> = 1. \tag{2.6}$$

The phase convention is chosen such that $\sigma(t) = \int_0^t <H> dt'$ where $<H>$ is the total energy of the system. This convention enables a direct comparison to the exact two-dimensional propagation. The choice of the r_1 and r_2 coordinates is critical to the success of the TDSCF as well as static SCF methods [16]. The choice of equation (2.3) reflects the physical intuition that the He degree of freedom is separated from the I$_2$ motion. The equations of motion are generated by the Hamiltonian. In order to save storage in a computer code, the sum of products form for the interaction potential is used

$$\hat{H}(r_1, r_2) = \hat{H}_1(r_1) + \hat{H}_2(r_2) + \sum_i \hat{V}_1^i(r_1)\hat{V}_2^i(r_2) \tag{2.7}$$

Inserting (2.5) into the Schrödinger equation (2.1):

$$i\frac{\partial\psi(r_1, r_2)}{\partial t} = i\frac{\partial\phi(r_1)}{\partial t}\chi(r_2) + i\phi(r_1)\frac{\partial\chi(r_2)}{\partial t} \tag{2.8}$$

$$= \chi(r_2)\hat{H}_1(r_1)\phi(r_1) + \phi(r_1)\hat{H}_2(r_2)\chi(r_2) + \sum_i \hat{V}_1^i(r_1)\hat{V}_2^i(r_2)\phi(r_1)\chi(r_2)$$

Multipling by $\chi^*(r_2)$ and integrating on the r_2 variable using the normalization condition $\frac{\partial}{\partial t}<\phi|\phi> = 0$ and the fact that the r.h.s. is Hermitian one obtains the TDSCF equations:

$$i\frac{\partial\phi(r_1)}{\partial t} = \hat{H}_1^{SCF}(r_1)\phi(r_1) \tag{2.9}$$

and:

$$i\frac{\partial\chi(r_2)}{\partial t} = \hat{H}_2^{SCF}(r_2)\chi(r_2)$$

where:

$$\hat{H}_1^{SCF}(r_1) = \hat{H}_1(r_1) + <\chi(r_2)|\hat{H}_2(r_2)|\chi(r_2)> + \sum_i <\chi(r_2)|\hat{V}_2^i(r_2)|\chi(r_2)>\hat{V}_1^i(r_1)$$

and:

$$\hat{H}_2^{SCF}(r_2) = \hat{H}_2(r_2) + <\phi(r_1)|\hat{H}_1(r_1)|\phi(r_1)> + \sum_i <\phi(r_1)|\hat{V}_1^i(r_1)|\phi(r_1)>\hat{V}_2^i(r_2)$$

$$H_i(r_i) = \frac{\hat{P}_{r_i}^2}{2} + \hat{V}(r_i) \qquad i = 1, 2$$

The equation is solved by constructing a grid for the r_1 and r_2 coordinates. On this grid the operations of equation (2.9) are calculated. The $\hat{H}_i$ operator in equation (2.9) is calculated by the Fourier method for one degree of freedom. The second term is the average energy of the complementary degree of freedom. The third operation in equation (2.9) is a sum of potential terms multiplied by averages of potential terms from the other degree of freedom. Summing up these contributions completes the Hamiltonian operation. The propagation in time is done simultaneously in both degrees of freedom by the second order differencing scheme (SOD) [6][7].

$$\phi(t+\Delta t) \approx \phi(t-\Delta t) - 2i\,\Delta t\,\hat{H}_{SCF}\phi(t) \qquad (2.10)$$

where $\hat{H}_{SCF}$ is the operator on the r.h.s. of equation (2.9). A similar propagation scheme exists for χ. Use of the SOD propagating method is employed because the operators of equation (2.9) become time dependent and the Chebychev scheme cannot handle the time ordering operation involved. This scheme preserves the norm of both wavefunctions since $\hat{H}_{SCF}$ is Hermitian. The total energy becomes the expectation value

$$E = <\phi|\hat{H}_{SCF}(r_1)|\phi> = <\chi|\hat{H}_{SCF}(r_2)|\chi> \qquad (2.11)$$

and is conserved in time. The second order propagation scheme requires two initial conditions $\phi(0\cdot\Delta t)$ and $\phi(1\cdot\Delta t)$. The scheme is therefore started by a second order Runga-Kutta propagation.

Results.

A typical calculation starts by constructing an initial wavefunction on the grid. The initial wavefunction was a product of the ground state of a Morse in the He direction and an excited vibrational state in the I_2 direction calculated by a relaxation method [13]. The potential was a Morse potential in the R_{He-I} and in the R_{I-I} coordinates, the parameters were adopted from reference [3], other details of the calculation as well as a more extensive discussion of the results can be found in reference [11]. The same initial wavefunction was propagated by the direct Fourier method with the Chebychev scheme and by the TDSCF method. Figure 1 compares propagation of both schemes for $v=5$, $v=11$, and $v=22$ vibrational initial states of I_2 by displaying the overlap:

$$b(t) = \left| <\psi(r_1,r_2)|\phi(r_1)\chi(r_2)> \right|^2 \qquad (2.12)$$

as a function of time.

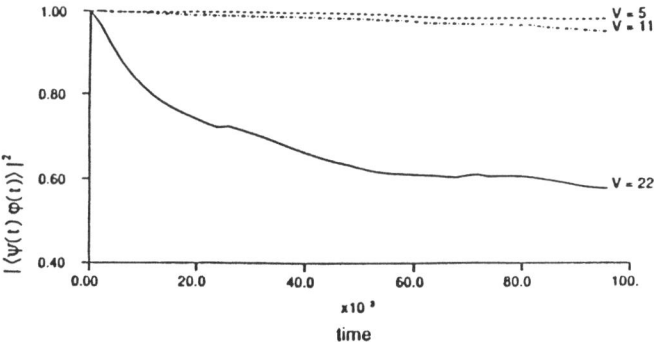

Figure 1: The overlap between the TDSCF wavefunction and the exact wavefunction as a function of time for the initial vibration of the I_2, v=5, v=11, and v=22.

The overlap criteria for comparison of wavefunctions is a more strict test of the approximation than commonly used average quantaities such as the dissociation rate or average energy.

Lifetimes of the different states can be estimated by the autocorrelation function [17][18]:

$$a(t) = |<\psi(0)|\psi(t)>|^2 \tag{2.13}$$

Figure 2 displays the autocorrelation function as a function of time for both the exact and TDSCF wavefunctions.

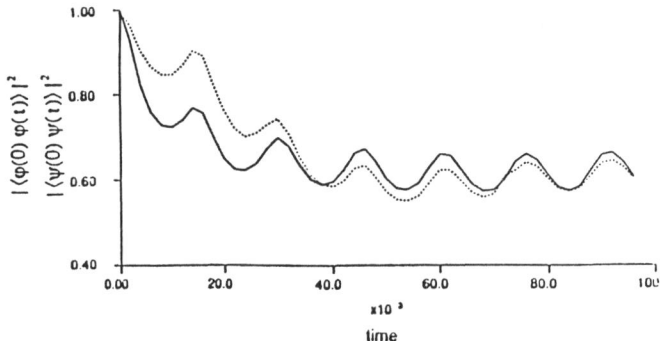

Figure 2: The autocorrelation function as a function of time for the solid line is the exact result and the dashed line is the TDSCF approximation. the initial state is v=22.

Figure 3 shows the exact vs the TDSCF wavefunction superimposed at 40000 au for $v=5$.

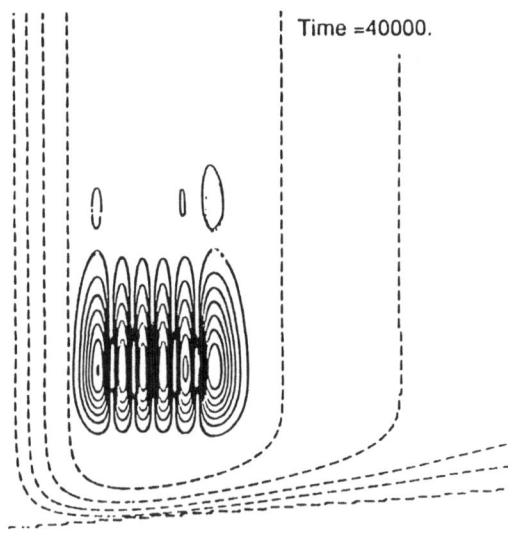

Figure 3: A contour plot of the wavefunction on the potential at t=40000 a.u. (approximately 1 p.s.) for v=5. The solid line is the exact result and the dotted line is the TDSCF approximation. The vertical direction is r_1 and the horizontal direction r_2. The potential is drawn as a dashed line

Upon examining the plots it can be concluded that the TDSCF wavefunction compares well with the exact evolution. Deviations are found for the $v=22$ I_2 initial state. The reason is that for this highly excited state part of the wavefunction penetrates into the classically forbidden region. The result is a fast initial dephasing of the wavefunction which is not well represented in the TDSCF wavefunction. The long time behavior manifested by the superposition of the I_2 vibrations on the He vibrations is represented quite well.

The success of the TDSCF approximation for the collinear He-I_2 is encouraging to the use of this approximation for larger clusters. It is expected that the quality of the method should improve due to the extra averaging when more particles are bunched around the central I_2.

III. The T-shaped and X-shaped He_n-I_2 clusters.

Experimental evidence suggests [1] that the He$-I_2$ molecule has a T shape. It is speculated that additional He atoms will stay on the same plane demonstrated by the X shape of He_2-I_2 displayed in Figure 4.

The TDSCF equations of motion for the T-shaped molecule are identical to equation (2.9). The mass scaling in equation (2.3) is changed to account for the different geometry : $b=(\frac{1}{2}m_I)^{\frac{1}{2}}$. The He-$I_2$ potential is expanded to second order in $R_3 = R_{I-I} - (R_{I-I})_{eq}$ in order to preserve the product form of the potential in equation (2.7).

The equations of motion for the X-shaped cluster are solved in the set of coordinates represented in Figure 4.

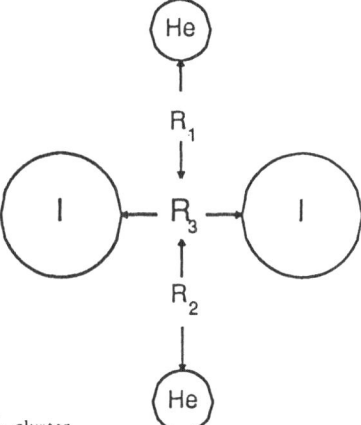

Figure 4: Coordinate set of the X-shaped He_2-I_2 cluster.

Using mass scaled coordinates,

$$r_1 = a R_{He^1-I_2} \qquad r_2 = a R_{He^2-I_2} \qquad r_3 = b R_{I\ I} \qquad (3.1)$$

where:

$$a = \left(\frac{m_1(m_2 + 2m_3)}{(m_1+m_2+2m_3)} \right)^{\frac{1}{2}} \qquad b = \left(\frac{1}{2}m_3 \right)^{\frac{1}{2}}$$

$m_1 = m_2 = m_{He}$ and $m_3 = m_I$,

the kinetic energy operator becomes:

$$\hat{T} = \frac{1}{2}\left(\hat{P}_1^2 + \hat{P}_2^2 + \delta\hat{P}_1\hat{P}_2 + \hat{P}_3^2\right) \tag{3.2}$$

and δ the kinetic energy coupling term is

$$\delta = \left(\frac{m_1 m_2}{(2m_3 + m_2)(2m_3 + m_1)}\right)^{\frac{1}{2}} \tag{3.3}$$

The potential energy is a sum of the I-I potential and the two He-I_2 potentials. The He-He potential is neglected because it is expected to be very small. The TDSCF wavefunction for this system becomes:

$$\psi(r_1, r_2, r_3) = \phi(r_1)\phi(r_2)\chi(r_3) \tag{3.4}$$

One should notice that the TDSCF wavefunction is symmetric to the exchange of the He atoms 1 and 2. Therefore only one TDSCF equations of motion for the He atoms has to be solved:

$$i\frac{\partial\phi(r)}{\partial t} = \hat{H}_{12}(r)^{SCF}\phi(r) \tag{3.5a}$$

where:

$$\hat{H}_{12}^{SCF}(r) = \hat{H}_{12}(r) + \langle\chi(r_3)|\hat{H}_3(r_3)|\chi(r_3)\rangle + \sum_i\langle\chi(r_3)|\hat{V}_3^i(r_3)|\chi(r_3)\rangle\hat{V}_{12}^i(r)$$

$$+ \frac{\delta}{2}\langle\phi(r)|\hat{P}_r(r)|\phi(r)\rangle\hat{P}_r(r) + \langle\phi(r)|\hat{H}_{12}(r)|\phi(r)\rangle$$

and:

$$\hat{H}_{12}(r) = \frac{\hat{P}_r^2}{2} + \hat{V}(r)$$

The equation of motion for the I_2 degree of freedom becomes:

$$i\frac{\partial\chi(r_3)}{\partial t} = \hat{H}_3^{SCF}(r_3)\chi(r_3) \tag{3.5b}$$

where

$$\hat{H}_3^{SCF}(r_3) = \hat{H}_3(r_3) + 2\langle\phi(r)|\hat{H}_{12}(r)|\phi(r)\rangle$$

$$+ 2\sum_i\langle\phi(r)|\hat{V}_{12}^i(r)|\phi(r)\rangle\hat{V}_3^i(r_3) + \frac{\delta}{2}\langle\phi(r)|\hat{P}_r(r)|\phi(r)\rangle^2$$

and

$$\hat{H}_3(r_3) = \frac{\hat{P}_3^2}{2} + \hat{V}(r_3)$$

Very similar equations are obtained for larger clusters where the mass scaling and δ parameter change. For the system under consideration the heavy I_2 atoms serve as a kinetic energy block similar to the situation in the light heavy light triatomic molecule. To a first approximation one can use $\delta = 0$ instead of $\delta = 2/131$. Using this approximation the main difference as a function of cluster size comes through the conservation of energy terms inherent in the TDSCF equations. The He atoms are only subject to the forces exerted by the central I_2 and no correlations between the different He atoms is assumed.

Results.

Calculations were performed for cluster sizes n=1 , n=2, and n=4, for $v=23$. Figure 5 displays the autocorrelation function as a function of time.

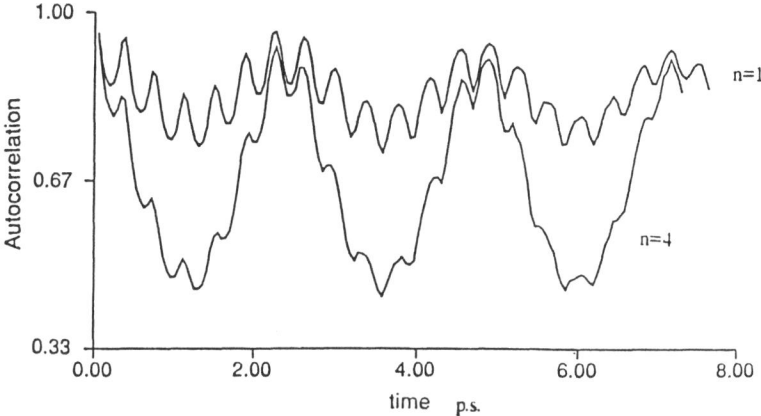

Figure 5: The autocorrelation function $\left|<\psi(t)\mid\psi(0)>\right|^{2}$ as a function of time for n=1 and n=4 size clusters.

The main feature seen in the autocorrelation function is the superposition of the He and I_2 vibrations. The larger amplitude of the He_4 cluster is only the result of the larger cluster size. When examining the autocorrelation functions of an individual He atom very small differences in the autocorrelation are found as a function of cluster size. This result demonstrates the almost perfect separability between the I_2 degree of freedom and the He degrees of freedom. One way to picture the problem is to think of the I_2 vibrations as a time varying force driving the He atoms out almost independently. The I_2 frequency is much higher than the He frequency with a frequency mismatch of approximately 19/3 and the high excitation of the I_2 molecule enables it to act as an energy sink. This situation is opposed to the situation explored by Brown and Wyatt [19] in which a Morse oscillator was driven by a constant frequency comparable to the vibrational spacings. In their study bottlenecks to the dissociation were found. Brown and Wyatt reasoned that these bottlenecks are produced while the molecule acquires energy and its frequency changes, causing a frequency mismatch with the driving frequency. This mismatch in the two frequencies has to go through the golden mean which is the most difficult irrational number to be approximated by a rational product [20]. Classical studies on this system [4-5] also found bottleneck-type restrictions in the classical phase space. Considering the evidence concerning the quantum calculations [3], which show a monotonic increase in the dissociation rate as a function of the initial vibrational excitation, no evidence for bottlenecks is found.

IV. Correlations and symmetry.

The approximation inherent in the TDSCF method is of a separate motion for each degree of freedom in an effective time dependent Hamiltonian. Correlations between degrees of freedom exist only as time correlations i.e. energy can flow back and forth between the individual degrees of freedom. Two types of spatial correlations should be considered: He - I_2 correlations and He - He correlations which are also associated with the symmetry to exchange. Although the He–I_2 correlations are weak some correlation still persists. As a demonstration of this effect one can consider the collinear He–I_2 molecule in the R_{He-I} and R_{I_2} coordinates. In these coordinates the TDSCF initial state is an eigenstate of the TDSCF Hamiltonian. Therefore the TDSCF wavefunction does not dissociate. Examining the exact dynamics one finds that dissociation occurs through a highly correlated motion which originates in the soft turning point of the I_2. This is an example of a rare correlated event which in this choice of coordinates is the only route to dissociation. It is therefore desirable to amend the simple TDSCF method by including the important correlations without increasing the dimensionality of the calculations. This task can be done by adding configurations to the simple product wavefunction.

The symmetry to exchange manifests itself by requiring that the He wavefunction be symmetric to exchange of two He atoms for He^4. For He^4 this symmetry can only manifest itself if the He motion is correlated with other He atoms. Considering the He - He correlations two types of interactions can lead to correlations: a kinetic energy coupling, equations (3.2)-(3.3), and a potential coupling term which is dominantly an excluded volume term which is present in a full 3-D description. The question one wants to answer is the influence of symmetry on the dissociation rate. Considering the picture manifested by the calculations where the He motion is uncorrelated and the dissociation is impulsive, symmetry plays no role. For larger clusters ($n \geq 3$) excluded volume interactions can lead to a correlated He motion. It is expected that this correlated motion should resist the I_2 driving, resulting in a longer lifetime. An opposite extreme is an RRKM picture where phase space has to fill up before dissociation can start. For this case the symmetry reduces the density of states and therefore increases the dissociation rate.

A more involved picture can be found for He^3 clusters because He^3 are fermions. Therefore the total wavefunction has to be antisymmetric including also the nuclear spin contribution. The situation is in analogy to the electronic structure of the He atom where the He^3 atoms play the role of the electrons and the I_2 the role of the positive charge. This picture leads to the conclusion that when a third He^3 atom is added it can only occupy an excited state, and as a result the system is already partially excited, with a higher dissociation rate. Another kinematic isotopic effect which also enhances the He^3 cluster dissociation is that being lighter than He^4, its vibrational frequency is higher by a factor of 1.15. This will lead to a more efficient energy transfer from the I_2 vibration to the He motion. One can conclude that the main effects of symmetry are present only for larger He clusters.

The addition of configurations to the wavefunction is the most simple way to include some correlations. This addition increases the number of coupled equations that have to be solved without increasing the dimensionality. The Multi-Configuration TDSCF (MC-TDSCF) approach is demonstrated for the symmetric He^4 cluster. A similar approach can be used to include He-I_2 correlations. A simple TDSCF approximation is assumed for the separation of the I_2 motion.

A multi-configuration symmetric wavefunction is chosen which has the form

$$\psi(r_1, r_2, r_3) = \chi(r_3) \left(N_1 \phi_a(r_1)\phi_a(r_2) + N_2 \phi_b(r_1)\phi_b(r_2) + \right.$$

$$\left. N_3(\phi_a(r_1)\phi_b(r_2) + \phi_a(r_2)\phi_b(r_1)) \right) \tag{4.1}$$

where the χ and ϕ wavefunctions are normalized:

$$<\chi|\chi> = 1 \quad , \quad <\phi_i|\phi_j> = \delta_{ij}$$

where i, $j = a$, b and N_1, N_2 and N_3 are normalization constants. The indices a and b can be interpreted as channel indices. The choice of the wavefunctions is related to a projection operator $\hat{P}$ which defines the desired correlation. The projection operator is symmetric to particle exchange and operates in the tensor product Hilbert space of the two particles:

$$\hat{P} = \hat{P}_1 \hat{P}_2 \tag{4.2}$$

and

$$\hat{P}_1\phi_a(r_1) = \phi_a(r_1) \quad , \quad \hat{P}_1\phi_b(r_1) = 0$$

with a similar relation for $\hat{P}_2$. Defining $\hat{Q} = \hat{I} - \hat{P}$ one obtains from the Schrödinger equation (2.1) the coupled equations

$$i\hat{P}\frac{\partial\psi}{\partial t} = \hat{P}\hat{H}\left(\hat{P} + \hat{Q}\right)\psi \tag{4.3}$$

$$i\hat{Q}\frac{\partial\psi}{\partial t} = \hat{Q}\hat{H}\left(\hat{P} + \hat{Q}\right)\psi$$

The following Hamiltonian is now used:

$$\hat{H} = \hat{H}^s(r_1) + \hat{H}^s(r_2) + \hat{H}^\delta(r_1)\hat{H}^\delta(r_2) + \hat{V}(r_1)\hat{V}(r_2) \tag{4.4}$$

where $\hat{H}^s(r_i)$ is the single particle Hamiltonian of particle i. $\hat{H}^\delta(r_i)$ is the kinetic energy coupling term, and $\hat{V}(r_i)$ is the potential energy coupling term. Using the time dependent variational principle and the norm conservation one obtains the equations of motion for ϕ_a and ϕ_b (for simplicity the dependence on r_3 is omitted from the equations).

$$i\frac{\partial\phi_a}{\partial t} = \hat{H}_{aa}^{SCF}\phi_a + \hat{H}_{ab}^{SCF}\phi_b \tag{4.5}$$

$$i\frac{\partial\phi_b}{\partial t} = \hat{H}_{bb}^{SCF}\phi_b + \hat{H}_{ba}^{SCF}\phi_a$$

where

$$(4.6)$$

$$
\begin{aligned}
\hat{H}_{aa}^{SCF} =\ & -\frac{N_2}{N_3}<\phi_a|\hat{H}^s|\phi_b> - <\phi_a|\hat{H}^s|\phi_a> + \hat{P}\hat{H}^s \\
& + \frac{N_1}{N_3}<\phi_b|\hat{H}^\delta|\phi_a>\hat{P}\hat{H}^\delta + <\phi_b|\hat{H}^\delta|\phi_b>\hat{P}\hat{H}^\delta + \frac{N_1}{N_3}<\phi_b|\hat{V}|\phi_a>\hat{P}\hat{V} \\
& + <\phi_b|\hat{V}|\phi_b>\hat{P}\hat{V} - \frac{N_2}{N_3}<\phi_a|\hat{H}^\delta|\phi_b><\phi_b|\hat{H}^\delta|\phi_b> \\
& - <\phi_a|\hat{H}^\delta|\phi_a><\phi_b|\hat{H}^\delta|\phi_b> - \frac{N_2}{N_3}<\phi_a|\hat{V}|\phi_b><\phi_b|\hat{V}|\phi_b> \\
& - <\phi_a|\hat{V}|\phi_a><\phi_b|\hat{V}|\phi_b> - \frac{N_1}{N_3}<\phi_a|\hat{H}^\delta|\phi_a><\phi_b|\hat{H}^\delta|\phi_a> \\
& - <\phi_a|\hat{H}^\delta|\phi_b>^2 - \frac{N_1}{N_3}<\phi_a|\hat{V}|\phi_a><\phi_b|\hat{V}|\phi_a> - <\phi_a|\hat{V}|\phi_b>^2
\end{aligned}
$$

and

$$(4.7)$$

$$
\begin{aligned}
\hat{H}_{ab}^{SCF} =\ & \frac{N_2}{N_3}\hat{P}\hat{H}^s + \frac{N_2}{N_3}<\phi_b|\hat{H}^\delta|\phi_b>\hat{P}\hat{H}^\delta + <\phi_b|\hat{H}^\delta|\phi_a>\hat{P}\hat{H}^\delta \\
& + \frac{N_2}{N_3}<\phi_b|\hat{V}|\phi_b>\hat{P}\hat{V} + <\phi_b|\hat{V}|\phi_a>\hat{P}\hat{V}
\end{aligned}
$$

A similar equation is found for $\hat{H}_{bb}^{SCF}$. Simultaneously with equation (4.5) one solves the equations of motion for the terms N_1, N_2, and N_3

$$(4.8)$$

$$
\begin{aligned}
i\frac{\partial N_1}{\partial t} =\ & 2N_1<\phi_a|\hat{H}^s|\phi_a> + 2N_3<\phi_a|\hat{H}^s|\phi_b> + N_1<\phi_a|\hat{H}^\delta|\phi_a>^2 \\
& + 2N_3<\phi_a|\hat{H}^\delta|\phi_b><\phi_a|\hat{H}^\delta|\phi_a> + N_1<\phi_a|\hat{V}|\phi_a>^2 \\
& + 2N_3<\phi_a|\hat{V}|\phi_a><\phi_a|\hat{V}|\phi_b> + N_2<\phi_a|\hat{H}^\delta|\phi_b>^2 + N_2<\phi_a|\hat{V}|\phi_b>^2
\end{aligned}
$$

A similar equation exists for N_2. N_3 can then be calculated from the total normalization condition.

The set of equations (4.5)-(4.8) can be readily implemented. Only two one-dimensional partial differential equations (4.5) need be solved. This formalism allows flexibility in the choice of projected states and should prove useful for many different scattering phenomena.

V. Conclusions.

This paper presents a framework for calculating the dissociation dynamics of van der Waals clusters. Considering quantum calculations for large systems with many degrees of freedom a mean field approach seems the only alternative. The approach adopted here of a TDSCF approximation relies on the physical picture in which a time scale separation exists between the He motion and the I_2 degrees of freedom. This basic approximation can be amended by including part of the correlations as more configurations without increasing the dimensionality of the problem. The procedure is not limited by the number of configurations. By a proper choice of projection operators all important correlations can be included while still maintaining the feasiblity of the calculation, even for relatively large systems.

From a numerical point of view the TDSCF method saves much computation time by decomposing one n-dimensional problem into n 1-dimensional problems. For the collinear He-I_2 calculation a reduction of a factor of ten was found. An exact calculation for larger clusters would not be feasible on the minicomputer used for this calculation. The TDSCF algorithm is also very well suited for parallel computation. By assigning one processor to each degree of freedom the processors carry out a lot of computation independently, with only a very small amount of communication between them.

The simple TDSCF calculation shows that the dissociation rate depends very little on cluster size. An impulsive model of the dissociation therefore results. This is in contrast to an RRKM picture which would predict a decrease in the dissociation rate. The emerging picture is of a He atom driven out of its well by a periodic fast perturbation.

Considering the role of symmetry in He clusters it is expected that different dissociation rates can be found for He^4 than He^3 for clusters with more than three He atoms.

Acknowledgment:

We want to thank the organizers of the Jerusalem Symposium for stimulating this work, and to A. Ben-Shaul, R. S. Berry, J. A. Beswick, and J. Jortner for helpful discussions. The Fritz Haber Research Center is supported by the Minerva Gesellschaft für die Forschung, GmbH München, BRD. R. K. acknowledges support by grants from the U.S. - Israel Binational Science Foundation.

Reference List

[1]   R. E. Smalley, D. H. Levi, and L. Wharton, *J. Chem. Phys.* **64** 3266 (1976);
        M. S. Kim, R. E. Smalley, L. Wharton, and D. H. Levi, *J. Chem. Phys.* **65**
        1216 (1976); R. E. Smalley, L. Wharton, and D. H. Levi, *J. Chem. Phys.* **68**
        671 (1978); G. Kubiak, P. S. H. Fitch, L. Wharton, and D. H. Levi, *J. Chem.*
        Phys. **68** 4477 (1978); K. E. Johnson, L. Wharton, and D. H. Levi, *J. Chem.*
        Phys. **69** 2719 (1978).

[2]   S. Beatty Woodruff and D. L. Thompson, *J. Chem. Phys.* **71** 376 (1979).

[3]   J. A. Beswick and J. Jortner, *J. Chem. Phys.* **68** 2277 (1978); J. A. Beswick
        and J. Jortner, *J. Chem. Phys.* **69** 512 (1978); J. A. Beswick, G. Delgado-
        Barrio, and J. Jortner, *J. Chem. Phys.* **70** 3895 (1979).

[4]   S. K. Gray, S. A. Rice, and D. W. Noid, *J. Chem. Phys.* **84** 3745 (1986).

[5]   M. J. Davis and S. K. Gray, *J. Chem. Phys.* **84** 5389 (1986).

[6]   R. Kosloff and D. Kosloff, *J. Comp. Phys.* **52** *35 (1983).*

[7]   R. Kosloff and D. Kosloff, *J. Chem. Phys.* **79** *1823 (1983).*

[8]   P. A. M. Dirac, *Proc. Camb. Phil. Soc.* **26** 376 (1930).

[9]   R. B. Gerber, V. Buch, and M. A. Ratner, *J. Chem. Phys.* **77** 3022 (1982).

[10]   G. C. Shatz, V. Buch, M. A. Ratner, and R. B. Gerber, *J. Chem. Phys.* **79**
        1808 (1983).

[11]   R. H. Bisseling, R. Kosloff, R. B. Gerber, M. A. Ratner, L. Gibson, and C.
        Cerjan, *submitted to J. Chem. Phys.* (1987).

[12]   H. Tal-Ezer and R. Kosloff, *J. Chem. Phys.* **81** *3967 (1984).*

[13]   R. Kosloff and H. Tal-Ezer, *Chem. Phys. Lett.* **127** 223 (1986).

[14]   P. M. Morse, *Phys. Rev.* **34** *57 (1929).*

[15]   H. Kobeissi, *J. Comp. Phys.* **61** *351 (1985).*

[16]   R. B. Gerber and M. A. Ratner, *Adv. Chem. Phys.* in press (1987).

[17]   R. Bisseling, R. Kosloff, and J. Manz, *J. Chem. Phys.* **89** *993 (1985).*

[18]   R. H. Bisseling, R. Kosloff, J. Manz, F. Mrugala, and J. Romelt, *J. Chem.*
        Phys. **86** 2626 (1986).

[19]   R. C. Brown and R. E. Wyatt, *Phys. Rev. Lett.* **57** 1 (1986).

[20]   D. Bensimon and L. P. Kadanoff, *Physica,* **13D** 82 (1986).

PROPERTIES OF LARGE FINITE SYSTEMS: CONSTRAINTS FROM ASTROPHYSICAL
OBSERVATIONS

W.W. Duley
Physics Dept.
York University
4700 Keele St.
Toronto, Ontario
CANADA M3J 1P3

ABSTRACT. Analysis of conditions under which carbon clusters form and
are stable under interstellar conditions are shown to offer insight
into the stability of PAH-like structures in UV rich and reactive
environments.

1. INTRODUCTION

The observation of a set of discrete infrared emission lines in the
spectra of a variety of objects (see Allamandola 1984) whose wavelengths match those of vibrational bands in organic materials has provided astronomers with an important new clue to the chemical nature of
interstellar matter. Assignment of these transitions to equilibrium
thermal emission from CH and other groups in amorphous carbon grains
was proposed by Duley and Williams (1981). Subsequent observations by
Sellgren, Werner and Dinerstein (1983) suggested that the amplitude of
emission within the 3.38 and 3.4µm features together with the existence
of an extended continuum in the ~ 2µm region in certain objects could
only be accounted for on the basis of a non-equilibrium emission process in very small grains (Sellgren 1984). The mechanism proposed
involved transient heating with UV photons to produce temperature spikes
and excess IR emission (Duley 1973, Purcell 1976). The composition of
these particles has been discussed by Léger and Puget (1984) and
Allamandola, Tielens and Barker (1985) leading to the conclusion that
the emitters were similar in structure to polycyclic aromatic hydrocarbon (PAH) molecules. A survey of the properties of these interstellar PAH molecules has recently been given by Omont (1986). Duley
and Williams (1986) and Duley (1986) have discussed the relation
between HAC and PAH molecules and conclude that HAC can be considered
to be a collection of PAH molecules and molecular groups weakly bonded
to form larger particles.
  In this paper I discuss the formation and growth of these PAH
particles under interstellar conditions and show how this information
can be used to constrain the behaviour of small carbon clusters.

67

J. Jortner et al. (eds.), Large Finite Systems, 67–73.
© 1987 *by D. Reidel Publishing Company.*

2. CONDITIONS IN INTERSTELLAR CLOUDS

While there are a hierarchy of interstellar objects (cf. Duley and Williams 1984), we will limit this discussion to those existing in diffuse clouds. PAH emission is observed from such clouds adjacent to energetic regions (Léger and Puget 1984, Allamandola et al 1985).

Diffuse clouds are characterized by densities of n ~ 10^2 hydrogen nuclei cm^{-3}. Hydrogen is in both atomic and molecular form. H_2 molecules are formed by H atom recombination on the surface of dust grains. These grains are small (0.01 – 0.1 μm) but are still larger than the PAH clusters responsible for several IR emission features. The space density of dust grains ranges from 10^{-9}n for 0.01 μm particles to 10^{-12}n for 0.1 μm particles. These particles consist of amorphous silicates (both Mg and Fe-rich) coated to a greater or lesser degree with hydrogenated amorphous carbon (α C:H or HAC)

The ambient gas in diffuse clouds consists of mainly atomic and ionic species. Dominant components include C, C^+, O and N as well as electrons. The space density of atomic carbon in all forms is ~ 3.7 × 10^{-4}n although this value is reduced when some carbon has accreted on dust. Generally C << C^+ under diffuse cloud conditions because C can be ionized by interstellar UV radiation. The spectrum of this UV light (which is the diluted light from hot stars) extends to 13.6 eV, the ionization limit of H. Between 130 and 300 nm the interstellar radiation flux F ~ 10^6 photons cm^{-2} eV^{-1} $ster^{-1}$ (Duley and Williams 1984). The gas kinetic temperature is ~ 100 K. A summary of collision rates for a 0.01 μm radius particle is given in table 1.

TABLE 1. Representative collision rates
with a 0.01 μm radius particle under
diffuse cloud conditions

Species	Rate (sec^{-1})
UV photon	2×10^{-4}
H	6×10^{-5}
O	1×10^{-8}
C, C^+	6×10^{-9}

3. GROWTH OF CLUSTERS UNDER IS CONDITIONS

We will limit this discussion to the growth of carbon clusters. These clusters will consist of C atoms bonded in sp^2 (graphite-like) and sp^3 (diamond-like) configurations. Many C atoms will also be bonded to H, O, N, etc. The growth of a molecule MX by addition of C can be represented in terms of the reaction

$$MX + C \xrightarrow{k} MC + X \tag{1}$$

For small MX, k is typically 10^{-11} cm^3 sec^{-1} for neutral-neutral reactions rising to $k \sim 10^{-9}$ cm^3 sec^{-1} when one of the reactants is a positive ion. For example, the reaction

$$C + C_2H^+ \longrightarrow C_3^+ + H \qquad (2)$$

has a rate constant $= 1.1 \times 10^{-9}$ cm^3 sec^{-1} (Duley and Williams 1984).

The accretion of carbon on larger particles can be described similarly

$$MX + C \longrightarrow MXC \qquad (3)$$

The rate constant for this reaction will be

$$k_3 = \gamma A\ v_C \qquad (4)$$

where A is the projected surface area of the particle, v_C is the speed of a C atom and γ is a probability factor for thermal accomodation and adsorption.

The volume of a graphite-like carbon particle containing N carbon atoms is $V = N(1.1 \times 10^{23})^{-1}$, where the numerical factor is the density of C atoms in graphite. A can then be approximated as

$$A \quad \frac{N^{2/3}}{2.33 \times 10^{15}}\ cm^2 \qquad (5)$$

With $v_C \sim 5 \times 10^4$ cm sec^{-1} and $\gamma \sim 0.1$, k_3 can now be related to N. The result is shown in figure 1 where it is compared with k for simple neutral-neutral (N-N) and ion-molecule (I-M) reactions. As expected, k_3 extrapolated to small N correlates with k_{N-N} since both rate constants are essentially based on geometrical cross-sections. The cross-section for ion-molecule reactions is enhanced due to polarization effects.

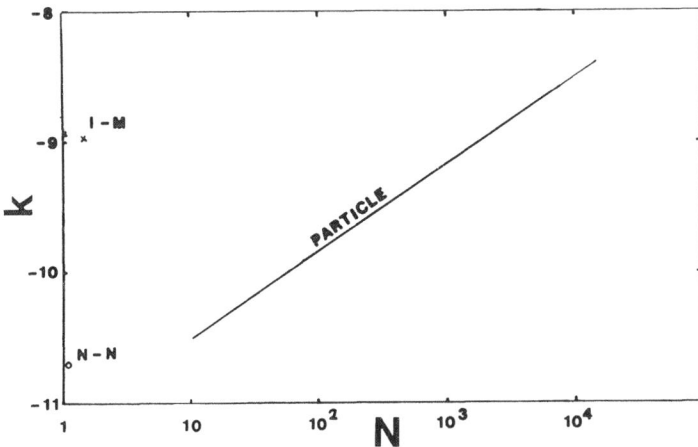

Fig. 1. Rate constant k(cm^{3}sec^{-1}) vs. number of atoms in cluster for reactions (1) and (3).

4. PHOTON IMPACT

The lifetime of these carbon clusters under IS conditions will be
limited by photon induced reaction or photochemical dissociation. The
latter process is most important for small clusters ($\lesssim 10$ atoms Léger
and Puget 1984, Allamandola et al. 1985, Duley 1986) and provides a
lower limit to the size of clusters that would be stable over extended
periods (10^8y) in diffuse clouds when subject only to photodissociation. Photon-induced reaction has also been identified (Duley and
Williams 1986) as a possible destruction channel for PAH under IS
conditions.

   The photon absorption rate for a molecule of projected area A is
$P = 6.3 \times 10^7 \sigma = 6.3 \times 10^7 \eta A$ where σ is the absorption cross-section
and η is an efficiency factor for absorption. Generally $\eta \sim 0.1 - 1$.
The numerical factor comes from the photon flux at ca. 8eV, assuming
a bandwidth of 5eV. P can be compared to the growth rate G from equ. 3

$$\frac{P}{G} = \frac{6.3 \times 10^7 \eta A}{\gamma A v_c n_c} \tag{6}$$

where n_c = space density of C, C^+. With the numerical values $\gamma = 0.1$,
$\eta = 0.1$, and $n_c = 1.85 \times 10^{-4}n$ (half the available carbon) equation 6
becomes

$$\frac{P}{G} = 6.8 \times 10^4 \tag{7}$$

For a molecule to be stable against photo-induced decomposition the
probability per event for decomposition must then be $< (6.8 \times 10^4)^{-1} =$
1.5×10^{-5}

5. CONSTRAINTS ON SIZE OF STABLE CLUSTER

Since H, N and O atoms are relatively abundant in diffuse IS clouds
while all these species are reactive with respect to carbon surfaces,
at least at higher temperature (Vietze Flaskamp and Phillips 1982)
reactions with these atoms may limit the size of carbon clusters. The
primary stage of attack is likely at peripheral carbon atoms - for
example

$$\diagdown\!\!\!\diagup - C + 3H \longrightarrow \quad \diagdown\!\!\!\diagup -CH_3 \tag{8}$$

or

$$\diagdown\!\!\!\diagup + 0,H \longrightarrow \quad \diagdown\!\!\!\diagup -OH \tag{9}$$

Subsequent reactions may result in the liberation of the surface group

$$\left. \begin{array}{c} \text{(structure)} \end{array} \right\rangle \!\!-CH_3 \;+\; H \;\longrightarrow\; \left. \begin{array}{c} \text{(structure)} \end{array} \right\rangle \;+\; CH_4\,(\text{gas}) \tag{10}$$

or

$$\left. \begin{array}{c} \text{(structure)} \end{array} \right\rangle \!\!-NH \;+\; H \;\longrightarrow\; \left. \begin{array}{c} \text{(structure)} \end{array} \right\rangle \;+\; NH_3\,(\text{gas}) \tag{11}$$

Such reactions will tend to keep the surface clear of surface functional groups, but will not in general, result in the destruction of the substrate. On the other hand oxidation reactions may result in ring disruption - for example

$$+\; HCO\,(\text{gas}) \tag{12}$$

The activation energy for such a reaction is expected to be appreciable since C-C bonds must be broken. It would certainly not occur under equilibrium conditions at low temperature.

However, as evidenced by IR data (Léger and Puget 1984, Allamandola et al. 1985) interstellar PAH clusters are unlikely to be in equilibrium with the gas kinetic temperature. Instead these molecules are subjected to rapid rises in vibrational temperature whenever a UV photon is absorbed. The average energy per mode on absorption of an 8eV photon will be

$$\varepsilon = \frac{8}{3N-6} \quad eV \tag{13}$$

where $N \sim$ number of carbon atoms in the cluster. This excitation persists for the IR radiative lifetime of the molecule - typically $\tau = 0.1$ - 0.01 sec. If E_a is the activation energy for a reaction such as 12 then

$$\beta = 10^{12}\,\tau\,\exp[-E_a/\varepsilon] \tag{14}$$

is the probability per absorbed photon that bond breaking will occur resulting in the liberation of HCO. The factor 10^{12} in this equation is a vibrational frequency.

As we have seen in section 4, $\beta \lesssim 1.5 \times 10^{-5}$ for stable clusters to exist. Then

$$10^{12}\,\tau\,\exp[-E_a/\varepsilon] \sim 1.5 \times 10^{-5} \tag{15}$$

defines the minimum size of these clusters. Taking $\tau = 0.1$ sec, $E_a = 3.5eV$ (Blyholder and Eyring 1957) and ε from equation 13 one obtains

$$N_{min} \sim 30 \tag{16}$$

This result is in agreement with the observations of Barker et al. (1987) and supports the conclusions arrived at in that paper.

6. CLUSTER GROWTH

For $N \tilde{>} N_{min}$ the rate constant for growth of a cluster by addition of
C is $k_3 = \gamma A v_c$. The number of C atoms accreted per second is then

$$\frac{dN}{dt} = A v_c n_c \tag{17}$$

$$= 4.3 \times 10^{-16} \gamma v_c n_c N^{2/3} \tag{18}$$

Equation 18 implies a doubling time

$$\tau = \frac{2.3 \times 10^{1/3}}{\gamma v_c n_c} \ N \qquad \text{sec} \tag{19}$$

Taking $\gamma = 0.1$, $v_c = 5 \times 10^4$ cm sec^{-1} and $n_c = 1.8 \times 10^{-4} n = 1.8 \times 10^{-2}$
cm^{-3} one obtains

$$\tau = 2.56 \times 10^{13} \ N^{1/3} \text{ sec}$$

$$= 8.1 \times 10^5 \ N^{1/3} \text{ y} \tag{20}$$

The first doubling time for growth of a cluster with $N = N_{min} = 30$ is
then $\tau = 2.5 \times 10^6$y. This timescale is somewhat lengthened if the
clusters are positively ionized. However, calculations indicate that
in diffuse clouds [PAH$^+$]/[PAH] $\sim$ 1 so that neutral clusters will
always be abundant (Omont 1986). Clusters with $N_{min} = 30$ will grow
kinetically (ie. by accreting 10% of incident C, C$^+$) to $N \sim 10^5$ over a
timescale of 1.5×10^8y - a typical timescale for quiescent conditions
in diffuse clouds (Duley and Williams 1984). Small clusters, therefore, grow very rapidly under IS conditions once they are large enough
to be stable against photo oxidation reactions. This then suggests
the interesting question - why is IS PAH emission observed at all?
     A way out of this conundrum is likely provided by reactions 8 and
10. Such reactions would keep PAH surfaces clear of new C atoms,
thereby inhibiting growth. There is some evidence from laboratory data
that the activation energy for reaction 10 is small for edge atoms
(Bar-Nun 1975, Bar-Nun et al. 1980) so that this reaction may proceed
even at low temperatures (< 100 K).
     Since there are $\sim 10^4$ more H atom collisions than C atom colli-
sions, the probability that reaction (10) proceeds need only be $\gtrsim 10^{-4}$
to inhibit subsequent cluster growth past $N \sim N_{min}$. If this were the
case, then only clusters with $N_{min} \sim 30$ would be observed.
     This then gives rise to the additional question of how it is that
clusters of this size could form initially under interstellar conditions. Duley (1986) concludes that direct synthesis via ion-molecule
or other chemical routes is unlikely given the timescales and the size
of molecules involved. It is perhaps more probable that PAH-like molecules appear as the degradation products of amorphous carbon layers
accreted on silicate dust particles in denser interstellar objects
where few H atoms exist. Chemical erosion of these solids in regions
of higher excitation where H atoms are abundant then results in the

transient appearance of clusters in the correct size range. These
clusters may still be attached to larger dust grains while emitting in
the 3.4 μm region.

ACKNOWLEDGEMENTS

This research has been supported by grants from the NSERCC. WD
acknowledges many useful discussions with Professor D.A. Williams.

REFERENCES

Allamandola, L.J. 1984. In "Galactic and Extra Galactic IR Spectros-
   copy" eds. M.F. Kessler and J.P. Phillips, Reidel, Holland p.5.
Allamandola, L.J., Tielens, A.G.G.M. and Barker, J.R. 1985. Ap.J.
   290, L25.
Barker, J.R., Allamandola, L.J. and Tielens, A.G.G.M. 1987. Ap.J.
   315, L61.
Bar-Nun, A. 1975. Ap.J. 197, 341.
Bar-Nun, A., Litman, M. and Rappaport, M.L. 1980. Astr. Ap. 85, 197.
Blyholder, G. and Eyring, H. 1957. J. Phys. Chem. 61, 682.
Duley, W.W. 1973. Nature Phys. Sci. 244, 57.
Duley, W.W. 1986. In "Polycyclic Aromatic Hydrocarbons and Astro-
   physics" eds. A. Léger, L. d'Hendecourt and N. Boccara, Reidel,
   Holland, p.373.
Duley, W.W. and Williams, D.A. 1981. Mon. Not. Roy. Astr. Soc. 196,
   269.
Duley, W.W. and Williams, D.A. 1984. "Interstellar Chemistry" Academic
   Press, London.
Duley, W.W. and Williams, D.A. 1986. Mon. Not. Roy. Astr. Soc. 219,
   859.
Léger, A. and Puget, J.L. 1984. Astr. Ap. 137, L5.
Omont, A. 1986. Astr. Ap. 164, 159.
Purcell, E.M. 1976. Ap.J. 206, 685.
Sellgren, K. 1984. Ap.J. 277, 623.
Sellgren, K., Werner, M.W. and Dinerstein, H.L. 1983. Ap.J. 271, L13.
Vietzke, E., Flaskamp, K. and Phillips, V. 1982, 111/112, 763.

PORPHYRIN DIMERS: THE PHOTOEXCITED TRIPLET STATE AS A PROBE FOR STRUCTURE AND DYNAMICS

H. Levanon
Department of Physical Chemistry
and the Fritz Haber Research Center for Molecular Dynamics
The Hebrew University
Jerusalem, 91904
Israel

ABSTRACT. In this work we monitor the photoexcited triplet state of several covalently linked porphyrin dimers considered as model compounds related to the primary photosynthetic constituents. Two spectroscopies are applied: i. selective laser excitation-time resolved EPR; ii. optical-pulse radiolysis. The former method, utilized on the dimers dissolved in a nematic liquid crystal, enables to determine: i. intramolecular singlet energy transfer rates; ii. mutual orientation of the chromophores constructing the dimer; iii. triplet spin dynamics. The latter method allows, via triplet sensitization of the dimers in isotropic media, the determination of intramolecular triplet energy transfer. The mechanisms of the singlet and triplet energy transfer are also discussed.

1. INTRODUCTION

The substantial progress in the general field of photosynthesis, in particular regarding the crystallization of the reaction-center protein in bacteria [1-3], should be considered as a major breakthrough in this field. Despite this latest progress, the exact mechanism of energy absorption and redistribution, which is followed by charge separation and subsequent vectorial electron transfer process, is still obscured and is under extensive experimental and theoretical studies [4-10]. It is clear that the complexity of the *in vivo* systems has triggered the search for appropriate model systems [11] on which the mechanism of energy and electron transfer reactions can be studied, understood and further extended to the *in vivo* systems.

In choosing model systems one should consider several parameters which bear upon the basic photophysical and photochemical processes which are relevant to primary photosynthesis. These should consist of suitable probe compounds, which are related to the reaction-center photosynthetic constituents, together with the matrices in which the chromophores are embedded. As to the experimental methods, several time-resolved spectroscopies should be mentioned: ultrafast (picosecond) optical laser spectroscopy [12-15] ; fast (microsecond) and very fast (nanosecond) optical-magnetic

75

J. Jortner et al. (eds.), Large Finite Systems, 75–81.

resonance spectroscopies in several modes [5,11]; optical-pulse radiolysis [16,17]; laser photolysis and fluorescence spectroscopies.

Although the photoexcited triplet state does not participate in the main pathway of photosynthesis, it may be considered as a major diagnostic tool in providing information on [5,11,18]:

i. molecular structure;

ii. dynamics of electron and energy transfer processes;

iii. spin interactions and dynamics associated with it.

The work described herein will focus on the triplet state of several covalently linked hybrid porphyrin dimers embedded in anisotropic (liquid crystals) and isotropic environments. Two experimental methods will be discussed:

i. selective laser excitation-time resolved EPR spectroscopy [11];

ii. triplet sensitization by optical-pulse radiolysis [19].

2. EXPERIMENTAL

2.1 Materials and Sample Preparation

The porphyrin dimers were synthesized in a method described by Little and co-workers [20]. Three dimers, $Zn(-CH_2-)_nH_2$, were employed: one para-para dimer (n=3, Pa3Pa) and two ortho-ortho dimers (n=3, 6, Or3Or and Or6Or). The Zn and H_2 on each side of the methylene group symbolize the zinc and free base tetratosyl porphyrin chromophores, respectively. For EPR measurements the dimers were dissolved in a liquid crystal, E-7 [11], in its isotropic phase and were oriented in its nematic phase in the presence of a strong magnetic field (9 kG) and cooled down to the desired temperature (~100K).

2.2 Time-Resolved EPR Experiments

The experimental setup we employ in our studies is capable of monitoring, on-line, the temporal changes of the transverse magnetization, $M_x(t)$, produced by light excitation together with the corresponding two-dimensinal CW spectra with respect to time and magnetic field [21]. The oriented sample in the microwave cavity is slectively excited into its triplet state by laser pulses of 10 ns width, at a repetition rate of 10 Hz and the signal output is digitized and further processed to its final form as shown in Fig. 1. Two orientations of the dimer, distributed in the nematic phase were examined:

i. the liquid crystal director, L , is parallel to the external magnetic field;

ii. the director is perpendicular to the external field (for details see reference 11).

2.3 Optical-Pulse Radiolysis Experiments

Pulse radiolysis of benzene solutions containig the dimers and a suitable sensitizer (β-acetonaphthone) produces directly the photoexcited triplet state of the chromophore [19]. This experimental procedure is well documented in previous studies [22-24].

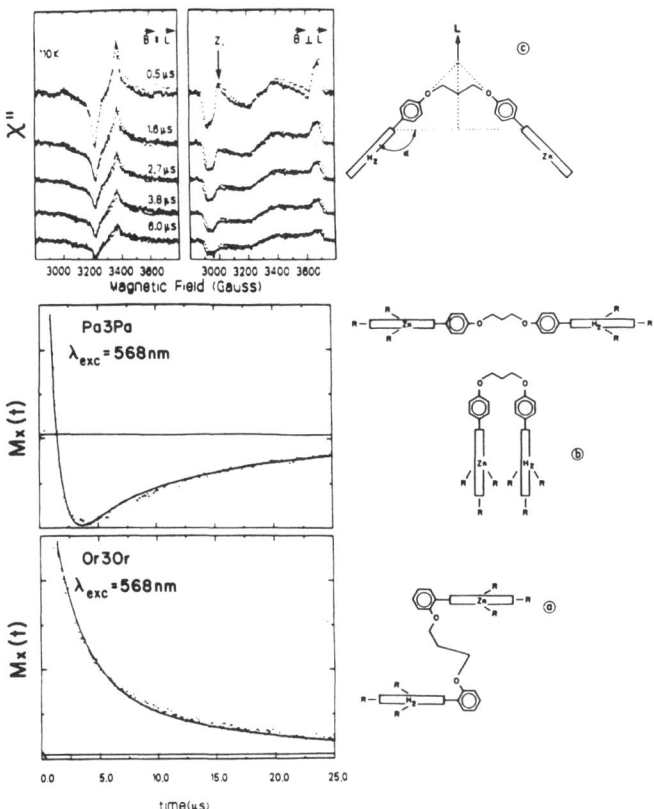

Fig. 1. Right: Schematic structure of the dimers: (a) Or3Or; (b) Pa3Pa in the extended and folded configurations: (c) Pa3Pa in its partial folded conformation in a nematic uniaxial LC (E-7). The angle α is calculated to be ~120°. For details see reference 11. Left: Time-resolved triplet CW spectra of Pa3Pa oriented in a unlaxial LC (parallel and perpendicular configuration) at progressive times after the 568 nm laser pulse. Z_1 indicates the magnetic field at which the kinetic curve was taken as shown in the middle trace. The lower trace is the kinetic profile taken at the emission peak of the spectrum (not shown) of Or3Or. Notice the mixed out-of phase kinetics for Pa3Pa which is attributed to a mixture of ZnTPP^T and H$_2$TPP^T which is reflected by the early time spectrum. The dotted lines on the CW spectra and the solid lines on the kinetic trace are best fit lines. For details see reference 11.

3. RESULTS AND DISCUSSION

3.1 Time-Resolved EPR

Schematic representation of the dimers studied is shown in Fig. 1. Similar to a previous study [25] employing a covalently linked dimer with n=2, selective laser excitation at 568 nm will preferably promote the system into an excited triplet attributed to the zinc part, Zn. Photoexcitation at 650 nm will exclusively promote the free base component, H_2, into its triplet state [25,26]. Time-resolved triplet spectra of the Or3Or and Or6Or are identical to that of free base monomer, H_2TPP, indicating that intramolecular singlet energy transfer from the Zn to H_2 chromophore in the dimer is complete within the time scale of our measurements ($k_{et} < 1.5 \times 10^9$ s^{-1} [26]). On the other hand, the Pa3Pa does shows a strong dependence on the selective excitation (Fig. 1); whereas the 650 nm excitation gives rise to the triplet state of H_2TPP, the 568 nm excitation indicates that the early spectrum is different from its successive ones which are identical with that of H_2TPP [26]. The most likely interpretation of this behavior, as reflected by the mixed kinetics (Fig 1.), is due to an admixture of both triplet components of the two porphyrin moieties constructing the Pa3Pa dimer [26]. Analysis of the data results in a relatively low rate constant for the intramolecular singlet energy transfer which is calculated to be 4.7×10^8 s^{-1} [26]. Such a low rate is attributed, by employing the dipole-dipole type mechanism, to a partially folded configuration imposed by the liquid crystalline structure (Fig. 1). Non- linear configurations have already been suggested in mixed-metal dimers of porphyrins [27] and other bichromophoric systems bridged together by n=3 methylene groups [28,29] ("n=3 Hirayama rule" [30]). As to the other two dimers, their structure favors an extended configuration ("Z-shaped" as shown in Fig. 1) resulting in an efficient route for intramolecular singlet energy transfer.

3.2 Optical-Pulse Radiolysis

In the time scale of the above time-resolved EPR experiments (1-2 μs) we did not notice any evidence for triplet intramolecular energy transfer. Several reasons may accoumt for it:
i. the energy transfer is beyond time resolution (too fast or too slow);
ii. the participation of the singlet dynamics, as shown above, interferes with possible triplet dynamics.

To circumvent the latter difficulty we have employed the method of direct triplet sensitization via a sensitizer whose triplet state is produced under pulse radiolytic conditions [19]. We have chosen for this purpose a sensitizer, namely β-acetonaphthone, having its first excited triplet far above the triplet energy levels of the dimers (58 Kcal/mol [31]). Such a large energy gap and, on the other hand, a gap of about 3 Kcal/mol [32,33], between H_2TPP and ZnTPP ensures an efficient route of populating the triplet states of the separate chromophores, linked together, with equal probability:

$$D \quad \rightarrow \quad D^T \tag{1}$$

$$D^T \quad + \quad Zn(\text{-CH}_2\text{-})H_2 \quad \rightarrow Zn^T(\text{-CH}_2\text{-})H_2 \tag{2}$$

$$D^T \quad + \quad Zn(\text{-CH}_2\text{-})H_2 \rightarrow Zn(\text{- CH}_2\text{-})H_2{}^T \tag{3}$$

Thus, if there is an intramolecular triplet energy transfer within the instrumental time resolution it should show up in the time evolved spectra. Indeed, all the dimers studied show triplet spectra which are different from a 1:1 admixture of the separate monomers [19]. Moreover, the Pa3Pa differs from the Or3Or and Or6Or. The data analysis results in the following conclusions [19]:

i. intramolecular triplet energy transfer occur in all dimers studied giving rise to a final triplet spectrum typical of H_2TPP:

$$Zn^T(-CH_2-)H_2 \quad \xrightarrow{k_{et}} \quad Zn(-CH_2-)H_2^T \qquad (4)$$

ii. process 4 is complete in both compounds but with different rates, i.e.,
$k_{et}(Or3Or,Or6Or) > k_{et}(Pa3Pa)$.

The former rates are too fast to be monitored via the spectral time evolution, whereas that of the latter could be estimated from the observed time evolved spectra. The rate constant, k_{et}, for the para-compounds is estimated to be $(2.6-10)x10^5$ s^{-1}, and that of the ortho-compounds exceeds $10x10^5$ s^{-1} [19].

The dipole-dipole mechanism for the intramolecular triplet energy transfer should be excluded as the triplet radiative lifetimes of the porphyrins are extremely low [34,35]. Certainly this is not the case for the intramolecular singlet energy transfer as discussed above. On the other hand, the large distance between the chromophores building up the dimers (~17 A of the "Z-shaped" and ~27 A of the fully extended conformation [32,36], cf. Fig. 1) does not allow for an efficient static mechanism [37]. We thus propose a dynamical exchange mechanism as suggested earlier for this process [37,38].

4. CONCLUSIONS

Although the probe compounds described above are far from being the most appropriate model systems for primary photosynthesis, the examples presented clearly demonstrate the linking between the photoexcited triplet state and structural problems as well as dynamical processes associated with it. A crucial point for further investigations is to search for dimers where both ends are of the same type and with short interchromophore distance ("super molecules") where ground state interactions are noticeable. These points are currently being encountered by synthesizing new type of covalently linked compounds as well as studies on *in vivo* systems.

5. ACKNOWLEDGEMENTS

This work was supported by a U.S.-Israel BSF, Grant No. 3443. The Fritz Haber Research Center is supported by the Minerva Gessellschaft fur die Forschung, GmbH, Munchen, BRD. Part of this work (pulse radiolysis) was also supported by the Office of Basic Energy Sciences of the Department of Energy, U.S.A.

6. REFERENCES

1. H. Michel and J. Deisenhofer in: *Photosynthesis III. Photosynthetic Membranes and Light Harvesting Systems* (L.A. Staehelin and C.J. Arntzen eds.). Encyclopedia of Plant Physiology, New Series, Springer Verlag, Berlin, 1986, Vol. **19**, p. 371.
2. J. Deisenhofer, O. Epp, K. Miki, R. Huber and H. Michel, *Nature* **318**, 618 (1985).
3. C.-H. Chang, D. Tiede, J. Tang, U. Smith, J.R. Norris and M. Schiffer, *FEBS Lett.* **205**, 82 (1986).
4. For recent reviews and list of references see e.g., *Antennas and Reaction Centers of Photosynthetic Bacteria. Structure, Interactions, and Dynamics.* (M.E. Michel-Beyerle ed.) Springer Series in Chemical Physics 42, Springer-Verlag, Berlin 1985.
5. J.R. Norris, C.P. Lin and D.E. Budil, *J. Chem. Soc., Faraday Trans. I* **83**, 13 (1987).
6. R. Haberkorn, M.E. Michel-Beyerle and R.A. Marcus, *Proc. Natl. Acad. Sci. USA* **76**, 4185 (1979).
7. J. Jortner, *J. Am. Chem. Soc.* **102**, 6676 (1980).
8. M. Redi and J.J. Hopfield, *J. Chem. Phys.* **72**, 6651 (1980).
9. R.J. Cave, P. Siders and R.A. Marcus, *J. Phys. Chem.* **90**, 1436 (1986).
10. J.W. Warner and R.S. Berry, *Proc. Natl. Acad. Sci. USA* (in press, 1987).
11. For a recent review and list of references see e.g., H. Levanon, *Rev. Chem. Int.* (in press, 1987).
12. D. Holten, C. Hoganson, M.W. Windsor, C.C. Shenk, W.W. Parson, A. Migus, R.L. Fork and C.V. Shank, *Biochim. Biophys. Acta* **592**, 461 (1980)
13. T.L. Netzel, M.A. Bergkamp, C.-K. Chang and J. Dalton, *J. Photochem.* **17**, 451 (1981).
14. M.R. Wasielewski, M.P. Niemczyk, W.A. Svec and E.B. Pewitt, *J. Am. Chem. Soc.* **107**, 1080 (1985); and ibid. 107, 5562 (1985).
15. T. Gilbro, V. Sundstrom, A. Sandstrom, M. Spangfort and B. Andersson, *FEBS* **193**, 267 (1985).
16. H. Levanon and P. Neta, *J. Phys. Chem.* **86**, 4532 (1982).
17. H. Levanon and O. Gonen, *Chem. Phys. Lett.* **104**, 363 (1984).
18. H. Levanon and J.R. Norris in: *Light Reaction Path of Photosynthesis* (F.K. Fong ed.) Springer Verlag, Berlin, 1982, Vol. 35, p. 35.
19. H. Levanon, A. Regev and P.K. Das, *J. Phys. Chem.* **91**, 14 (1987).
20. R.G. Little, J.A. Anton, P.A. Loach and J.B. Ibers, *J. Heterocycl. Chem.* **12**, 343 (1975).
21. O. Gonen and H. Levanon, *J. Phys. Chem.* **89**, 1637 (1985).
22. O. Brede, R. Mehnert, W. Naumann and J. Teple, *Ber. Bunnsenges. Phys. Chem.* **89**, 1036 (1985).
23. E.J. Land, A. Sykes and T.G. Truscott, *Photochem. Photobiol.* **13**, 311 (1971).
24. N.-H. Jensen, R. Wilbrandt and P.B. Pagsberg, *Photochem. Photobiol.* **32**, 718 (1980).
25. O. Gonen and H. Levanon, *J. Chem. Phys.* **84**, 4132 (1986)
26. A. Regev, T. Galili, H. Levanon and A. Harriman, *Chem. Phys. Lett.* **131**, 140 (1986).
27. R.A. Brookfield, H. Ellul, A. Harriman and G. Porter, *J. Chem. Soc. Faraday Trans.* II **82**, 219 (1986), and references therein.
28. F.C. de Schyver, N. Boens and J. Put, in: *Advances in Photochemistry* (J.N. Pitts Jr., G.S. Hammond and K. Collnick eds.), Wiley, New York, Vol. **10**, p. 359.

29. V.C. Anderson and R.G. Weiss, *J. Am. Chem. Soc.* **106,** 6630 (1984).

30. F. Hirayama, *J. Chem. Phys.* **42,** 3163 (1965).

31. S.L. Murov, in: *Handbook of Photochemistry,* Dekker, New York, 1973, p. 5.

32. R.A. Brookfield, H. Ellul, A. Harriman, *J. Chem. Soc. Faraday Trans. II,* **81** 1837 (1985).

33. A. Harriman, *J. Chem. Soc. Faraday Trans. I* **76,** 1978 (1980).

34. R.L. Ake and M. Gouterman, *Theoret. Chim. Acta* **15,** 20 (1969).

35. M. Gouterman and G.-E. Khalil, *Mol. Spectrosc.* **53,** 88 (1974).

36. L. Benthem, *Ph.D. Thesis,* Wageningen Agricultural University, The Netherlands, 1984.

37. D. Gust, T.A. Moore, R.V. Bensasson, P. Mathis, E.J. Land, C. Chachaty, A.L. Moore, P.A. Liddel and G.A. Nemeth, *J. Am. Chem. Soc.* **107,** 3631 (1985).

38. J.C. Mialocq, C. Giannotti, P. Maillard and M. Momenteau, *Chem. Phys. Lett.* **112,** 87 (1984).

COUPLING BETWEEN LOCALLY EXCITED AND CHARGE TRANSFER STATES OF JET COOLED MOLECULAR COMPLEXES.

M. Castella[a], P. Claverie[b], J. Langlet[b],
Ph. Millié[a], F. Piuzzi[a] and A. Tramer[c].
[a]CEN Saclay, Département de Chimie Physique,
91191-Gif-sur-Yvette (France)
[b]Laboratoire de Dynamique des Interactions Moléculaires
Université Paris VI, 4,Place Jussieu, 75005-Paris (France)
[c]Laboratoire de Photophysique Moléculaire, CNRS,
Université Paris-Sud, 91405-Orsay, (France).

ABSTRACT. Fluorescence and fluorescence-excitation spectra were recorded for a series of anthracene and perylene complexes with aromatic ligands (electron donors). Perturbation calculations were carried out for some model systems in order to evaluate potential energy surfaces in their ground, locally excited and charge-transfer electronic states. The vibrational structure of excitation (emission) spectra may be qualitatively explained by the amount of perturbation between both excited states.

1. INTRODUCTION

It is well known since the pioneer Mulliken's work [1] that the excited states of molecular complexes with weakly bound, van der Waals ground state (G) − A.B may be divided into two groups :
(i) locally excited (LE) states − $A^*.B$ (or $A.B^*$) correlated at $R_{AB} \to \infty$ with excited state (A^*) and ground state (B) of neutral molecules and stabilized by a van der Waals interaction, usually a little stronger than in the ground state and
(ii) charge-transfer (CT) − $A^- B^+$ states correlated to the ion pair and strongly bound by a Coulomb interaction.

Since the Coulombic energy does not depend on details of molecular structures in a series of complexes with aromatic donors (B) and acceptors (A), the relative positions of the first LE and CT states depend mainly on the value of the parameter :

$$\Delta = I_B - E_A - \epsilon_A^* \dots\dots\dots\dots\dots\dots\dots\dots\dots(1)$$

where I_B is the ionization potential of the donor, E_A − the electron affinity of the acceptor and ϵ_A^* − the energy of the lowest excited

83

J. Jortner et al. (eds.), Large Finite Systems, 83–99.

singlet state of A (we suppose $\epsilon_A^* < \epsilon_B^*$). For $\Delta \geq 4.5eV$, the lowest excited state of the complex (responsible for the fluorescence) is the LE state, while for $\Delta < 4eV$, the CT state is the lowest one [2].

This simple model was confirmed by extensive spectroscopic studies is condensed phases (solvent effects, charge transfer complexes, exciplexes etc...). More recently, the studies of van der Waals complexes formed in supersonic expansion have been extended to the A.B systems with the CT state as the lowest excited state[3-6].

In this work, we are mainly interested by the intermediate case, where LE and CT states are nearly resonant. As previously shown [5] the spectra of this type of complexes are complicated and suggest a strong coupling between LE and CT states. We will discuss here new data concerning the electronic spectra of some model molecular systems and preliminary results of computations of their potential energy surfaces. More complete data will be published elsewhere[7,8].

2. EXPERIMENT

Experimental techniques

Fluorescence and fluorescence-excitation spectra of molecular complexes were recorded using a free-jet expansion. He carrier gas was saturated with A and B compounds by passing through two, independently heated reservoirs. The jet was crossed at $x/d = 20$ to 40 with the light beam from a Lambdaphysik FL 2002 dye laser (bandwidth of ca. 0.3 cm^{-1}) pumped by a Lambdaphysik EMG 102 excimer laser. The fluorescence was recorded either through a low-resolution Bausch & Lamb monochromator (excitation spectra) or through a .6m Jobin-Yvon HRS3 monochromator with spectral slits of 4 to 25 cm^{-1} (fluorescence spectra).

Results

The spectra were recorded for anthracene and perylene complexes of benzene, phenol, anisole, ortho-dimethoxy-benzene (1,2-DMB) meta-dimethoxy-benzene (1,3-DMB), aniline, N-methyl-aniline (MA) and N-dimethyl-aniline (DMA). Only the anthracene-MA complex is not fluorescent.

The complexes may be divided-following their spectral properties schematically described in Table I — into three groups.

The perylene-phenol complex is a good representative of the Group I. A part of its excitation spectrum (including the $\overline{o}_0^0$ origin) and a part of the emission spectrum are represented in Fig.1. The $\overline{o}_0^0$ band is red shifted by ~455 cm^{-1} with respect to the $\overline{o}_0^0$ band of the bare perylene, what indicates a stronger intermolecular bond in the excited than in the ground state. The spectrum is composed of narrow bands (as narrow, as in the spectrum of the bare molecule with the widths limited by the rotational population distribution).

TABLE I.

Group	I	II	III
Excitation spectra			
Bandwidths	narrow	narrow	broad unresolved.
0→v' progressions	short	long	short
Intensity distribution	$I_0^0 > I_0^{v'}$	$I_0^{v'} > I_0^0$	–
Perturbations	no (or a few Fermi resonances)	strong (irregular multiplet structure)	–
Fluorescence			
Spectra	resonant (below the IVR onset)	red-shifted	strongly red-shifted.
Vibrational structure	simple, narrow band	complex (unresolved) + exciplex emission (?)	diffuse exciplex emission.
Lifetimes	$\tau_{(AB)^*} \simeq \tau_{A^*}$	$\tau_{(AB)^*} > \tau_{A^*}$	$\tau_{A^-B^+} \tau_{A^*}$
Perylene complexes	phenol $\Delta=4.7eV$ anisole $=4.4$	aniline $\Delta=3.9eV$ MA $=3.5$	DMA $\Delta=3.3eV$
Anthracene complexes	benzene $=5.25$ phenol $=4.5$	1,2 DMB $=4$ 1,3 DMB $=3.9$	aniline $=3.7$ DMA $=3.1$
Conclusion :	$E_{CT} > E_{LE}$	$E_{CT} \simeq E_{LE}$	$E_{LE} > E_{CT}$

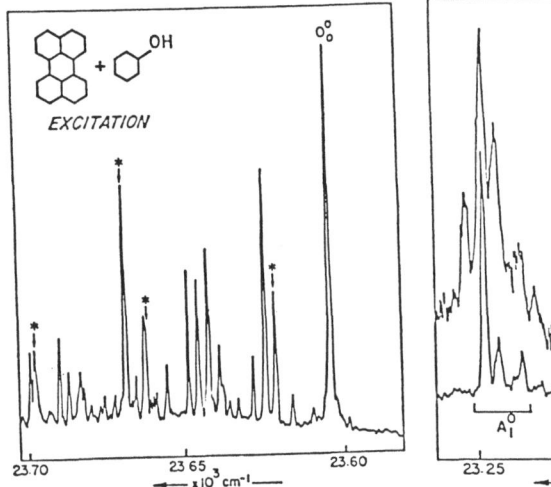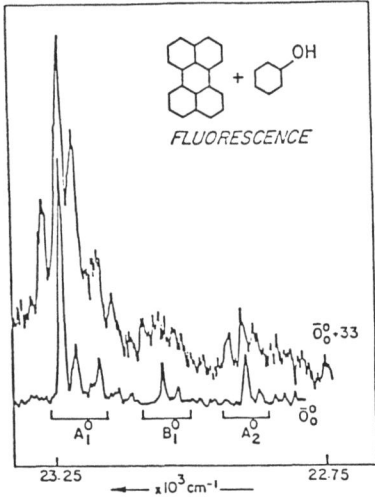

Fig.1. Spectra of the perylene—phenol complex : (a) fluorescence-excitation spectrum; bands coinciding with perylene hot bands are marked by asterixes. (b) fluorescence spectrum under $\bar{o}_0^0 = 23618$ cm^{-1} and 23652 cm^{-1} excitation. A and C indicate 351 cm^{-1} and 547 cm^{-1} intramolecular modes of perylene.

All strong bands may be assigned to overtones and combinations of three low-frequency intermolecular modes : $\omega_1 = 71$ cm^{-1}, $\omega_2 = 33$ cm^{-1} and $\omega_3 = 27.5$ cm^{-1} and to their combinations with intramolecular perylene modes. The fluorescence spectrum when excited below the onset of the intramolecular vibrational redistribution (IVR) corresponds to the resonant emission from initially excited level (Fig. 1b). In spite of a lower resolution, the structure due to ~30 and ~70 cm^{-1} modes combining with perylene frequencies may be easily seen. Vibrational progressions — in excitation as well, as in emission — are short with a maximum intensity in the $\bar{o}_0^0$ band, what corresponds to a similar equilibrium geometry in the ground and in the excited state. At last, the fluorescence lifetime is pratically identical to that of the bare perylene.

The spectra of complexes : perylene—anisole, anthracene—phenol as well, as those of as perylene—isoprene[9] and anthracene—benzene[10] reported by other authors are similar. They seem to correspond to transitions between two states with similar potential-energy surface and electronic excitation localized on the acceptor molecule. As expected from high values of the Δ parameter (Eq.1) the first excited state is a LE state unperturbed (or only weakly perturbed) by a higher lying CT state.

The DMA–perylene, aniline–anthracene and DMA–anthracene systems classed in the groupe III are classical examples of exciplexes, extensively studied in solutions. The structureless, broad–band fluorescence with lifetime much longer than that of the hydrocarbon[5,6] corresponds obviously to the "exciplex" emission from the CT lowest state of the complex. The same transition in absorption cannot be detected in the excitation spectrum (probably because of its low oscillator strength and diffuse structure, cf. Ref.3) or appears only as a continuous background. The main feature of the excitation spectrum is a broad, asymmetric band in the same frequency range, as the LE absorption in the Group I complexes. In the case of the perylene–DMA complex, the shape of this band may be approximatively fitted as an envelope of the band system with similar spacing and intensity distribution as in the phenol complex but homogeneously broadened ($\delta\nu$ of the order of 20 cm^{-1}. This picture is consistent with a "statistical limit" coupling of the low vibronic levels of the LE state to a dense quasi–continuum of the CT–state levels.

The supposed bandwidths correspond to the electronic relaxation rate 10^{12} s^{-1} i.e. of the same order of magnitude as for the $S_2 \rightarrow S_1$ relaxation in large isolated molecules.

TABLE II.

Fluorescence excitation spectrum of the anthracene –
1,2-DMB complex (27054.2 cm^{-1} band tentatively assigned as σ_0^0).

v'	$\nu - \nu_{00}$		$\langle \Delta \nu \rangle$
"hot" ?	−27.5		
0	0	ω	0
1	23.8	ω	24.5
	24.5	m	
2	48.1	s	48.7
	49.4	m	
3	71.2	vω	73.2
	72.5	ω	
	73.6	vs	
4	88.9	m	95.2
	95.7	vs	
5	110.5	vω	117.0
	112.6	ω	
	117.9	vs	
	122.2	vω	
6	136.6	s	139.0
	141.0	s	
7	158.4	vω	161.0
	160.8	m	
	161.1	m	
8	182.7	ω (b)	182.7

M. CASTELLA ET AL.

The spectra of the complexes classed in the second group have – in spite of some differences – a number of common features :

(i) The fluorescence excitation spectra are composed of bands as narrow, as in the first group but with an irregular spacing ; some of the bands are obviously split into two or more components with a ramdom intensity distribution. In the simplest case of the anthracene – 1,2–DMB complex (Fig.2),

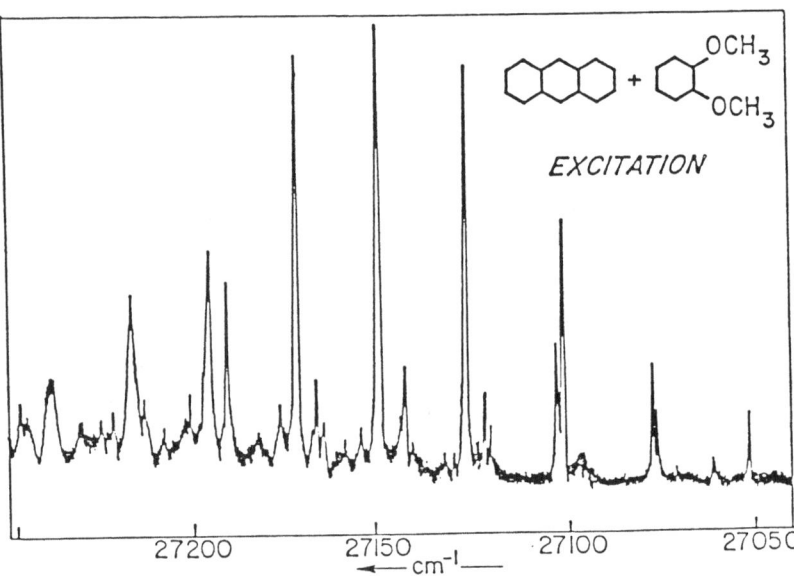

Fig.2. Fluorescence excitation spectrum of the anthracene – 1,2–DMB complex.

all strong lines may be assigned to a single progression if centers of gravity of each multiplet are considered as frequencies of the unperturbed bands (Table II). If the first relatively strong band is taken as the $\bar{\sigma}_0^0$ transition, a good fit is obtained by assuming $\omega_e = 25.0$ cm^{-1} and $\omega_e x_e = 0.2$ cm^{-1}. The characteristic feature of the more complex spectrum of the perylene–aniline system (Fig.3) is its doublet structure with irregular frequency intervals. By considering –as previously– gravity centers of each doublet, one can describe the spectrum by assuming two modes $\omega_1 = 70$ cm^{-1} and $\omega_2 = 8.5$ cm^{-1}. Surprisingly, the bands are narrow in the first and third clump ($v_1' = 0$ and 2) and strongly broadened in the second one ($v_1' = 1$). At last, the excitation spectrum of the anthracene–1,3–DMB complex is so complex and irregular that no vibrational analysis was still attempted.It seems impossible to explain the observed structure by intra-state pertur-bations (Fermi resonances). They suggest rather the coupling between two different electronic states.

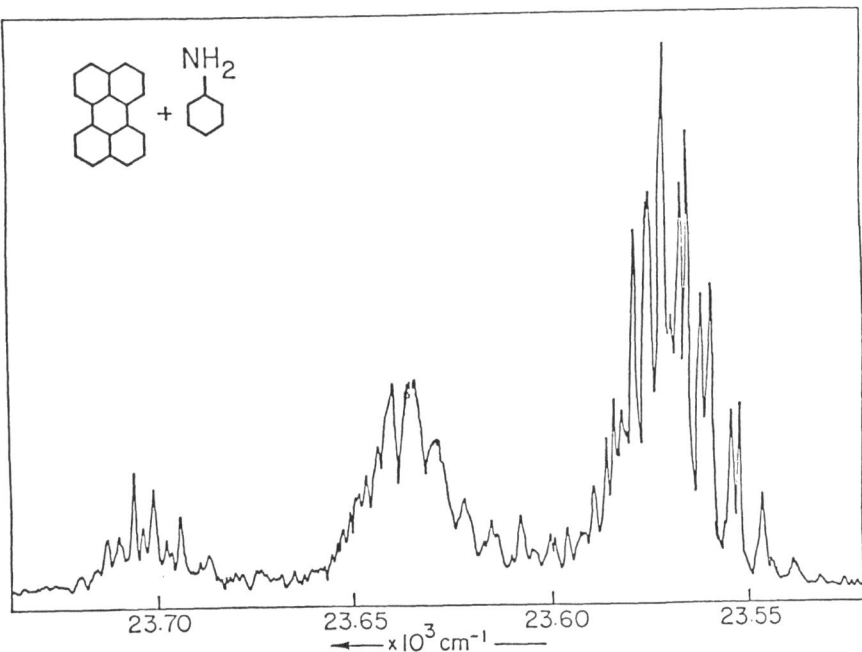

Fig.3. Fluorescence excitation spectrum of the perylene–aniline complex.

(ii) in all cases, the intensity distribution (at least for some of the intermolecular modes) is very different from that observed in the first group. The σ_0^0 band is very weak or missing. If the first observed band is considering as origin, the strongest band in the anthracene-1,2-DMB complex corresponds to v'=4 and in perylene-aniline complex to $v_2' = 6$

(iii) the fluorescence spectra are not resonant but more or less red shifted with respect to the excitation frequency and show only a diffuse broad-band structure, even in the case of excitation of relatively low levels (Fig.4).

Note that the resolution in fluorescence detection does not exceed 10–15 cm^{-1} and that the assignment of origin being uncertain we ignore the vibrational energy of initially excited level. The apparent absence of the fine structure in the fluorescence spectrum may be due either to the long progressions of closely spaced (and unresolved) bands or to a low onset of the IVR.

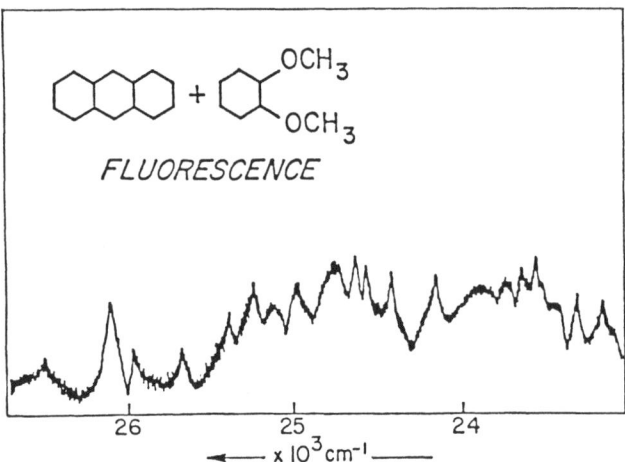

Fig.4. Fluorescence spectrum of the anthracene – 1,2-DMB complex excited in the 27078 cm^{-1} (tentatively assigned to σ_0^0 + 25 cm^{-1}).

(iv) the fluorescence lifetime is longer than that of the parent hydrocarbon (i.e. than that of the pure LE state) but much shorter than that of the "exciplex" fluorescence from a pure CT state. This effect is strongly pronounced in the perylene-aniline complex (5b).

All these features may be considered as resulting from a relatively strong, local perturbation of the LE state, bearing the major part of the oscillator strength by the "dark" CT state. Long progressions with weak or missing σ_0^0 band indicate a much more pronounced difference between equilibrium geometries of the ground and excited states than in the group I. The appearance of very low-frequency modes in aniline-perylene complex may be also considered as due to a deformation of potential energy surfaces. The vibrational analysis of the excitation spectra shows that almost all levels are perturbed. At last, the increase of the fluorescence decay time is a necessary consequence of the strong mixing of "radiant" and "dark" discrete states.

Such a behaviour corresponds to the "strong-coupling case" of the theory of radiationless transitions[11] in contrast to the "statistical limit" attained for the group III complexes. This implies either low density of CT levels in quasi-resonance with the low levels of the LE state or a very selective coupling to some specific CT levels.

The case of the perylene–MA complex must be considered separately. Its fluorescence–excitation spectrum is intermediate between very rich, narrow–band spectrum of perylene–aniline complex and almost completely structureless spectrum of perylene–DMA [5b]. The spectrum is much more structured than that of perylene–DMA but still composed of apparently broad bands. The spectral resolution being limited by the rotational envelope of bands, it is impossible to say, whether this width is due to a congestion of narrow bands closer spaced than in perylene–aniline spectrum or to the homogeneous broadening. The fluorescence spectrum and lifetime are also intermediate between those of aniline and DMA complexes.

POTENTIAL ENERGY SURFACES

1. Methods

In previous publications, binding energies of homo– and heterodimers of aromatic molecules in their ground state have been calculated using either empirical atom–atom potentials [6,10] or a localized multipole expansion neglecting polarization and dispersion effects [12] or by the SCF treatment of the "supermolecule" with empirical corrections for dispersion energy [13]. Here, we apply for evaluation of potential energy surfaces of the complex in its ground ($G = A.B$), locally excited ($LE = A^*.B$) and charge–transfer ($CT = A^-B^+$) zero–order electronic states the simplified formulae elaborated by Claverie et al.[14b] based upon the "exchange perturbation" theory (see Ref.14b and references therein). In this model, the total energy of the complex is a sum of independently calculated terms corresponding to electrostatic, polarization, dispersion and repulsive interactions :

$$E_{tot} = E_{el} + E_{pol} + E_{disp} + E_{rep} \cdots\cdots\cdots(2)$$

At first, the charge distribution in all the concerned states of acceptor (A, A^*, A^-) and donor (B, B^+) molecules is approximated by a set of multipoles (charges, dipole and quadrupole moments) localized on atoms and centers of intramolecular bonds. The multipoles are evaluated by an ab initio SCF treatment using a minimum set of gaussian orbitals (1s for H, 1s, 2s and 2p for C, N and O atoms). The polarizibility of each molecule is represented as a set of isotropic atomic polarizibilities assuming additivity of molecular refractions.

– the electrostatic energy is then calculated as a sum of interactions between all local multipoles using well known classical formulae.
– the polarization effects are calculated in the same way as sum of interactions between all permanent and induced moments.
– the dispersion and repulsion effects are described by the formula :

$$E = \sum_i^{(A)} \sum_j^{(B)} k_i \, k_j \left[-A \left(\frac{1}{\rho_{ij}^6} + \frac{c'}{\rho_{ij}^8} + \frac{c''}{\rho_{ij}^{10}} \right) + C \, \sigma_{ij} \, e^{\alpha \rho_{ij}} \right] \dots (3)$$

where i and j number the atoms of A and B molecules ρ_{ij} are related to interatomic distances ρ_{ij} by :

$$\rho_{ij} = \frac{R_{ij}}{2(W_i W_j)^{1/2}} \quad \dots\dots\dots\dots\dots\dots \quad (4)$$

where W_i, W_j are van der Waals radii. Atomic parameters k_i, k_j and A, C, C', C" and α constants are deduced from sublimation energies and from model calculations for simple molecules [14] and applied without any changes. The factor σ_{ij}

$$\sigma_{ij} = (1-q_i/N_i)(1-q_j/N_j) \dots\dots\dots\dots\dots\dots(5)$$

(where q are effective charges and N — numbers of valence electrons for i and j atoms) accounts for the dependence of repulsion forces on the electronic populations of atomic orbitals.

In order to explore the potential energy surface, the molecule B is placed in arbitrary position with respect to the fixed molecule A and allowed to translate and rotate until it gets trapped in an energy minimum. By repeating this operation for different initial positions we obtain in principle a complete map of the surface. The replacement of continuous charge distribution by localized interactions may be responsible for minor artifacts : a number of pseudominima corresponding to nearly the same energy and only slightly different configuration in the region where the surface is flat. On the other hand, the assumption of identical polarizibilities in all states of the acceptor leads to an under-estimation of the binding energy in LE (A.B) state. An empirical correction to E_{disp} was thus applied in order to bring $E_{A^*.B} - E_{A.B}$ into agreement with the observed red shift in the excitation spectra of perylene-phenol and anthracene-phenol complexes. The same corrections were then used to all complexes of the same acceptor.

2. Results

1. In the G and LE states with a van der Waals bond, the stability of the complexes is essentially due to dispersion forces, the electrostatic and polarization terms being small and, in most cases, of opposite signs (Table III). The same conclusion about a predominating role of dispersion was derived in a quite different way in the case of benzene and tetrazine dimers[13]. As can be seen, the ground-state binding

TABLE III.

Bonding energies and configurations (in kcal/mole) of perylene
complexes in their ground (G) and charge transfer (CT) states
(average values for a few deep "minima").

Donor state[a]		E_{el}	E_{pol}	E_{disp}	E_{rep}	E_{tot}	z_{CM}[b]	z_N[c]
phenol	G - ρ	0.2	- 0.2	- 9.4	4.1	- 5.3	3.45	3.42
	CT - ρ	-79.1	-11.9	-14.4	-13.4	-92	3.21	3.16
	CT - t	-80.4	-12.3	-14.8	14.3	-93	3.24	2.95
aniline	G - ρ	0.9	- 0.3	- 9.2	3.8	- 4.8	3.5	3.5
	CT - ρ	-78	-11.6	-15	14	-90	3.17	3.05
	CT - t	-84	-15	-14.6	17	-96	3.65	2.5
N-methyl aniline	G - ρ	0.8	- 0.33	-11.3	4.9	- 6.2	3.52	3.39
	CT - ρ	-77.2	-14.1	-16.5	15.0	-92.7	3.27	3.21
	CT - t	-80.1	-15.4	-17.4	17.0	-95.9	3.43	2.97
N-dimethyl aniline	G - ρ	- 0.2	- 0.2	-11.4	4.8	- 7.0	3.55	3.54
	CT- ρ	-76.9	-12.3	-17.9	16.0	-91.4	3.30	3.26

[a] ρ - almost parallel configurations, t - strongly tilted configurations.

[b] distance of the center of mass of the donor from the acceptor
plane (in Å).

[c] the same for the N (or O) atom of the donor.

energy is not very different for different complexes, of the order of −4 to −7 kcal/mole in qualitative agreement with experiment : the dissociation threshold of the perylene–DMA complex in the LE state is not attained with vibrational energy excess of ca. 2000 cm^{-1}, what corresponds to the ground–state binding energy E_G > +4.5 kcal/mole [7].

2. The most stable configurations correspond to stacked structures with almost parallel aromatic rings but T–shaped structures are also relatively stable. The most detailed study of the perylene–MA complex yields E_G = −6.23 kcal/mole for the most stable stacked structure and E_G = −5.31 kcal/mole for a T–shaped one with CN bond parallel to the perylene plane and the methyl group pointing to this plane. This agrees with previous observations of isomeric forms for perylene–MA and anthracene–DMA complexes[5b].

3. Potential–energy surfaces are relatively flat for rotations and translations of the B molecule in the polane parallel to the A–molecule plane. We cannot decide on this stage whether we have to do with a single, broad minimum or with several shallow minima separated by low energy barriers. In any way, all stable configurations correspond to the substituent group (−OH, −NR$_2$) of the donor located above the central aromatic ring of perylene or anthracene while its benzene ring is displaced towards the edge (or even out of it) of the acceptor molecule. This tendency is enhanced when the positive charge on the substituent group increases (i.e. from phenol to DMA).

4. In strongly bound CT states electrostatic interactions become predominant. In the case of amine donor ions, a large part of the positive charge is localized in the −NR$_2$ group which is strongly attracted by the negative charge distributed over the whole aromatic system of the acceptor. This attraction is, however, counterbalanced by rapidly increasing repulsion forces, when the donor benzene ring and methyl groups (if present) approach closely the acceptor plane. The compromise between two forces will lead, when possible, to deviations from the parallell configurations : strong in aniline complex, less pronounced in MA complex because of the steric hindrance by the methyl group and negligible in strongly hindered DMA (Fig.5). Since aniline, MA and DMA differ not only by steric effects of methyl groups but also by their ionization potentials (I_B = 7.69, 7.32 and 7.12 eV, respectively), the interaction between LE and CT states is significantly different for their perylene complexes.

In perylene–DMA the energy minimum in the CT state corresponds to a parallel configuration (the angle between the long axis of DMA and perylene plane : $\simeq$ = 0°). Because of the low ionization potential of DMA, the lowest excited state of the complex is the CT state, ca. 5.5 kcal/mole below the LE state. The both surfaces cross at a relatively large intermolecular distance.

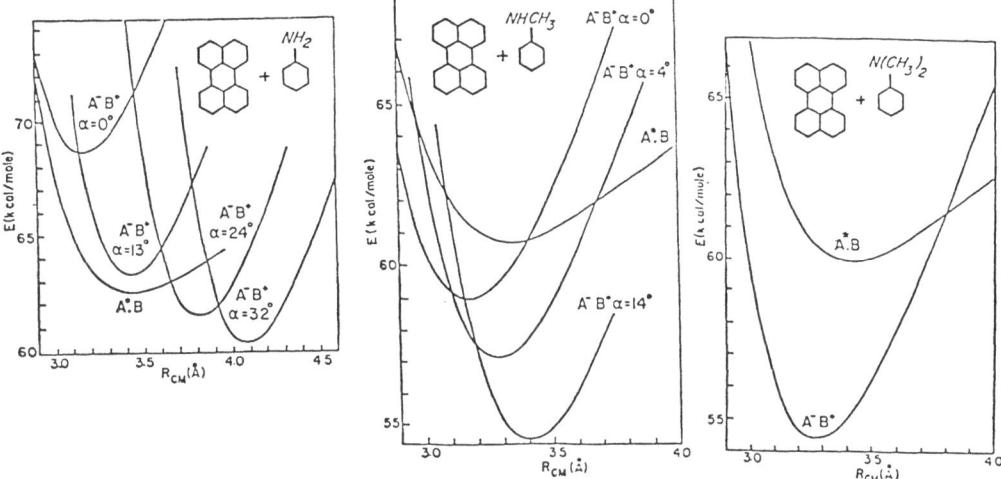

Fig.5. Calculated potential energy curves as a function of the distance
of the center of gravity of the donor molecule to the acceptor plane $-R_{CM}$
for different tilting angles (cf. insert in Fig.6) for perylene
complexes with aniline, MA and DMA.

In aniline, we find the energy minimum for a parallel CT
configuration above that of the LE state. In order to minimize the energy
of the CT state the donor molecule must be tilted by about 30°. In this
way, the distance of amino H and N atoms and of the center of the donor
benzene ring from the perylene plane are of 2.23, 2.49 and 4.1 Å instead
of 3.15 Å in the parallel configuration. The CT and LE potential energy
surfaces cross at an intermediate α angle giving an energy barrier
between both states (Fig.6).

In MA, the minimum of the parallel CT state lies slightly below that
of the LE state but still we obtain an important energy gain for a tilted
configuration : α = 10° plus a rotation around a long axis of the donor by
ca. 19°. The distance from the perylene ring for $-CH_3$ group, amino H
atom, N atom and the center of the ring are now of 3.23, 2.64, 2.97 and
3.41 Å, respectively.

Such a close approach of the hydrogen to the acceptor aromatic
system suggests a kind of hydrogen bonding to π–electrons. This is
consistent with experimental data in solutions : while anthracene and
pyrene complexes with DMA are strongly fluorescent, the fluorescence is
completely quenched in aniline and MA complexes[15]. This effect is
tentatively explained by the proton transfer in the excited state

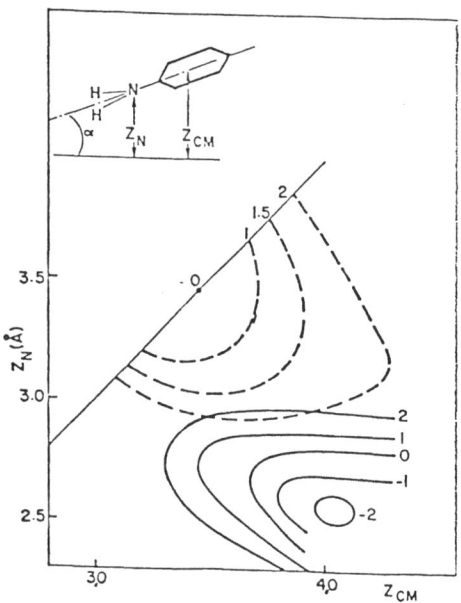

Fig.6. Simplified picture of potential surfaces in R_{CM} and R_N coordinates (defined in the insert) for LE (broken lines) and CT (solid lines) for the perylene-aniline complex. The energy dependence on other coordinates is neglected.

followed by recombination after the electronic relaxation ; this explanation is corroborated by an efficient photochemical hydrogen-deuterium exchange when $C_6H_5ND_2$ is used as donor[16]. The complexes formed in supersonic nozzles are, however, fluorescent excepted for anthracene-MA complex, where no emission was detected : in this case the amino hydrogen atom comes closely to the reactive C (9) site of the anthracene molecule and the proton transfer may take place without activation energy. It is more difficult to explain why such a process is not so efficient in the aniline complex.

DISCUSSION

Let us take as an example the series of complexes of perylene with aniline, MA and DMA.

In the DMA complex, the energy of the most stable (parallel) configuration of the CT state is lower by about 1500 cm^{-1} than that of the first LE state. This energy gap is large as compared to the average frequency of intermolecular modes ; the density of the vibronic levels of the CT state isoenergetic with the lowest levels of the LE state is high enough to form a dense, dissipative quasi-continuum. Such a level pattern is consistent with a rapid electronic relaxation and with a large red shift between $| G > \rightarrow | LE >$ absorption and $| CT > \rightarrow | G >$ emission. On the other hand, the surfaces of LE and CT states cross at a relatively large intermolecular distance, so far from the LE state energy minimum (i.e. from the Franck-Condon region in absorption) that the intensity distribution in the excitation spectrum is not very different from the case of an unperturbed LE state.

For the aniline complex in parallel configuration we find the CT state above the LE state (note that energy gap may be overestimated in our model by neglect of reverse charge transfer). The interaction between two states may induce a significant deformation of its energy surface : A relatively long progression in the ~ 70 cm^{-1} mode (tentatively assigned to the stretching vibration) and appearance of a long progression with an exceptionnally low frequency of 8.5 cm^{-1} may be considered as an evidence of such a deformation.

The energy minimum of the CT state corresponds to the tilted configuration. The potential energy surfaces cross at an intermediate tilting angle forming a kind of energy barrier. (The existence of such a potential energy barrier for exciplex formation was already postulated by Saigusa et al. in the case of 1-cyanonaphthalene-triethylamine complex[11]). It may explain a relatively weak coupling (level splitting of 1-4 cm^{-1}) between low levels of the LE state and a small number of quasi-resonant levels of the CT state. Franck-Condon factors are unfavorable excepted for levels involving large-amplitude bending vibrations. From this point of view, the broadening of individual bands in the $v_1' = 1$ clump in the excitation spectrum may correspond to a more efficient coupling.

It is interesting to note, that the red shift of the complex σ_0^0. band with respect to the o_0^0 transition in the parent hydrocarbon is significantly smaller in the strongly bonded DMA complex than in the whole series of weaker bonded systems including benzene-perylene complex[10]. A possible reason of this anomaly may be level reversal occuring when we move from the Group II to the Group III complexes. As long as the CT state is located above the LE state, their interaction will push the LE state to lower energies (Groupe I and II), while in Groupe III systems the interaction with the lowest CT state will increase the energy of the LE state.

The case of the perylene-MA complex is more difficult. Our calculations give the potential energy surfaces of two states crossing (in the parallel configuration) in the vicinity of the energy minimum of the LE state. Since slightly tilted CT configurations have still lower energies, one can expect strong perturbations of the LE state in the

Franck–Condon region and a relaxation from the LE to the CT state even more efficient than in the DMA complex what implies a broad–band, structureless excitation spectrum in the | G > → | LE > spectral range.

It is obviously not the case : the structure in the excitation spectrum of the MA complex is better resolved than in DMA complex and the fluorescence spectrum does not correspond to a pure exciplex emission.

We cannot propose any convincing interpretation of this discrepancy. It is however interesting to note that the case of the MA–perylene complex is not unique. A similar broad–band structure was observed in the fluorescence excitation spectra of DMA complexes with 9–methyl-anthracene and 9,10–dimethyl–anthracene, while that of the anthracene complex is completely structureless[6b]. The methyl substitution decreases probably the electron affinity of the acceptor; the effect is similar to that due to the increase of the donor ionization potential in the MA and DMA complexes of perylene.

CONCLUSIONS

The amount of experimental data for molecular complexes involving two aromatic molecules is still very limited and any attempt of genera-lization would be dangerous. it is nevertheless, interesting to note that inter–molecular modes – non-active or weakly active in complexes with rare–gas atoms or simple molecules – become optically active in complexes with aromatic ligands and their activity (length of progres-sions) increases with decreasing ionization potential of ligands (donors). This effect reflects an increasing difference between the shape of the potential energy surface in the ground and in the first locally excited state. It seems reasonable to describe this observation in terms of the coupling between the LE state and a higher lying CT state, non-negligible even in the case of a large energy gap. A correct description of complicated spectra of complexes where the surfaces of LE and CT states cross or anti–cross would necessitate more elaborated theoretical models including the configuration–dependent coupling between both states.

REFERENCES :

1. R.S. Mulliken, *J. Amer. Chem. Soc.* 74, 811, (1952).
2. R.S. Mulliken and W.B. Person, *Molecular Complexes* Wiley, New-York (1969).
3. T.D. Russell and D.H. Levy, *J. Phys. Chem.* 86, 2718, (1982).
4. (a)H. Saigusa and M. Itoh, *Chem. Phys. Letters* 106, 391 (1984). *J. Chem. Phys.* 81, 5692, (1984).
   (b)H. Saigusa, M. Itoh, M. Baba and I. Hanazaki, *J. Chem. Phys.* 86, 2588, (1987).

5. (a)M. Castella, J. Prochorov and A. Tramer,
       J. Chem. Phys. <u>81</u>, 2511, (1984).
   (b)M. Castella, F. Piuzzi and A. Tramer,
       Chem. Phys. Letters <u>129</u>, 105, 112, (1986).
6. O. Anner and Y. Haas,
   (a) *Chem. Phys. Letters* <u>119</u>, 199, (1985).
   (b) to be published.
7. M. Castella, F. Piuzzi and A. Tramer (to be published).
8. M. Castella, P. Claverie, J. Langlet and Ph. Millié
   (to be published).
9. M.M. Doxtader and M.R. Topp, *J. Phys. Chem.* <u>89</u>, 4291, (1985).
10. M.M. Doxtader, E.A. Mangle, A.K. Bhattacharya, S.M. Cohen
    and M.R. Topp, *Chem. Phys.* <u>101</u>, 413, (1986).
11. A. Nitzan, J. Jortner and P.M. Rentzepis
    Proc. Roy. Soc. London <u>A327</u>, 367, (1972).
12. S.L. Price and A.J. Stone *J. Chem. Phys.* <u>86</u>, 2859, (1987).
13. J. Pawliszyn, M.M. Szczesniak and S. Scheiner,
    J. Phys. Chem <u>88</u>, 1726, (1984).
14. (a)J. Langlet, P. Claverie, F. Caron and J.C. Boeuve
       Int. J. Quant. Chem. <u>20</u>, 299, (1981).
    (b)P. Claverie in : *Intermolecular Interactions (B. Pullman — Ed)*
       Wiley, New-York (1978).
15. T. Okada, T. Mori and N. Mataga,
    Bull. Chem. Soc. Japan <u>49</u>, 3398, (1976).
16. J. Gebicki, W. Reimschüssel and T. Nowicki
    Chem. Phys. Letters <u>59</u>, 197, (1978).

Intermolecular potential and dissociation of excited state mercury ammonia complexes.

R. van Zee, W.B. Bosma, T. Zwier,
Calvin College, Department of Chemistry
Gran Rapids, Michigan
M.C. Duval and B. Soep,
Laboratoire de Photophysique Moléculaire du
C.N.R.S.), Université de Paris Sud, France.

ABSTRACT. The excitation of mercury ammonia van der Waals complexes allows the direct observation and characterisation of the well known exciplex in a spectroscopic manner. The complex has been formed in a supersonic expansion and excited by a laser, two electronic states were observed very deep (>5000 cm^{-1}) and showing strikingly different behaviour. The A state is not oberved in fluorescence but dissociates into Hg($^3P_0$) and NH3, while the other state, B fluoresces. The vibrations have been assigned in both states through isotopic substitution (ND3) as well as the electronic nature of the A and B states through rotational contours. This gives a picture of the states and their decay modes helpful to interpret the formation of the related exciplex observed in collisions of Hg $^3P_1$ +NH3.

INTRODUCTION.

The correspondence between collision complexes and van der Waals complexes has been the focus of our attention over the past years. There is a more than formal relation ship inbetween the photodissociation of a van der Waals complex and the collision of the equivalent pair . We thus studied electronic relaxation (1) and reactive collisions (2) through excited state complexes. This has led us to study in detail the intermolecular potential within the complex in order to investigate the effect of the intermolecular movements upon the decay routes of the system. Metal-molecule complexes are attractive candidates for such experiments as their photoexcitation reveals many internal modes; among them mercury systems are easy to prepare and pertain to a large litterature relevant to collision

J. Jortner et al. (eds.), Large Finite Systems, 101–111.

induced effects,"the mercury sensitisation".

We report here some new results on mercury ammonia obtai
ned by the cooperation of two groups, one in Calvin college
the other at Orsay. The mercury ammonia system has long
been known to form exciplexes through collisions of
Hg($^3P_1$)+NH3 (3) , hence if the relevant potential can be
accessed by collisions, why not excite it directly via the
ground state Hg-NH3 complex (the direct excitation of aromatic charge transfer complexes is also reported in this
volume (4))? Hikida and Mori(5) as well as Callear et
al.(6), have found evidence of a multistep process in the
Hg-NH3 exciplex emission. The first step results in metastable Hg($^3P_0$) formation trough collisions while in the second a short lived excimer is produced by three body
collisions, emitting a broad fluorescence centered at λ
≈ 280 nm. From Hikida(4) this unrelaxed molecule correlates to Hg($^3P_1$)+NH3 and can be further relaxed to produce
the stabilised emission at 340 nm, while Callear (6) assumes that the stabilised excimer correlates to $^3P_0$. Recently
Tsuchiya (7) excited directly the ground state collision
pair Hg-NH3 and showed that the precursor could be attained
by optical excitation, thus showing that it pertained to a
potential correlating to Hg($^3P_1$)+NH3. Subsequent collisions
also lead to the the formation of the stabilised excimer.

We propose here the spectroscopic identification of the
precursor through the photoexcitation of the ground state
Hg-NH3 complex. Detailed informations have been obtained on
the molecular levels of the complex, their symmetry and
their dissociation routes, giving a convincing picture for
the nature of the excimer.

EXPERIMENTALLY, the complexes are formed in a supersonic
expansion (2,8) of a mixture of mercury, ammonia and argon
or helium. Argon expansion is more effective to form the
complexes, however as it was found necessary to identify
the various species, helium was used as a more progressive
clusterisation is observed. as a function of the backing
pressure, and careful NH3 concentration dependence was
thus undertaken.

Fluorescence excitation spectra as well as action spectra have been recorded with .15 or.3 cm^{-1} maximum resolution allowing rotationally resolved spectra to be taken.
The action spectra probe the dissociation products of the
complex by LIF, while a pump laser scans the the complex
potential thus measuring the efficiency of each level to
promote the dissociation into the observed channel. We probe here Hg($^3P_0$) formation using a laser fixed at 404.7 nm,
a metastable mercury transition.

THE MOLECULAR VIBRATIONS OF THE EXCITED STATE HG-NH3
COMPLEX.

    We shall endeavour to describe the observed spectra in
the simplest fashion; the complex will be considered as
composed of the mercury chromophrore attached to a rigid
ammonia molecule taken as a rigid ball and the geometry of
the ammonia moiety is not supposed to differ in the complex
excited state. This rigid rugby ball will exhibit stretching movements corresponding to a pure V(R $_{Hg-NH_3}$) potential and torsional movements corresponding to a NH3 twist,
off the molecular axis (Hg-NH3). The stretching is the
equivalent to the vibration of a diatomic composed of Hg
and NH3. In a good approximation these movements are uncoupled near the bottom of the torsional well in the usual separation of the radial and angular potentials (9).

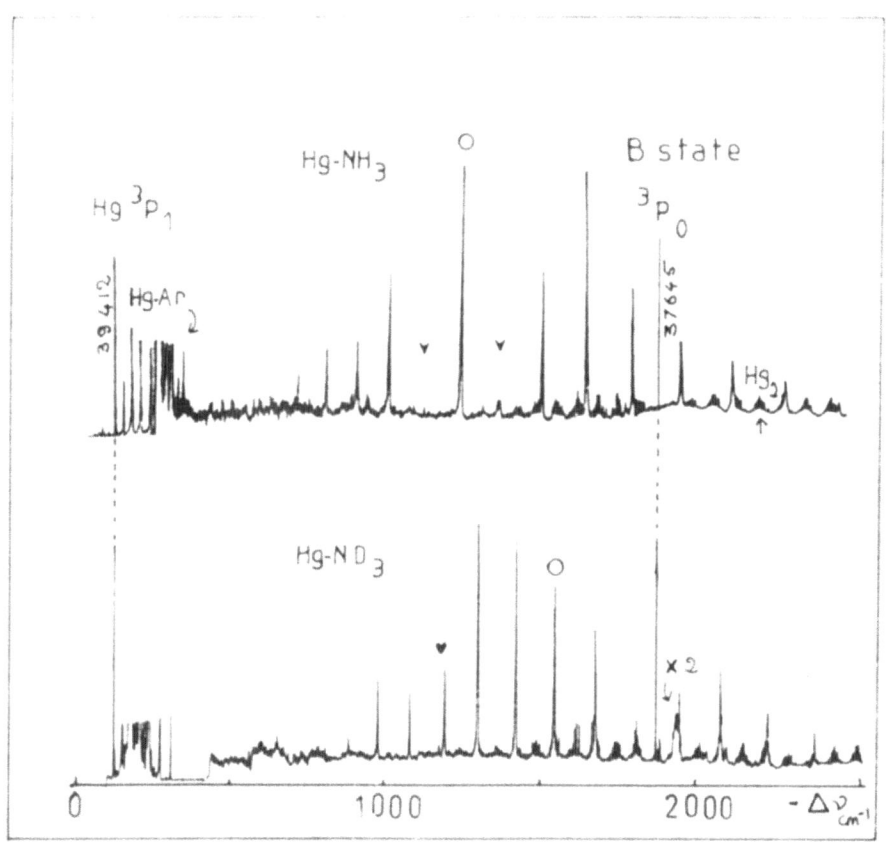

Figure1: fluorescence excitation spectra B state, upper
Hg-NH3, lower Hg-ND3. The atomic transitions are indicated,
the ♥ signals the perturbed intensities,the O,the corresponding quanta in Hg-NH3/ND3.

We shall refer to the potentials as pertaining to the $^3P_1$ and $^3P_0$ states, thus in the quasi diatomic approximation the J momentum of mercury is coupled to the molecular axis to yield two states $\Omega=1$ and $\Omega=0^+$ for $^3P_1$ and one $\Omega=0^-$ for $^3P_0$. These electronic states have been observed in Hg-Ar(10,11) and in Hg-N2 (12) while in the latter complex molecular motion is also observed.

The fluorescence excitation spectra.

The first spectrum displayed in figure 1a reveals a simple one membered progression ,diatomic-like, which we have assigned to the Hg-NH3 stretch. This assignment can be verified through isotopical substitution of NH3 with deuterium whose corresponding spectrum appears in figure 1b.We suppose that the stretch potentials are the same for both NH3 and ND3 molecules,and also that the potential fits a Morse function all the way down to the bottom of the well, which derives from linear Birge-Sponer plots (fig.2).

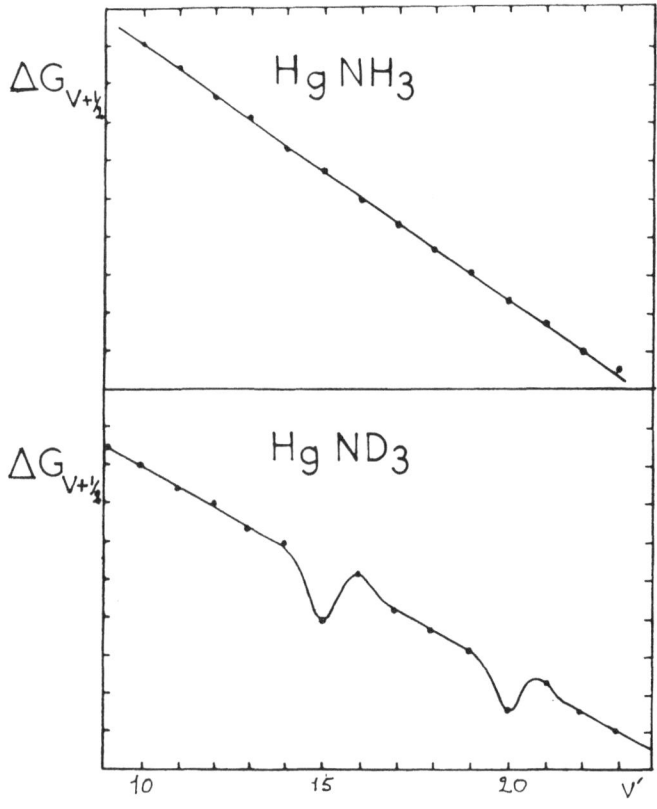

Figure2: Birge Sponer plots in the B state: upper Hg-NH3, lower HgND3.

The identity of the stretching progression is confirmed
when comparing the ratio of the anharmonicities
$w_e x_e$(Hg-ND3)/$w_e x_e$(Hg-NH3)=.84, with the ratio of the reduced   masses   of   the   quasi   diatomics   molecules
μ(Hg-ND3)/μ(Hg-NH3)=.85.
The spectra displayed in figure 1 are similar but they show
perturbations in position as seen in the B.S. plot (fig.2b
for ND3) and intensity in their bluer part. The fluorescence lifetime of the respective vibrational transitions does
not vary much off 200ns (a greater value than the bare
mercury $^3P_1$ decay time,120 ns) except for these perturbed
bands. Hence they must undergo a more efficient nonradiative process which may yield either:

$Hg(^3P_0)$+NH3   or

$Hg(^1S_0)$+H+NH2.

We investigated the first mechanism through the action
spectra detecting $3P_0$ metastable mercury, we also tried to
observe HgH by the same method but without success.

The $^3P_0$ action spectra.

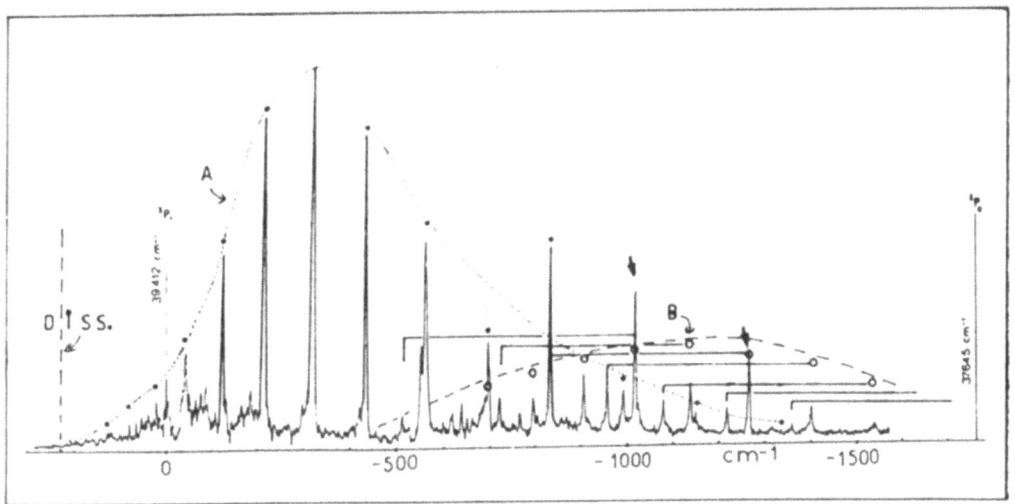

Figure3: action spectrum, A and B are observed. A=full
circles, B= open circles; the B torsions are represented by
horizontals and the arrows denote the perturbed quanta
v_s=19 and 21 in B.

The spectrum in figure 3 shows a completely different
picture as opposed to LIF spectra (fig.1). However all the
bands observed in the preceding also appear here but with
different intensities :the stronger here are the weaker in
the LIF spectra which means that $^3P_0$ is a preferred non radiative decay channel.

On the other hand two other progressions are noticeable, which could not be seen in the LIF spectrum. The first one is a combination of the B state stretch and another vibration that we have to assign to the NH3 torsion within the complex. The latter bands can be combined with the main progression with either $\approx 40, 180, 310, 460$ cm^{-1} spacings in Hg-NH3 .We chose the 460 cm^{-1} spacing in Hg-NH3 with 330 cm^{-1} in Hg-ND3 giving a ratio of 1.3 close to the expected one, $(B(NH3)/B(ND3))^{1/2}$, for the ammonia torsion within the complex. These frequencies seem rather high at first sight, but this well of the complex, B, will be found as high as 5096 cm^{-1} and can accept a large anisotropy. The anisotropy can be estimated also from the electostatic interaction of the quadrupole due to the Hg 6p orbital with the ammonia dipole,which amounts to 2000 cm^{-1} with a 2.3 Å separation we shall determine in the following. Such a barrier to torsion can, if compared to the ammonia inversion barrier, produce a large bending frequency.

The latter difference in the isotope torsion frequencies acts as a zero point energy shift upon the stretch levels and, as the torsion vibration is doubly degenerate (E type here in C_{3v}), the shift equals the vibrational frequency, w_T. Thus the isotope shift between the same stretch quanta v_s of the Hg-NH3/-ND3 molecules is:

$$\Delta v = (1-p)(v_s+\tfrac{1}{2}) - (1-p^2)(v_s+\tfrac{1}{2})^2 + w_T(\sqrt{2}-1) - \tfrac{1}{2}w_e x_{eT}(v_s+\tfrac{1}{2})$$

We thus find two possible combinations for the isotopically substituted B state stretch bands, yielding two assignments differing by 12 quanta, and giving well depths of 5096 and 3600cm^{-1}, ie very deep wells for van der waals complexes.

On the blue side of the action spectrum another progression is observed disconnected from the latter B state and with a larger spacing ; we assign it to a different electronic state of the complex, A, correlating to Hg $^3P_1$. Following the same line of arguments as in the preceding, this progression belongs to the stretch movement, but, to assign the quanta we have to assume that the torsion frequency is the same in this state as in B, as no combination bands are observed here.

Finally the onset of the spectrum on the red side yields the ground state energy, $D_o=250$ cm^{-1}, corresponding to the threshold of the dissociation into ^3P_o +NH3.

Thus we can describe up to now the potentials in the A and B states with the quasi-diatomic approximation, and we are going to verify that the electronic structure of these states can be related to the excitation of the mercury atom within the complex ,through rotationally resolved spectra.

ROTATIONAL STRUCTURE IN THE A AND B STATES.

The B state ,in fluorescence excitation exhibits a structure apparently simple under .1cm⁻¹ resolution. The spectrum can be analysed with a simple rigid rotor model as a superposition of rotational bands due to the six isotopes of mercury (204..198) combined with NH3, as was done recently by Tsuchiya on Hg-Ar(13).

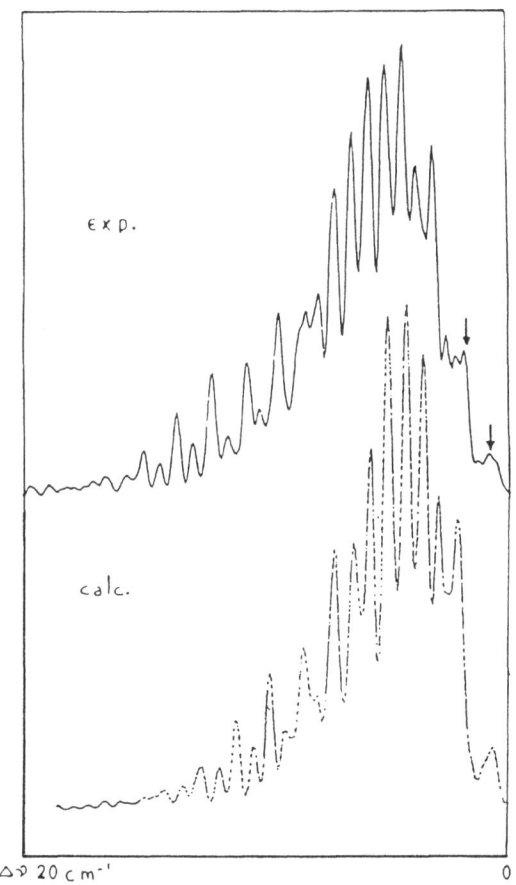

exp.

calc.

Δν 20 cm⁻¹   0

Figure4: rotational contour of the υ'=15 band in the B state, the arrows indicate the 202 and 204 (red) band heads (fluorescence excitation).

The B state bands can be readily assigned to parallel

bands from the existence of band heads on the red side,
corresponding to the Hg 202 and 204 isotopes. The typical
band contour appears in figure 4, each line in the spectrum
corresponds to the superposition of many rotational transitions and the regular appearance of some of the bands is
due to the accidental frequency match of the isotope shift
with the rotational line separation, yielding an accurate
measure of this shift. Therefrom the vibrational quantum v_s
can also be assigned, as in the preceding, even directly
from the spectra, using the formula for the 202- 204 isoto-
pe splitting: $\triangle v = (1-p)v_s \triangle G$, where $\triangle G$ is the experimental vibrational spacing and p is the ratio
$(\mu_{202}/\mu_{204})^{1/2}$. The resulting v_s values are in good agreement with the former.
Moreover the equilibrium distances in the A and X (ground)
states are obtained and listed in table 1, there is nevertheless some spread of the latter values as different couples of upper and lower constants fit the spectra.

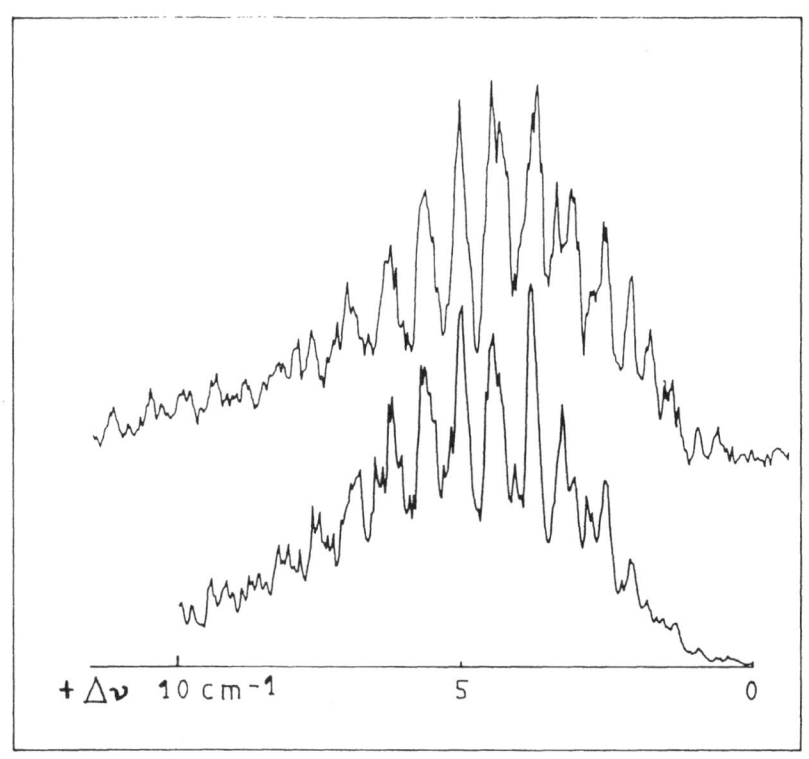

Figure5: rotational contour in the A state for $v'=17$, in
action.
    It was also found that the inversion doubling of the

ammonia   molecule need not to be taken into account for the
ground state,   meaning   that already here in the   G.S.   the
two   possible   orientations   of the ammonia with the mercury
are unequivalent,   in difference with   the equivalent   complex   Ar-NH3   which   must   arise from a shorter equilibrium
distance in Hg-NH3 (3.3Å compared to 3.8Å).
   The B state spectra were also recorded in action   giving a
similar contour for the stretch bands   but   very   different
for the bending transitions;   the fit was then obtained assuming a perpendicular transition and ξ close   to   1.   This
confirms   the   fact that these torsional bands,   pertaining
angular momentum (here vibrationnal) in difference with the
pure stretch bands,   correspond to a degenerate vibration in
a symmetric state (B).

   The A state bands had all the similar appearance of   perpendicular   transitions   and were all fitted with ξ =.92 in
figure 5.   This constant corresponds here   to   the   average
electronic   angular   momentum   ⟨Ω⟩   along the complex axis,
hence the diatomic description is a good one here and   this
A   state corresponds in this approximation to the Ω=1 state
correlating to Hg^{3}P$_1$+NH3,   while the B state   thus   corresponds to the Ω=0 component.

Table 1

	D$_e$	w$_e$	w$_e$x$_e$ cm^{-1}	R$_e$ Å
A :	5600	338	5.1	2.35
:				
B :	5100	263	3.38	2.35
:				
X :	250	(30)	–	3.35

     We   have   insofar evaluated with spectroscopic preci-
sion the potential parameters of the electronic states made
out of Hg^{3}P$_1$+NH3, which we have assigned to the Ω=0 and Ω=1
states, 5100 cm^{-1} deep and slightly different. We made the
assumption that the geometry of ammonia did not change upon
excitation   of the mercury,   there may exist some change as
the fit of the parallel bands could be improved when the
 N-H bonds are loosened by 0.9.

     The main deactivation paths of the corresponding ex-
cited levels are the fluorescence decay (B) or   the   dissociation to ^{3}P$_0$ mercury (A).
   The   latter   route   involves angular momentum along the
complex axis (electronic   or   vibrationnal),   which   former
studies   on   Hg-N2   (12,14) have shown to be determinant on
the coupling of the ^{3}P$_1$ to ^{3}P$_0$ states   via   the   upper ^{3}P$_2$

state. The $^3P_2$ states will be here deep enough to interact strongly with the A and B states.

The emission of the bottom of the two wells to the ground state will produce a bound free emission at $\approx$ 320nm, in the same domain as the excimer. On the other hand the resonant emission as observed in the B state lies at 280nm, the same as the precursor.This assigns the emitting states of the exciplex, moreover we can now infer its formation mechanism.The first step in collisions of excited state mercury is quenching to $^3P_0$, followed by three body stabilisation within the relevant well, $\tilde{a}$. Then this a state will communicate by collisions with the A and B states that we have observed to be very deep, through the A-$\tilde{a}$ coupling evidenced in the action spectra. If the pressure is low, "resonance" fluorescence will be emitted with the short 200ns lifetime at 280nm. We observed by exciting directly the complex in the B state this fluorescence. When the pressure is increased these levels can be relaxed, before emitting, to the excimer well. In proof to this relaxation, we excited bigger complexes which display a broad fluorescence excitation spectrum and have the same characteristic emission as the relaxed exciplex, either spectral or temporal (t_{fluo}> 500ns). If the dissociation energy of Hg-(NH3)$_2$< 2* the HgNH3 energy (3), as in strongly bound hydrogen complexes, the excitation of high levels of the bigger complex will yield dissociation to the bottom of the observed A or B states, ie the stabilised excimer emitters.

REFERENCES.

1. C. Jouvet and B. Soep J. Chem. Phys. 80 (1984) 2229.

2. C. Jouvet, W.H. Breckenridge and B. Soep J. Chem. Phys. 84 (1986) 1443.

3. A.B. Callear, Chem. Rev. 335 (1987) 335.

4. M. Castella and A. Tramer , companion paper in this volume.

5a. T. Hikida, T. Ishimura and Y. Mori, Chem. Phys. Lett. 27(1974) 548.
5b. T. Hikida, T. Ishihara and Y. Mori, Chem. Phys. 52(1977) 43.

6. A.B. Callear and C.G. Freeman, Chem. Phys. 23 (1977)
343.

7. H. Hiroguchi and S. Tsuchiya, Chem. Phys. 108 (1986)
153.

8. C.A. Taatjes, W.B. Bosma and T. Zwier Chem. Phys. lett.
128 (1986) 127.

9. S. Bratoz and M.L. Martin J. Chem Phys.42 (1965) 1051.

10. M.C. Duval, O. Benoist d'Azy, W.H. Breckenridge,
 C. Jouvet and B. Soep J. Chem. Phys. 85 (1986) 6324.

11. K. Fuke, T. Saito and K. Kaya, J. Chem. Phys.81 (1984)
2591.

12. W.H. Breckenridge, O. Benoist d'Azy, M.C. Duval,
C. Jouvet and B. Soep in the proc. of the workshop
"Stochasticity and intramolecular redistribution of
energy", R. Lefebvre and S. Mukamel editors ,Reidel (1987)
149.

13.S. Isogai,K. Yamanouchi, M. Okunishi and S. Tsuchiya, to
be published.

14. J.A. Beswick and C. Jouvet J. Chem. Phys. in press.

FROM FIVE-FOLD TO CRYSTALLINE SYMMETRY IN LARGE CLUSTERS

J. Farges, M.F. de Feraudy, B. Raoult and G. Torchet
Laboratoire de Physique des Solides - LA 02
Université Paris Sud
91405 ORSAY, France

ABSTRACT. When microclusters of a fcc material adopt the icosahedral
structure, strains appear which develop with increasing cluster size so
that above a given size, this structure is no longer able to grow
while keeping its symmetry. This is observed in argon clusters, the diffraction patterns of which show icosahedral features up to about 750
atoms i.e. 5.5 icosahedral layers, but anomalous features beyond this
size. Such anomalous features tend to give to the patterns a fcc character when a size of 2 or 3 thousand atoms is reached. In this paper,
arguments are given to show that the fcc character is not due to a structural transition experienced by the icosahedra but is due to the addition of uncomplete layers onto the fifth one.

INTRODUCTION : THE ICOSAHEDRAL STRUCTURE

Some twenty years ago, scientists having to represent a small piece
of solid material would have pictured a microcrystal showing particularly important effects of surface relaxation and possibly some defects
like stacking faults, twins or dislocations. Nowadays, they probably
would think of an icosahedron. So what happened in the mean time ?
Certainly the recent interest for cluster properties has questioned
most of generally accepted ideas. Moreover, crystallographers have been
fascinated by the icosahedron which in spite of its perfect geometry is
unable to depict any monocrystal due to its five-fold symmetry.
In 1962, Mackay [1] described a possible polycrystalline structure in
the form of an icosahedron built up with 20 tetrahedra connected together
through twinning planes. As shown in Fig. 1, each tetrahedron exhibits
a perfect rhombohedral [2] structure and therefore can be extended to infinity. The α angle of the unit cell is $63° 26'$. Fundamentally, Mackay
solved the incompatibility between the crystalline structure and the
five-fold symmetry by locating the latter not in the structure itself
but in the particular arrangement of the twins which in turn should not
be considered here as defects but as constitutive elements of the icosahedral structure.
One can easily understand that a compound is expected to adopt the

113

J. Jortner et al. (eds.), Large Finite Systems, 113–119.

icosahedral structure if it possesses a rhombohedral phase with α = 63°
and further a low twin energy. Such a compound does not exist, at least
among monoatomic species. However, one may notice that 63° is not far
from 60°, the α value of the fcc cell, so that disregarding these 3 degrees would make every fcc compound with low twin energy a candidate
for the icosahedral structure. It is in fact possible to change 60° into
63° by stretching elastically every layer parallel to the surface. Doing
this, layers in adjacent tetrahedra exert strains upon each other which
result in radial strains and finally in a compression of the central region. Effects of this compression can be observed in multilayer icosahedral models made of Lennard-Jones atoms if atoms are allowed to move
from their initial position, as in Fig. 1, to their equilibrium position
in a computer calculation [3]. This relaxation process is carried on to
the point where the force exerted on each atom is reduced to about
2×10^{-14} N. Then, the following observations can be made :
1. In each tetrahedron, layers are no longer planes as in Fig.1 but con-
vex surfaces which can be considered as spherical segments with identical curvature, the radius of curvature being large compared to the distance between adjacent players.
2. In a tetrahedron with a given number of layers, the distance between

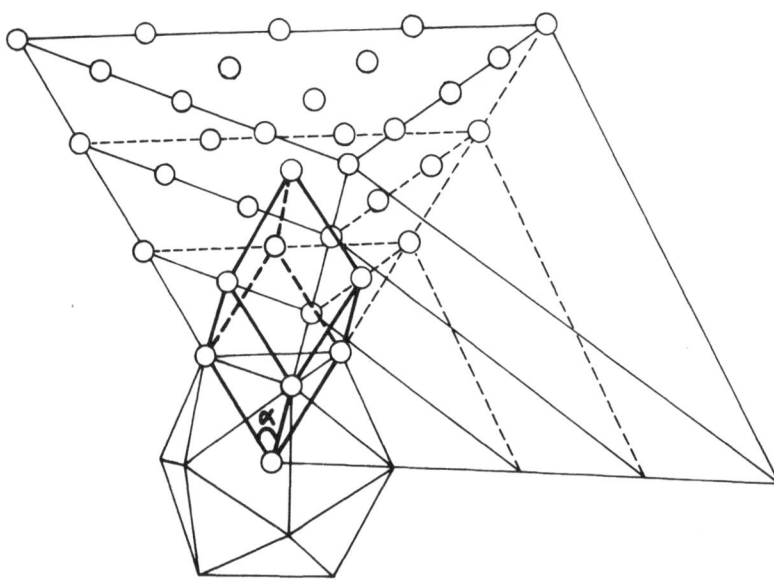

Fig. 1 Two tetrahedra among the twenty constituting the icosahedral
structure. In one of them, four layers of atoms are shown together with
the rhombohedral unit cell (α = 63° 26'). The number of atoms in the layer of index n is (10 n + 2).

adjacent layers decreases linearly when one moves from the surface towards the center. As an example, in a tetrahedron with 9 layers, the
distance decreases by 1% every time a layer has been crossed.

3. When a new layer is added onto a given icosahedron, every distance
between adjacent layers is decreased. Consequently, each layer comes
closer to the center while keeping the same curvature.

4. Distances between atoms belonging to the surface layer remain approximately unchanged whatever the total number of layers in the icosahedron.

5. The external layer experiences a normal relaxation which is positive.

PHYSICAL CLUSTERS WITH ICOSAHEDRAL STRUCTURE

In his paper, Mackay was very doubtful about the possibility that a
fcc compound could adopt the icosahedral structure. Then the question
should be asked whether clusters with icosahedral structure have even
been observed.Nowadays one would probably answer in the affirmative,
having in mind the beautiful micrographs on which nickel, gold or silver
clusters with icosahedral shapes are clearly visible. However, when diffraction experiments are carried out on these quite large clusters, made
of several ten thousands of atoms, the normal fcc structure is found[4].
Such large clusters do not show any strain but cracks, dislocations or
grain boundaries.The one case where diffraction has revealed the true
icosahedral structure was that of argon clusters produced in a free jet
expansion.

A free jet is obtained by expanding a gas with pressure p_0 through a
small hole into a chamber at low pressure. Clusters are formed by homogeneous nucleation during the expansion and grow by monoatomic addition.
Their mean size is easily increased by increasing p_0. One or two dia-
phragms are located on the jet axis in order to produce a cluster beam
downstream. Clusters travel into vacuum and the beam is crossed perpendicularly by a 50 kV electron beam. Scattered electrons are recorded
onto a photographic plate in the form of a powder pattern characteristic
of the cluster structure.

When clusters are small, patterns show that they have an amorphous
structure[6]. When the mean cluster size is increased, patterns are different and those recorded for p_0 = 3.3, 6 and 9 bar are shown in Fig. 2.
For higher inlet pressure (p_0 > 12 bar), patterns show crystalline lines
which can be indexed according to the fcc structure. Such lines are absent from the patterns p_0 = 3.3 and 6 bar. In front of these experimental patterns, diffraction functions have been reproduced in Fig. 2 calculated for a model of icosahedral structure with 147 and 420 atoms,
respectively. Diffraction functions are the Fourier inverses of the radial distribution function provided by a molecular dynamics calculation
on relaxed models, multiplied by an apparatus function[5].The almost perfect agreement between experimental and calculated curves proves that
for the considered number of atoms, Ar clusters have the relaxed icosahedral structure. This good agreement brings two more informations :

1. The size distribution in the cluster beam is certainly not very
broad otherwise the patterns would not show such a sharp resolution.

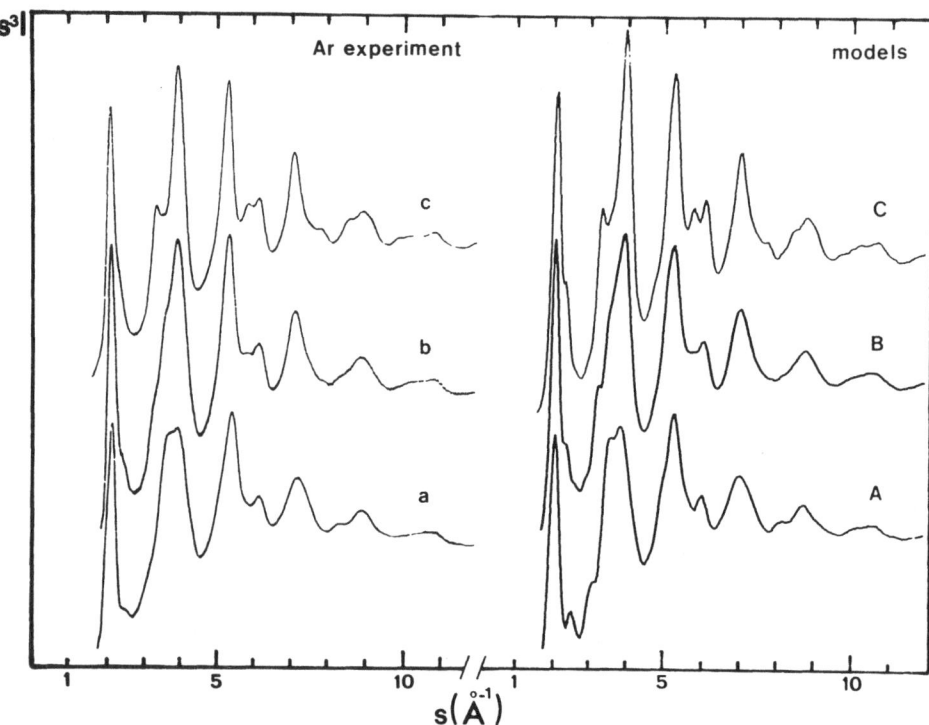

Fig. 2. Left side : Ar diffraction patterns recorded for several inlet
pressures p_0 = 3.3 (a), 6 (b) and 9 bar (c). Right side : diffraction
function calculated for several models with icosahedral structure
3 layers N = 147 (A), 4 and half layers N = 420 (B). Curve C is the superposition of two weighted functions of icosahedral (0.70) and fcc
(0.30) models.

2. The icosahedral structure must be the stable one, in this size range, otherwise a variety of structures would have appeared in the beam.
    The upper pattern p_0 = 9 bar in Fig. 2 corresponds to clusters with
an estimated mean size $\overline{N} \simeq$ 1000, as deduced from the extrapolation
of preceding results. It cannot be accounted for by any the icosahedral
models, whatever the considered size. On the contrary, the superposition
of both icosahedral and fcc structures provides a very good agreement as
shown in Fig. 2. As a matter of fact, the first signs of an fcc character
appear on the patterns as soon as $\overline{N}$ = 600 [5]. As these signs belong to
the largest clusters in the size distribution, the appearance of the fcc
structure is estimated to occur when N = 750 ± 150, the error being due
to the size distribution estimate.

FROM ICOSAHEDRAL TO FCC STRUCTURE

    In order to understand what happens in the clusters beyond 750 atoms,
we will study first the pattern p_0 = 15 bar shown on Fig. 3, which corresponds to an estimated number of atoms of about 3000. In this pattern,

one can notice that the angular positions of the maxima are those of
the Bragg lines in the fcc structure and that the width of the single
lines, which is related to the size of the fcc part of the cluster,gives
an estimate of about 600 atoms. This means roughly that clusters are made of five fcc parts connected to each other by some defects.

    It is possible to go further by calculating the diffraction function
of a fcc monocrystal of 600 atoms. This function, which is shown in
Fig. 3, does not depend on the model shape. When compared to the experimental pattern p_o = 15 bar, the Bragg lines in the calculated function
are correct in position and width, but several disagreements are found
in several line heights and in the resolution of several lines such as
(200) or (400). As an example, it is easily seen that the (111) and the
(642) lines are 40% too low. These discrepancies are caused by defects
or faults in the fcc cluster structure [7].

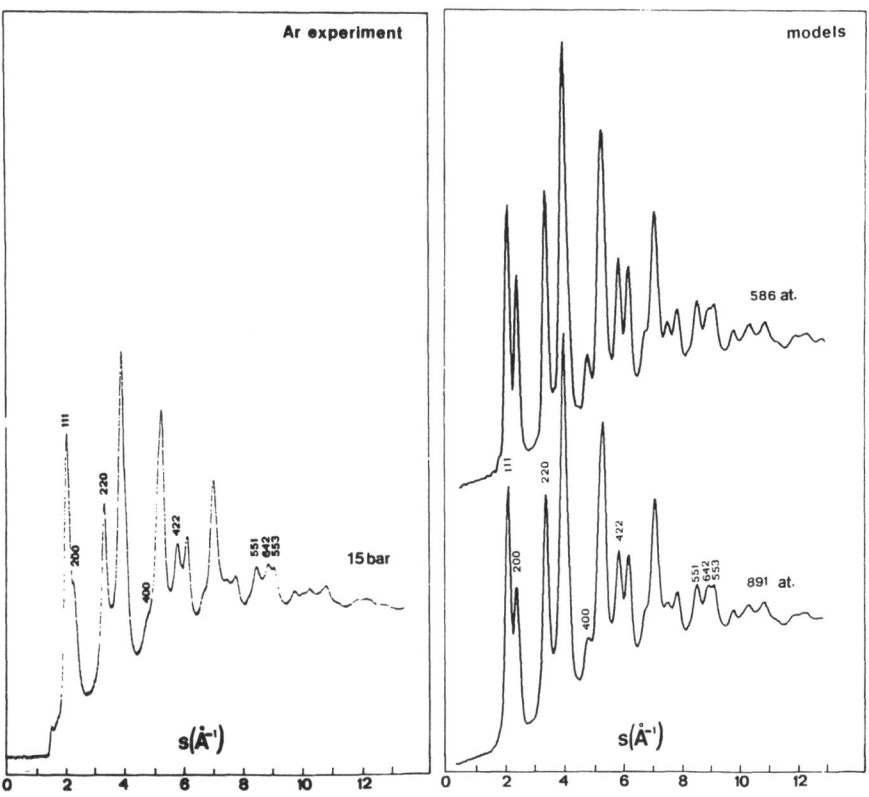

Fig. 3. Left side : Ar diffraction pattern for inlet pressure
p_o = 15 bar. Right side : diffraction function calculated for a pure
fcc model (upper curve), and for a model made of 2 tetraedra connected
through a twin plane (lower curve).

If we consider now a model made of 2 tetrahedra connected through a twin plane, the corresponding calculated function, shown in Fig. 3 is similar to the pure fcc function except in several line heights - (111), (642) - and resolution - (200, (400). It is remarkable that each of these differences which are due to the introduction of one twin in the model, brings the calculated function closer to the experimental pattern. One expects then that a better model would be made of fcc parts with 600 atoms connected to each other by more twins. A model with five fcc parts connected by five twins would contains 3000 atoms and would probably be convenient.

DISCUSSION

We may now come back to the question asked precedingly : what happens when the clusters while growing in the beam begin to produce a pattern different from that of an icosahedron, i.e. whenthey contain more than 750 atoms. Since the five layer icosahedron contains 561 atoms and the six layer one 923 atoms, one is led to guess that the sixth layer cannot grow up completely. The reason for this appears clearly : the existence of a sixth complete layer would result in a compression of the internal layers which would be unacceptable due to the resulting energy cost.

A first idea arises form the good fit obtained in Fig. 2c for the pattern p_0 = 9 bar ($\overline{N} \simeq$ 1000) by the superposition of two weighted calculated functions, 70% icosahedral and 30% fcc. This suggests that during the sixth layer growing, the icosahedra could be transformed into fcc cuboctahedra. Such a transformation is possible[1-3] without any drastic change since the atoms would have just to roll along each other while keeping the same nearest neighbors. Nevertheless, in order to account for the expected twinned structure of clusters with some 3000 atoms (p_0 = 15 bar), the fcc cuboctahedra should grow in such a way that every pure fcc part in them should never contain more than six or seven hundreds of atoms. This seems somewhat unrealistic and another growth process for the icosahedra is preferred. This process does not impose their destruction and would produce diffraction functions similar to the experimental patterns. One may then imagine that the five layer icosahedron keeps on growing by layer addition but that the successive layers cover only a segment of the icosahedral surface in order to avoid strains and compression effects. Thus atoms would gather round two or three vertices and, beyond a given cluster size, mainly round one of them. In this way, they would build up twinned tetrahedra epitaxially grown up upon several icosahedral faces. These tetrahedra would have roughly the fcc structure, giving to the pattern the expected feature. Constructing such models is now in progress in our group.

REFERENCES

1. A. Mackay, Acta Crystallogr., 15 : 916 (1962)

2. C.Y. Yang, J. Cryst. Growth, 47 : 274 (1979).
3. J. Farges, M.F. de Feraudy, B. Raoult and G. Torchet, Acta Crystal-
    logr., A38 : 656 (1982).
4. S. Ino and S. Ogawa, J. Phys. Soc. Jpn, 22 : 1365 (1967).
    B.D. DE Baer and G.D. Stein, Surf. Sci., 106 : 84 (1981).
    C. Solliard, thèse n° 497, Lausanne (Switzerland) (1983).
    M. Brieu, thèse n° 1260, Toulouse (France) (1986).
    M. Gillet, J. Cryst. Growth, 36 : 239 (1976).
5. J. Farges, M.F. de Feraudy, B. Raoult and G. Torchet, J.Chem. Phys.,
    78 : 5067 (1983)
6. J. Farges, M.F. de Feraudy, B. Raoult and G. Torchet, J. Chem. Phys.,
    84 : 3491 (1986).
7. J. Farges, B. Raoult and G. Torchet, J. Chem. Phys., 59 : 3454 (1973).

CLUSTER ENERGY SURFACES

T.P. Martin, T. Bergmann and B. Wassermann
Max-Planck-Institut für Festkörperforschung
Heisenbergstr. 1, 7000 Stuttgart 80, FRG

ABSTRACT. An accurate description of the structural and vibrational properties of small clusters can be achieved only through a detailed examination of multi-dimensional total energy surfaces. The importance of various properties of such a surface are illustrated for pure alcohol clusters and for alkali halide fragments dissolved in alcohol clusters.

INTRODUCTION

Although clusters are now routinely observed in mass spectrometers, their most fundamental property, their structure, is still essentially unknown /1/. This is a particularly frustrating state of affairs for the theorists, who, if given the structure, can use their sophisticated methods to calculate the electronic and vibrational properties of clusters. Since experimental investigations have not yet provided much structural information, the theorists have been forced to attack the problem themselves. However, it may be a task which is just beyond the present state of the art. This situation can be contrasted to that which exists in solid state physics where precise structural data is available. One can imagine the confusion which would exist today if the solid state physicist would still have to rely on total energy calculations for the determination of complicated crystal structures. It was the development of diffraction techniques that finally allowed rapid advances in solid state physics. Cluster science awaits a similar breakthrough. In the meantime, the cluster theorist faces a challenge even more difficult than that of the solid state physicist. Rather than examining a small set of structures, a vast multidimensional total energy surface must be mapped out and examined /2-4/.

Any attempt to determine the properties of a total energy surface encounters two fundamental difficulties. Suppose one tries to systematically define the energy surface at the points on a coordinate grid. Consider the seemingly modest goal of constructing a grid with only ten intersection points along each of the 3N-6

121

J. Jortner et al. (eds.), Large Finite Systems, 121–133.
© 1987 *by D. Reidel Publishing Company.*

axes. This means that for a 10 atom cluster the total energy must be evaluated 10^{24} times. Even if this were possible, which of course it is not, such a course mesh of points would leave many important features hidden. The sheer expanse of multidimensional space poses the first fundamental computational problem. The second difficulty is no less formidable. Each of the total energy calculations must be carried out with a high degree of precision because many features on the surface, e.g. well depth and barrier heights differ by only a small fraction of the total energy. An ab initio calculation of the total energy of a ten-atom metal cluster including the effects of correlation is a state-of-the-art task /5-7/ even if it is performed for only one point on the energy surface. A complete mapping of the surface using such techniques is unthinkable. Fortunately, this second difficulty can be avoided for certain materials. If the atoms in a cluster interact with one another isotropically, it is possible to define a pair potential that depends only on the distance between the atoms. This is the case for rare gas clusters /8-12/ and alkali halide clusters /13-15/. Pair potentials have also been defined and used for certain molecular clusters /16-21/.

MODEL

The clusters under consideration are too small to observe directly. We are forced to investigate their structure by means of total energy calculations. Mathematically, the problem can be stated very simply: find all minima on a multidimensional total energy surface. However, this entails computation of the energy for an enormous number of cluster configurations, a feat possible only if a simple interaction potential is used. But here we run into a second difficulty; the various minima often differ only slightly in energy. If the interaction potential is too simple, the calculated absolute minimum will not correspond to the observed clusters. Having these limitations always in mind, we have minimized the total energy of alkali halide-alcohol clusters using simple but tested interaction potentials. Experiments offer information about only limited regions of configuration space. That is, an empirically parameterized potential may be very good for the specific intermolecular distances and angles encountered in crystals, but the same potential will not adequately describe the large variety of configurations found in clusters. For this reason there is considerable justification for fitting potential parameters to quantum mechanical interaction energies calculated over a large region of configuration space. Such an approach has been used, for example, in a Monte Carlo study of fixed Li^+ and F^- ions surrounded by 200 water molecules /22-24/. The form of the potential we have used to describe alcohol was originally proposed by Snir et al. /25/. The molecule is composed of 13 centers of interaction, Fig. 1. Six of the centers are positively charged and are located at the known positions of the six atoms in the molecule. The remaining 7 centers are negatively charged and can be conveniently thought of as bond charge and lone pair charge. The bond charges are located on the axes between nuclei.

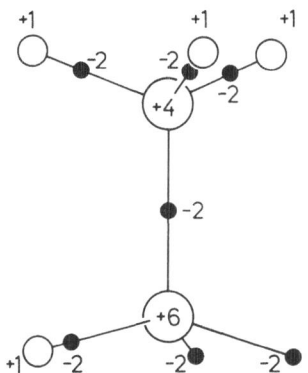

Fig. 1. The alcohol model used consists of 13 interaction centers (ref. 25 and 26). Six centers are located at atomic sites and 7 centers at lone pair sites.

All centers are allowed to interact with one another electrostatically, $q_i q_j / r_{ij}$. The negatively charged centers have an additional interaction with the form:

$$U_{ij} = q_i q_j (A_{ij} exp(-B_{ij} r_{ij}) - C_{ij} / r_{ij}^6) \tag{1}$$

where q_i and q_j are the charges and r_{ij} is the distance between the centers. Without any further assumptions this would lead to 43 fitting parameters, clearly an unmanageable number. Therefore, charges are assigned to each of the centers (Fig.1) and use is made of the combination rules:

$$A_{ij} = (A_{ii} A_{jj})^{1/2}, B_{ij} = (B_{ii} + B_{jj})/2, C_{ij} = (C_{ii} C_{jj})^{1/2} \tag{2}$$

thus reducing the number of fitting parameters to 17. We have used the parameters of Marchese et al/26/. The interaction potential between pairs containing either an alkali or a choride ion is assumed to contain an electrostatic Coulomb contribution and a repulsive Born-Mayer contribution/13/,

$$U(ions)_{ij} = q_i q_j / r_{ij} + A_{ij} exp(-B_{ij} r_{ij}). \tag{3}$$

The position and orientation of each rigid methanol molecule have been represented by the three cartesian coordinates defining the center of mass and by three Euler angles. To initiate a minimization run, the coordinates and angles were chosen at random. A fast but rough minimization was first carried out using the steepest descents method. This was followed by a refinement with the Davidson/Flecher/Powell method/27/. After establishing an energy minimum the vibrational frequencies were determined. Starting from a variety of initial configurations, the value of a given energy minimum was reproducible to six digits and the vibrational frequencies to four.

PURE ALCOHOL CLUSTERS

The bonding between alcohol molecules is similar to the bonding between water molecules. However, where water has a coordination four in the condensed phase, alcohol has coordination two. Each alcohol molecule can form one acceptor and one donor hydrogen bond with neighboring molecules. There is only one type of structure that can be constructed out of units with coordination two, and that is a ring. The most stable forms of small alcohol clusters are, in fact, rings. However, many stable isomers can be found. These isomers distinguish themselves from one another mainly in the orienation of the methyl groups with respect to the plane of the ring. A simple alternation (one methyl group pointing up, the next down, etc.) is energetically preferred.

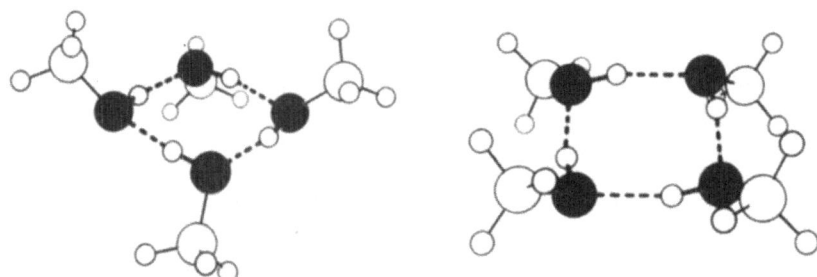

Fig. 2. Two stable forms of $(CH_3OH)_4$. For the global minimum the methyl groups alternate,up,down,etc.

As the alcohol cluster grows the global minimum can still be characterized as belonging to a simple ring structure. For n equal to 5 and 6 the rings are still almost planar. At n equal to 8, strong deformations begin to appear. At some point in the growth process the ring structure must cease to be the most favorable form. It is not plausible that a cluster containing, for example, 100 molecules should consist of one single enormous loop, no matter how contorted this loop might be. However Fig. 4 shows clearly that this stage has not yet been reached for 9- and 10-molecule clusters. These rings are certainly not circular or even planar. But all molecules do have a two-fold coordination resulting in closed chain structures at the global minima. The first global minimum for which the simple ring structure is not preferred occurs for n=11. For this cluster a six-molecule ring and a five-molecule ring are bonded together through van der Waals forces.

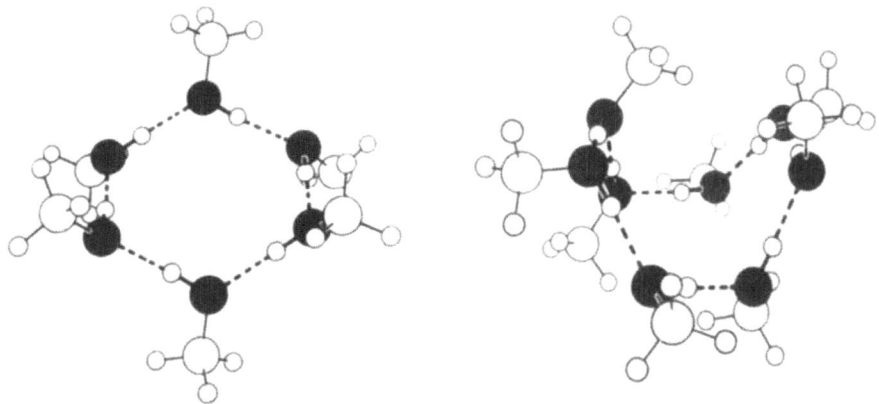

Fig.   3.   Configurations corresponding to the global minima for $(CH_3OH)_6$. Alcohol molecules with the coordination 2 tend to form small symmetric rings.

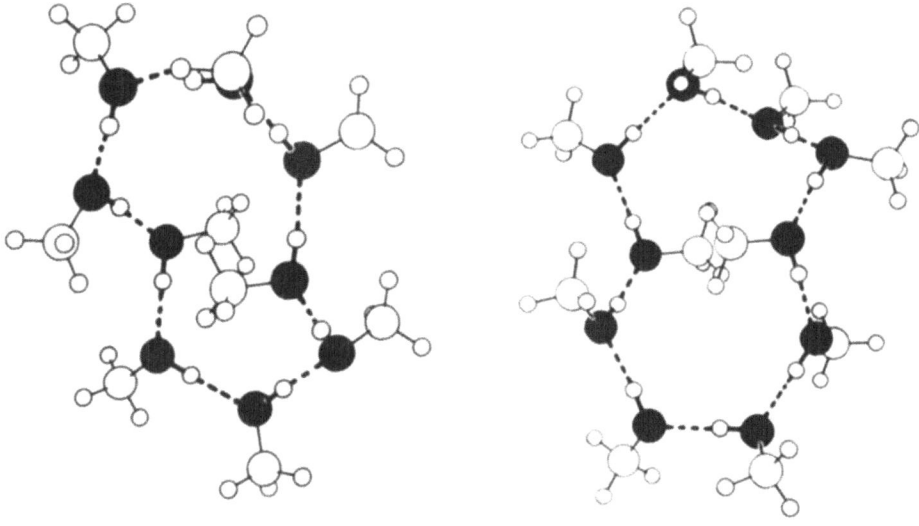

Fig. 4. The most stable forms of $(CH_3OH)_9$ and $(CH_3OH)_{10}$. As the rings grow larger they are strongly contorted by the van der Waals forces between adjacent molecules.

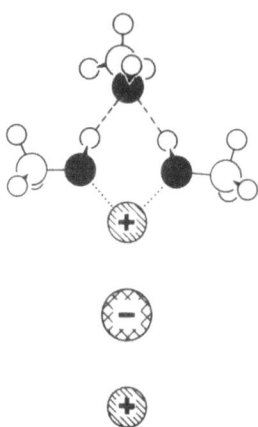

Fig. 5. Configuration for a local minimum.

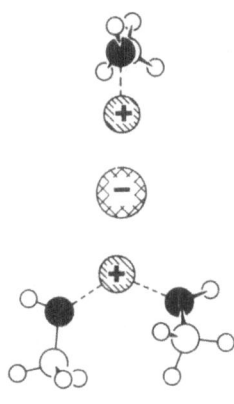

Fig. 6. Configurations for a global minimum.

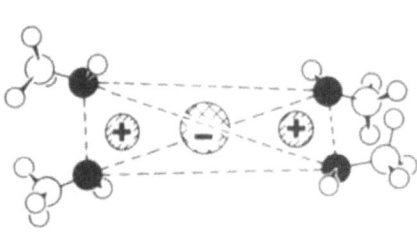

Fig. 7. Most stable configuration for $(Li_2I)^+(CH_3OH)_4$.

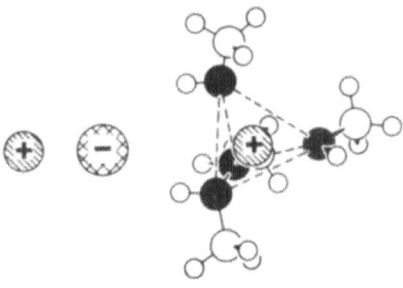

Fig. 8. Configuration for a local minimum.

CHARGED CLUSTERS

The stability of a LiI-methanol cluster is determined essentially by three important bonds. The strongest is the Li-I bond. This bond is so strong that the LiI fragment remains essentially undistorted until a large number of methanol molecules have been added to the cluster.

The Li-O bond provides the most important contribution to the solute-solvent interaction. This bond weakens as additional alcohol molecules are forced to share the same Li ion.

The solvent-solvent O-H hydrogen bond is relatively weak but becomes important with increasing methanol content, simply because there is no more room around the Li ions to allow the formation of an additional Li-O bond. The excess methanol molecules have no alternative but to form hydrogen bonds with existing solvent molecules.

In this paper, results will be presented for calculations on only one type of charged cluster, $(Li_2I)^+(CH_3OH)_n$. Even with this limitation the results must be summarized because of the large number of stable configurations exisiting for each cluster size; 5 configurations for n=2, 11 configurations for n=3, 30 configurations for n=4, etc. After analyzing all this information it has been possible to formulate three rules for constructing a cluster of high stability. It should be emphasized that these rules were not used in our calculations but are rather a qualitative summary of the results.

1) The $(Li_2I)^+$ fragment remains linear and undistorted for n less than 8.

2) The solvent molecules should be distributed as evenly as possible between the two Li ions.

3) Hydrogen bonding between methanol molecules should be avoided for n less than 8.

It will be instructive to apply these rules to a simple but nontrivial example, $(Li_2I)^+(CH_3OH)_3$. One local minimum on the energy surface of this cluster corresponds to the configuration shown in Fig. 5. This configuration is not energetically favorable because it does not obey rules 2 and 3. The molecules are bunched on one side of the LiI fragment and they form hydrogen bonds among themselves. The most favorable configuration is shown in Fig. 6.

Using these rules we can immediately identify which of the more than 30 stable configurations containing 4 methanol molecules is most stable. It is shown in Fig. 7. This cluster is particularly resistent to dissociation through loss of a solvent molecule, a property that may explain the strong mass peak of n=4.

A further examination of the local minima for $(Li_2I)^+(CH_3OH)_4$ reveals another structure which, although much less stable, is of considerable interest, Fig. 8. All the methanol molecules have grouped around one end of the alkali halide fragment and one of the Li-I bonds has almost doubled in length. It would appear that a minimum of four methanol molecules is necessary for the solvation of a Li^+ ion. Such solvated ions are found in the most stable configurations only for n greater than 8. The steep decline in the intensity of mass peaks for clusters containing more than eight molecules may very well reflect the onset of solvation.

NEUTRAL CLUSTERS

It should be emphasized that LiI is not a typical alkali halide. Its unusual properties derive from the large difference in the ion radii. If one tries to pack Li^+ and I^- ions together to form a rock-salt crystal structure or to form the usual cluster structures, the problem of I-I overlap immediately becomes apparent. The problem disappears if the coordination of each ion is lowered. Therefore, textbooks often state that LiI should crystallize into a four-fold coordinated zinc blende structure. The most stable forms of LiI clusters also tend to be less closely packed than for the other alkali halides. To give an example, consider the structure of the dimer $(LiI)_2$. In general, the dimer has two stable forms, a rhombus and a linear chain. Although the rhombus is the energetically favored form for all other alkali halides, for LiI, the rhombus does not even form a local minimum on the energy surface. The linear chain is the only stable form of $(LiI)_2$. Starting then with this linear form of the bare dimer, what happends as methanol molecules are added to the cluster one at a time?

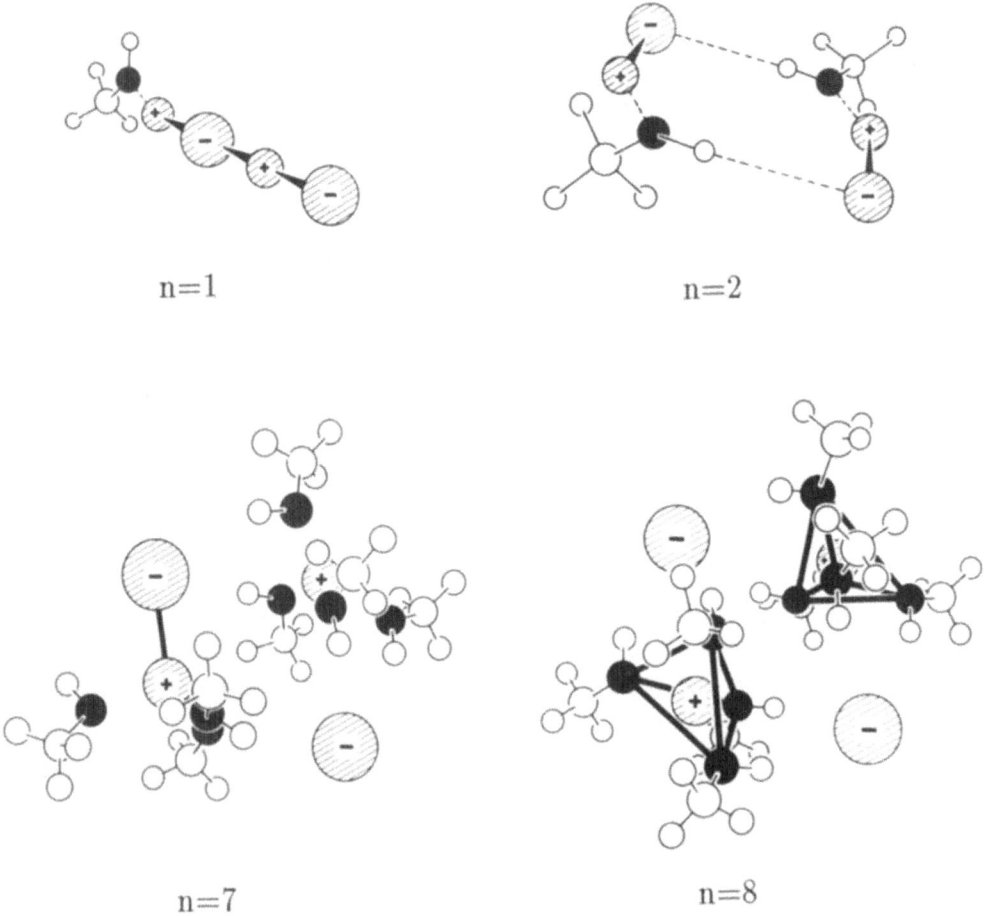

Fig. 9. Configurations for the global minima of $(LiI)_2(CH_3OH)_n$. Ion pair formation at n=2. Complete solvation at n=8.

One might ask the specific question, how many methanol molecules must be added to break a Li-I bond? However, it must be remembered that point-charge interactions are long range and that the ions are bound not merely to the nearest neighbors. The danger of using the concept of a local Li-I bond is best illustrated with an example. The linear $(LiI)_2$ dimer can be broken apart at two inequivalent points, either an end ion can be removed, or an ion pair can be removed. The energy required to remove an end ion is 3.6 eV while only 0.62 eV are necessary to symmetrically fragment the cluster into two ion pairs. Such large variations in a local Li-I bond energy indicate that such a bonding concept is not useful. The total energy surface of each $(LiI)_2(CH_3OH)_n$ cluster contains many local minima. In the following brief discussion the structure only at apparent global minima will be indicated for each value of n.

Two methanol molecules are sufficient to split the LiI dimer into ion pairs. The most stable configuration is a ring, Fig. 11 composed of alternating LiI and alcohol units. The 0.6 eV lost by fragmentation into ion pairs is more than compensated by the formation of two O-Li and two H-I bonds. In some sense the isolation of ion pairs represents the first stage of solvation. We will see that the next stage, the isolation of individual ions, requires a much larger number of alcohol molecules.

With the addition of 3,4 and 5 alcohol molecules to $(LiI)_2$ there is no qualitative change in the degree of solvations. Even six molecules are insufficient to dissociate the strongly bound ion pairs. Since the Li-O bond is the strongest solvent- solute interaction, it is not without meaning to describe the clusters in terms of the coordination of the Li ions only. In $(LiI)_2(CH_3OH)_6$ the six alcohol molecules distribute themselves evenly between the two Li ions, three molecules to each ion. Following this rule we would expect that for $(LiI)_2(CH_3OH)_7$ the alcohol molecules would be distributed 3 to one Li ion and 4 to the other. That this is indeed the case, can be seen in Fig. 9. However, now the second stage of solvation sets in. The ion with coordination 4 dissociates from its counter-ion, resulting in a cluster containing one ion pair and two fully solvated ions. Stated more generally, *for clusters with composition $(LiI)_n(CH_3OH)_m$, the second state of solvation starts at n/m greater than 3 and is complete for n/m greater than 4.* Just as this rule predicts, $(LiI)_2(CH_3OH)_8$ contains completely solvated ions, Fig. 9.

CATCHMENT AREA

Another important characteristic of a total energy surface is often called the catchment region. Imagine rain falling onto the surface. Every raindrop, no matter where it hits the surface, will eventually find its way to a minimum. That area of the surface which collects rain drops for a specific minimum is said to define a catchment area. All of configuration space can be uniquely divided up into a set of such areas. In this way each point in configuration space can be said to belong to a given minimum (except, of course, those points defining the boundry line between catchment areas). In the above discussion the concept of catchment area proved useful in a qualitative description of a total energy surface. We will now attempt to use the concept in a more quantitative way.

Catchment area has a particularly simple meaning for the case of a microcanonical ensemble. In such an ensemble each point in configuration space is equally probable, therefore, the probability of the cluster having a given configuration is simply proportional to the length of a constant energy contour in the corresponding catchment area. This interpretation must be only slightly modified when considering the canonical ensemble. In this case the probability of reaching a given point in phase space must be weighted by $exp(-H/KT)$. Specifically, the probability that a cluster occupies catchment region a relative to the probability that it occupies region b is just,

$$P_a/P_b = \int_a exp(-H/KT)dq^3\,dp^3 \Big/ \int_b exp(-H/KT)dq^3\,dp^3 \qquad (4)$$

For the purposes of this discussion we will assume that the integration over momentum coordinates is independent of the configuration since both configurations have the same mass. However, the configurations will not, in general, have the same moment of inertia.

The integration over configuration coordinates is easily carried out if it is further assumed that each catchment region has a parabolic form. This harmonic assumption is certainly valid at low temperatures.

$$V_a = E_a + \sum_i \omega_a^2(i)q_a^2(i)/2m \qquad (5)$$

where $\omega_a(i)$ is the ith vibrational frequency in the parabolic catchment region belonging to minimum a. The corresponding displacement amplitude has been denoted with $q_a(i)$. Using this expression the integral in Eq. 4 is easily evaluated to give:

$$P_a/P_b = exp[(E_b - E_a)/KT] \prod_i \omega_b(i)/\omega_a(i) \qquad (6)$$

Here we have used the classical expression for the energy of a harmonic oscillator in order to retain a simple geometrical description of the probabilities. The quantum mechanical result is obtained if each ω in the product in Eq. 6 is replaced with a corresponding $1 - exp(-h\omega/KT)$. Notice that the relative probability of a cluster being in a given minimum is dependent on the depth of the minimum (through the Boltzmann factor) and on the extend of the catchment region (through the product of inverse frequencies). Just one very low frequency vibration can, at non-zero temperature, stabilize an energetically unfavorable configuration.

Now it is possbile to state and understand a rather unusual observation. *Cluster configurations corresponding to shallow minima tend to become more favorable at high temperatures. Cluster configurations corresponding to deep minima are less likely to be found at high temperatures.* The reason for this behaviour is the following. Energetically unfavorable clusters sitting in shallow minima usually have open structures which possess incipient instabilities. Such clusters resist only weakly certain types of deformations, i.e. they have low frequency vibrations. The entropy generating, low frequency vibrations tend to make the configuration more favorable as the temperature increases. In principal, that configuration with the lowest frequencies will always become the most favorable if the temperature is made high enough. We will now present a specific example to illustrate this point.

Consider once again $(LiI)_2(CH_3OH)_2$. The global minimum on the energy surface of this cluster corresponds to a ring structure. However, this ring was the end result only 5 times in 95 minimizations from random starting configurations. It has a small catchment area. The local minimum corresponds to a chain structure, Fig.10. This local minimum has a larger catchment area, smaller curvatures, lower frequencies. Therefore, the probability of finding it in the vapor will increase with temperature, Fig. 10.

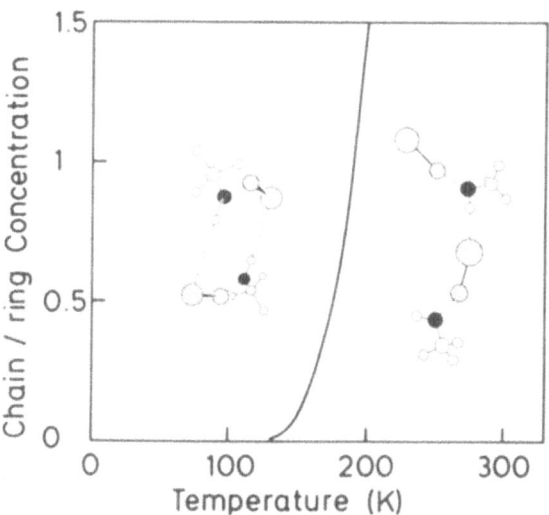

Fig. 10. Temperature dependence of the relative concentrations of two different $(LiI)_2(CH_3OH)_2$ cluster configurations. The low frequency chain vibrations generate large amounts of entropy at high temperatures.

CONCLUDING REMARKS

The solvation of LiI in alcohol clusters seems to proceed in two stages with increasing dilution; a) the isolation of ion pairs, b) followed by the complete solvation of ions. Complete solvation is achieved for dilutions allowing four alcohol molecules to surround each Li ion. Direct experimental verification of this conclusion would be difficult. Diffraction techniques are best suited to crystalline samples. If it were possible to grow and to determine the structure of LiI-alcohol crystals, we would have unambiguous experimental data for comparison. Recently, such experiments have actually been performed/57/ on $Li(MeOH)_4I$. This material turns out to have the highest ionic conductivity of any known Li compound, a fact which could make it of great interest for future applications. However, it is the structure of this material that interests us here. Each Li ion is tetrahedrally surrounded by four alcohol molecules.

It is, without a doubt, convenient to think of a cluster of a given size as having a specific geometry. Unfortunately, this simplified way of thinking does not lead to an accurate description of a collection of clusters at finite temperature. The total energy surface of a cluster contains many local minima, most of which play an important role in determining cluster geometry. The calculational effort required to characterize all of these minima is enormous because of the multidimensionality of configuration space. Numerical procedures which are simple in 3 dimensions become unthinkable in 100 dimensional space. However, some progress can be made through the use of two-body interaction potentials applied to clusters of small size.

REFERENCES

1. The proceedings of three International Conferences on Clusters published in: J. Phys. (Paris), Suppl. C2 (1977); Surf. Sci. **106** (1981); **156** (1985).
2. M.R. Hoare: Adv. Chem. Phys. **40**, 49 (1979).
3. T.P. Martin: Phys. Rep. **95**, 167 (1983).
4. F.H. Stillinger and T.A. Weber: Science **225** 983 (1984); Phys. Rev. A **25**, 978 (1982); **28**, 2408 (1983).
5. J. Koutecky and P. Fantucci: Chem. Rev. (1986).
6. J. Flad, G. Igel-Mann, H. Preuss and H. Stoll: Surf. Sci.**156**, 379 (1985).
7. P.S. Bagus, C.J. Nelin and C.W. Bauschlicher, Jr.: Surf. Sci. **156**, 615 (1985).
8. C.L. Briant and J.J. Burton: J. Chem. Phys. **63**, 2045 (1975).
9. R.D. Etters and J. Kaelberer: J. Chem. Phys. **66**, 5112 (1977).
10. E.E. Polymeropoulos and J. Brickmann: Chem. Phys. Letters **92** 59 (1982).
11. J. Farges, M.F. Feraudy, B. Raoult and G. Torchet, Surf. Sci.**106**, 95 (1981).
12. E. Blaisten-Barojas and H.C. Andersen: Surf. Sci. **156**, 548 (1985).
13. T.P. Martin: J. Chem. Phys. **67**, 5207 (1977).

14. D.O. Welch, O.W. Lazareth, G.J. Dienes, R.D. Hatcher: J. Chem. Phys. **68**, 2159 (1978).

15. B.I. Dunlap: J. Chem. Phys. **84**, 5611 (1986).

16. F.H. Stillinger and A. Rahman: J. Chem. Phys. **60**, 1545 (1974).

17. J.R. Reimers and R.O. Watts: Chem. Phys. **85**, 83 (1984).

18. V. Carravetta and E. Clementi: J. Chem. Phys. **81**, 2646 (1984).

19. I.P. Buffey and W. Byers Brown: Chem. Phys. Lett. **109**, 59 (1984).

20. M.J. Ondrechen, Z. Berkovitch-Yellin and J. Jortner: J. Am. Chem. Soc. **103**, 6586 (1981).

21. B.W. van de Waal: J. Chem. Phys. **79**, 3948 (1983).

22. H. Kistenmacher, H. Popkie, and E. Clementi, J. Chem. Phys. **58**: 5627 (1973).

23. R.O. Watts, E. Clementi and J. Fromm, J. Chem. Phys. **61**, 2550 (1974).

24. J. Fromm, E. Clementi, R.O. Watts, J. Chem. Phys. **62**, 1388 (1975).

25. J. Snir, R.A. Nemenoff and H.A. Scheraga, J. Phys. Chem. **82**, 2497 (1978).

26. F.T. Marchese, P.K. Mehrotra and D.L. Beveridge, J. Chem. Phys. **86** 2592 (1982).

27. J. Stoer: *Einführung in die Numerische Mathematik I*; 3rd ed. (Springer, Berlin, Heidelberg, 1979).

28. W. Weppner, W. Welzel, R. Kniep and A. Rabenau, to be published in Angew. Chem.

PHASE CHANGES: THE INTERPLAY OF DYNAMICS AND EQUILIBRIUM IN FINITE SYSTEMS

R. Stephen Berry
Department of Chemistry and the James Franck Institute
The University of Chicago
5735 South Ellis Avenue
Chicago, Illinois 60637
U. S. A.

ABSTRACT. Finite systems may exhibit solid-like and liquid-like "phases", whose conditions for mutual equilibrium do not correspond to those of any traditional bulk phase transition, whether first-order, second-order or higher. Rather, a finite system may display either a finite range of temperature and energy throughout which the different "phases" coexist, or a range in which the system appears to be a slush, with properties intermediate between those of a solid and a liquid. Which of these occurs is determined by the relationships of several time scales, including the time characterizing the observation. The passage from solid to liquid may be characterizable by the fractal dimension of the space on which the classical phase point of the system moves.

1. BACKGROUND

1.1. Early simulations

In 1973, four publications appeared[1] announcing, independently, evidence for distinguishable solid-like and liquid-like forms of finite clusters. Soon thereafter more elaborate simulations showed in detail some of the ways these two phase-like forms do indeed have characteristics we associate with bulk solids and liquids[2,3,4]: the solid-like clusters show spatial order, very slow diffusion, mean displacements less than 10% of a mean nearest-neighbor distance, and no soft modes (i.e. modes with frequencies close to zero) in the vibrational spectrum. By contrast, the liquid-like clusters show spatial disorder, rapid self-diffusion, large mean displacements and notable densities of soft modes. Furthermore the caloric curves, the curves of mean temperature or kinetic energy as a function of total energy per particle (from isoergic simulations) or of mean energy as a function of temperature (for isothermal simulations) showed transitional character, at least as a change of slope, in a transitional region.

There was some controversy over the shapes of the caloric curves because the Monte Carlo calculations of Etters an Kaelberer for isothermal clusters indicated that these are monotonic, but the molecular dynamics calculations by Briant and Burton for isoergic clusters

J. Jortner et al. (eds.), Large Finite Systems, 135–143.

of argon showed regions of negative slope for clusters with 7 or more atoms. These "loops" were suspected by Burton[5] to be the result of computer runs too short to give properly equilibrated behavior, and indeed this eventually proved correct[2,6,7]. However the comparison did open a question whose answer is suspected but not proven, namely "Do the Monte Carlo and molecular dynamics calculations generate the same caloric curves?" We shall return to this point.

1.2. An Analytic Theory

The behavior of simulated clusters of argon or "Lennard-Jonesium" turned out to be somewhat different from naive expectations based on current dogma. The notion of distinct phase-like forms of small, finite clusters seemed quite unlike the smooth change one might postulate as a cluster or molecule is given more energy and more of its highly-excited vibrational states are populated. The possibility of a first-order phase transition, considered by Briant and Burton as one interpretation of their the caloric curves, seemed altogether outlandish, inconsistent with the extant theory which requires that a system be infinite to show a sharp phase transition. (In fact, in contrast to critical phenomena, one can certainly not say that we understand the melting and freezing transition of even the simplest substances.)

By identifying solid-like clusters with ordinary, nearly rigid molecules exhibiting small-amplitude vibrations and near-rigid rotations, and liquid-like clusters with very nonrigid, floppy molecules, it proved possible to rationalize the quantitative results of the simulations[6]. With suitable models for the solid and liquid forms, one could construct approximate densities of states, partition functions and free energies for both forms.

At low temperatures, the solid form has the larger partition function and the lower free energy and is therefore the more stable. The partition function of the liquid form grows faster with increasing temperature than that of the solid form, so that at high temperatures, the liquid has the larger partition function, the lower free energy and the greater stability. At some temperature $T_e(N)$, dependent on N, the number of atoms in the cluster, the two forms have the same values for their partition functions and free energies.

The properties that can be estimated from the partition functions $Q(T,N)$ include $T_e(N)$, the temperature range $\Delta T(N)$ over which both forms might be present in detectable amounts (taken in Ref.6 as 10% or more of the less prevalent form), the surface free energy and the surface tension, the latter requiring Nishioka's relation[7] and analysis of the data of Ref.3 to make a comparison). The behavior of these properties as derived from the simulations was reasonably well reproduced by the very simple analytic approach derived from the densities of states of all but the crudest of representations of the solid and liquid forms of the clusters. The conclusion of that work was that only very simple, general properties are sufficient to account for at least some of the characteristics of the solid and liquid forms of clusters.

One postulate was required in order to carry out the calculations of Ref.6; namely, that the two forms, solid-like and liquid-like, can, under some circumstances, coexist in equilibrium. The second stage of the analysis was finding necessary and sufficient conditions for the validity of that assumption, and the exploration of the implications of those conditions.

To find conditions for the coexistence of solid-like and liquid-like forms of clusters, it was necessary to put the free energy of the system on a scale on which a minimum would correspond to a thermodynamically stable system--much as is done in the Landau theory of second-order phase transitions. This can be done for the rigid-to-nonrigid case by using the energy levels of the two forms as extreme limits in a correlation diagram. As with the correlation diagrams relating the separated-atom and united-atom limits for diatomic molecules, the diagrams relating prolate and oblate limits of the symmetric top, and the diagrams connecting the energies of complex ions in the weak- and strong-field limits, we choose limiting cases that we can treat with some exactitude. Then we assume which properties are conserved along the scale between the two limits and connect the states at the two limits with linking lines, starting at the bottom of the scale, so as to link states with the same spin and symmetry and avoid the crossing of any two links involving states of the same spin and symmetry (here is where the assumption about what is conserved is used). Every state at each limit is linked to another state at the opposite limit, the lowest available of the same spin and symmetry. This procedure gives us a qualitative correlation diagram for the passage between rigid and nonrigid limits of a cluster or molecule.

The next step was to give a quantitative meaning to the abscissa of the correlation diagram. This was done by extending the idea used by Yamada and Winnewisser[8] to quantify the degree of nonrigidity of linear molecules. The quantity used to define a general nonrigidity scale is the ratio of two excitation energies. The numerator is the energy required to excite the cluster to the lowest state with one unit of angular momentum; in the rigid-body or solid-like limit, this is the lowest rigid-rotor excitation. The denominator is the excitation energy to the lowest vibrationally excited state with no angular momentum. In the limit of an infinitely stiff vibrator this ratio becomes zero; in the nonrigid model we chose, the Gartenhaus-Schwartz pairwise harmonic attraction[9], the ratio becomes $1/2$. Multiplying by 2, we have a parameter γ running from 0 to 1, from the rigid to the nonrigid limit.

It is now possible to define the partition function Q and the free energy F as functions of *both* the physical variable of temperature T and the mathematical parameter γ measuring the degree of nonrigidity[10]. The condition for thermodynamic stability of the system at a particular value of γ and a temperature T is of course that $F(T,\gamma)$ be a minimum as a function of γ for that value of T. This may be satisfied by $F(T,\gamma)$ being a boundary minimum or by meeting the conditions $[\partial F(T,\gamma)/\partial \gamma]_T = 0$ and $[\partial^2 F(T,\gamma)/\partial \gamma^2]_T > 0$. If $F(T,\gamma)$ has a single minimum in the range of γ from 0 to 1 at temperature T, then only one form is stable at that temperature; if $F(T,\gamma)$ has more than one minimum, then there is a stable form at each of the minima.

It is worthwhile to comment here on the meaning of the minima in $F(T,\gamma)$ with respect to the nonrigidity parameter γ. A minimum in the free energy at a particular value of γ can be interpreted to mean that the full, exact Hamiltonian has a *stable* approximation corresponding to that value of γ; this usually means that there is a natural expansion of the Hamiltonian there which, when truncated, corresponds to a particular physical model. One example is the Born-Oppenheimer approximation, appropriate for solids, which is an expansion procedure based on the existence of a stable, near-rigid form of the cluster and the expansion of the Hamiltonian about the point of equilibrium in configuration space. Another, appropriate for liquids, is mean field theory, which is not an expansion about a physical point in configuration space but about an approximate, self-consistent effective potential.

Whether the energy level patterns and densities of states of clusters meet the conditions for the stability of multiple forms can only be answered by examining the energy level correlation diagrams. For any reasonable set of parameters in the interaction potentials, even if we neglect any promotion energy from the lowest state of the rigid form to the lowest state of the nonrigid form, we find the following behavior[11]:

1) all the lowest levels must rise in energy as γ goes from 0 to 1, but

2) almost all the high levels must drop in energy as γ goes from 0 to 1

This is a consequence of the fact that the density of states of the solid-like limiting form is the larger at low energies but that of the liquid-like form is the larger at high energies. We now assume that most of the energy levels are monotonic functions of γ. It is not necessary to make the much stronger assumption that the energy levels vary linearly with γ, but this assumption was made in order to make it possible to compute $F(T,\gamma)$ in a series of test cases[10].

With the assumption that most energy levels are monotonic functions of γ, the implication of the structure of the correlation diagram is that at sufficiently low temperatures, the free energy is a monotonic increasing function of γ, and at sufficiently high temperatures, it is a monotonic decreasing function of γ. These imply, not surprisingly, that at low temperatures, the one stable form of a finite system is solid-like and nearly rigid, and that at high temperatures the one stable form is very nonrigid and liquid-like. To go beyond these trivial conclusions, we need only recognize that because the density of states increases with increasing energy, and faster near the nonrigid limit than near the rigid limit. Hence as the temperature increases, the slope $[\partial F(T,\gamma)/\partial\gamma]_T$ becomes less for γ near 1 than near 0, the curve of free energy bends, becoming more concave toward the γ axis. This means that a temperature is reached at which the slope of the free energy becomes zero at or near the point $\gamma=1$. Above this temperature, the free energy has a minimum near the nonrigid limit as well as near the rigid limit. As the temperature increases still further, the curve of $F(T,\gamma)$ bends more, reaching a temperature at which the values of the free energy at its two minima are equal. This temperature, T_e, is the temperature at which an observer would see the two forms of the cluster present in equal amounts in a large sample. At any temperature for which F has two minima, the ratio of the amounts of the two forms is given by the chemical equilibrium constant $K\equiv[solid]/[liquid]=\exp[(F_{solid} - F_{liquid})/kT]$. A schematic representation of the behavior of $F(T,\gamma)$ is shown in Fig. 1.

At this point it seems that we have found a general phenomenon peculiar to finite clusters of atoms or molecules. The discussion just presented disconnects the freezing temperature from the melting temperature and seems to imply that all finite clusters should have *sharp but unequal freezing and melting points*. We shall see shortly that the situation is actually more complicated than this argument suggests; some clusters do indeed exhibit discrete ranges of temperature and energy within which solid and liquid phase-like forms coexist, but to do so, a system must satisfy a criterion regarding separability of certain time scales, in addition to the criteria that have no connection with dynamics. Those systems that meet all the criteria do indeed exhibit the special kind of phase transition just described-- neither a first order nor a second or higher order transition of the kind exhibited by bulk matter, but a special kind of transition that only certain systems can show. The following are *necessary* criteria for this special kind of phase transition:

Criterion 1: the system must be solid-like at sufficiently low temperatures, so that it cannot pass among potential wells in its lowest vibrational state like Li_3 does.

Criterion 2: the system must be finite, so that at temperatures different from T_e only a finite amount of free energy separates the two forms of the cluster.

The second of these criteria is of course the basis for the sharpness of a first order transition in bulk matter. If the two phases have the same free energy per atom, i.e. the same chemical potential μ, then the system is indifferent to which phase is present; this occurs

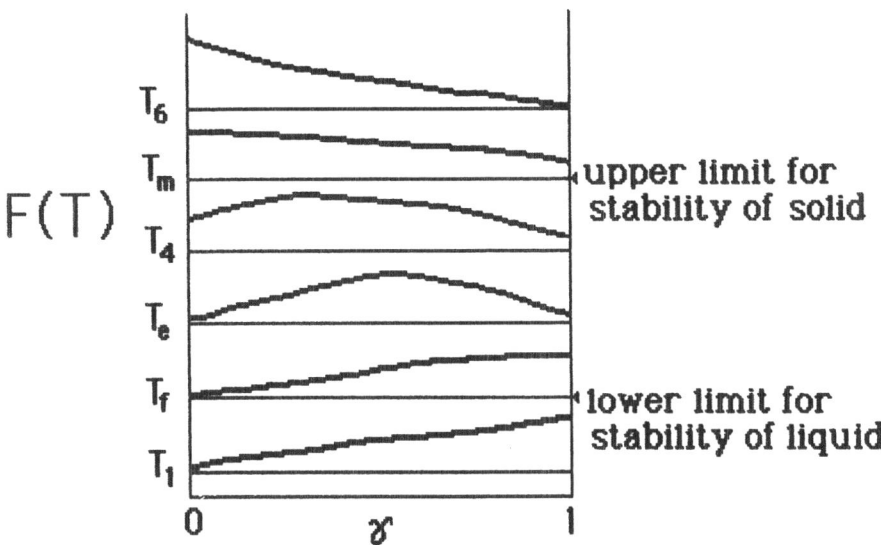

$F(T)$

upper limit for
stability of solid

lower limit for
stability of liquid

Figure 1. A schematic representation of the free energy $F(T,\gamma)$ as a function of γ for a series of temperatures. At T_1, F is a monotonic, increasing function of γ; at T_f, F has a zero derivative at the nonrigid limit; at T_e, the value of the free energy is the same at the two minima. The next temperature, T_4, is an arbitrary temperature for which F has two unequal minima; at T_m, F has a zero derivative at the rigid limit, and at the high, arbitrary temperature T_6, F is monotonic and decreasing as a function of γ. More complicated kinds of behavior of F would also be consistent with the description given here.

only at T_e. At any other temperature, in an ensemble of systems the ratio of populations in the two phases is, as stated earlier, given by the equilibrium constant. To emphasize the difference between small--microscopic and mesoscopic--systems and macroscopic systems, we write that constant as $K = \exp[N(\mu_s - \mu_l)/kT]$. This expression tells us that if N is not very large, K may vary over a range yielding two detectable phases, e.g. between 0.1 and 10, within a temperature band wide enough to be controlled and measured in the laboratory. However if N is more than a few powers of 10, K becomes a switching function, going from extremely large, effectively infinite, to very close to zero, as the temperature is raised through a very narrow band around T_e. The finiteness of the cluster makes the two forms appear as much like isomers as like phases, and one cannot apply the phase rule in its usual form to this situation.

We anticipate the other criterion, which is the most demanding, for clusters to exhibit a phase change with a finite band of coexistence temperatures:

Criterion 3. The individual clusters or molecules must pass between the two phase-like forms at an average rate that is slow relative to the times required for the establishment of well defined equilibrium properties of each phase, and also slow relative to the times required for the measurement of those equilibrium properties.

The importance of this criterion becomes clear when we examine the results of recent simulations. We turn to these now, because no calculations have yet been done from the analytic model with realistic parameters.

2. RECENT SIMULATIONS

When the analytic results emerged, it of course became crucial to test them. Experimental tests are far more expensive than simulations, and it is not a trivial matter to devise experiments that can give unambiguous confirmation or denial of a finite range of coexistence, and if so, address the second question, of the sharpness of its boundaries. Hence doing simulations first seemed the more effective course. The relevant simulations thus far include simulations of isoergic Ar$_{13}$ and Ar$_7$ by molecular dynamics[12,13], similar isoergic simulations of a variety of sizes of argon clusters[14], and several sets of simulations of isothermal argon clusters by Monte Carlo methods[15,16] and by both Monte Carlo and molecular dynamics methods[17] done by the method of Nosé[18]. Another relevant Monte Carlo simulation, seemingly from a very different starting point, has been done recently for the 10-state Potts model on a finite lattice[19]. Still others are in progress and may be discussed elsewhere in this volume[20].

2.1. Isoergic Simulations

The results of the molecular dynamics simulations of isoergic clusters have been reviewed recently[2,21] so we describe them only very briefly here. For N=6, 7, 9, 11, 13, 15 and 19, the simulations show distinct regions of coexistence. The most straightforward diagnostic for this is a range of energies for which the mean temperature--i.e. the mean, over a suitable but short number of steps of the simulation (500 in the cases cited), of the internal kinetic energy--exhibits a bimodal distribution. This bimodality is only seen for a band of energies whose bounds depend on the number N of atoms in the cluster; the band is as sharp as the resolution of the simulations, so the results are consistent with the analytic theory but of course can be no real proof. The properties of a cluster may be computed either from averages over the entire distribution of short term averages, or separately, for each of its two peaks, both Gaussian within the fluctuations of the calculation. Below and above the region of coexistence, there is no ambiguity about how to determine averages.

The properties of particular interest are, first, those distinguishing the two modes and identifying them with solid-like and liquid-like forms of the clusters. The two peaks of the bimodal region of the temperature distribution do indeed yield simple continuations of the low-temperature, solid-like branch and the high-temperature, liquid-like branch of the caloric curve. In this region of energy and temperature, the caloric function T(E) or E(T) behaves as a double-valued function. By averaging over all the points of the distribution, one of course obtains a smooth, single-valued curve. This curve is *monotonic* for all cases we have examined, provided the runs are long enough; short runs gave the S-shaped behavior reported by Briant and Burton[3]: Burton's skepticism[5] was well founded.

Other properties important for the diagnosis of the phenomenon are: the mean square displacement as a function of time or number of steps, whose slope is the diffusion coefficient, the velocity autocorrelation function and its Fourier transform, the spectral density; and the mean fluctuations of the nearest neighbor distances ("bond lengths"). All these criteria indicate unambiguously that at low energies, clusters of the sizes just listed are quite solid-like and the *high-temperature* branch of the bimodal region represents a smooth extension of all these properties, while the *low-temperature* branch in the coexistence region extends all these properties smoothly up to higher energies along the liquid-like branch. For example the hot solid shows a much smaller diffusion coefficient than the cold liquid at the same energy.

For all the clusters exhibiting bimodal distributions, we have found that the time spent in each form is indeed long enough to permit the establishment of well-defined average properties characteristic of each "phase". This is especially important for liquids because the very essence of being a liquid is one of ready compliance, and hence of response and therefore of time- or rate-dependent properties. If one strikes a water surface fast enough and over a large enough area, one realizes painfully the stiffness of a liquid when it is probed in a way that delivers a large amount of momentum in a very short interval. However this conclusion must in principle depend on the length of time used for the short-time averages. If these were taken to be 50,000 steps instead of 500, it would be much more difficult to pick out the bimodal structure of the distribution. The time interval chosen for the short-time averaging corresponds to the time constant of a measurement. It may be that some of the clusters, such as those for N of 8 or 14 might show bimodal or multimodal distributions if their short-time averages were taken over much shorter intervals, e.g. 100 steps. As it is, with the 500-step averages, clusters of several sizes exhibit unimodal, albeit in some cases asymmetrical, distributions of mean temperatures, and would appear to an observer as slush.

2.2. Isothermal Simulations

Isothermal simulations of Ar_{13}, whether by Monte Carlo or molecular dynamics, almost duplicate the caloric curve generated by isoergic molecular dynamics[17]. The fully averaged curves are superimposable except for a small but seemingly systematic deviation in which the isothermal system yields slightly higher energies, or corresponds to slightly lower temperatures, in the low-energy portion of the coexistence region. The overall agreement implies nearly ergodic behavior everywhere. The deviations can be attributed to the inability of an isoergic cluster to climb over a high pass into a spread-out, alpine valley, whereas the isothermal cluster occasionally has enough energy to cross and therefore can explore both the deep, narrow valleys corresponding to stiff solids and the broad, high valleys corresponding to nonrigid liquids. The isothermal systems can reach the regions of high potential and low kinetic energy inaccessible to the isoergic clusters and therefore have a slightly lower mean temperature at a given (low) energy in the coexistence region.

Extracting the bimodal distributions from Monte Carlo simulations of isothermal clusters is a delicate matter requiring very long runs and careful choice of averaging intervals. It can be done, however[17,19], and the same kind of double valued behavior is seen here as was found for molecular dynamics calculations of isoergic clusters. If the separation is not made, the transition can look deceptively gradual and lead to interpretating the results to mean that sharply distinct "phases" do not occur. Incidentally, one property that has been used in connection with the isothermal simulations but not the isoergic is the angular correlation function, which is a particularly useful diagnostic for polyhedra of known geometry[15,17]. Monte Carlo methods cannot be used to determine the velocity correlation function but the Nosé isothermal molecular dynamics method can, and does give a spectral density of states virtually identical to that of the isoergic calculation under corresponding conditions

2.3. Fractal Dimension

One other approach to interpreting melting and freezing of finite systems is still very much in progress but deserves mention here. This is the exploration of the dimension of the space in which the phase point of the system moves when the system is held at constant energy. If a system were composed of N particles which made up a solid with separable, harmonic vibrations, then in the center-of-mass system, the phase point would move on a hypersurface of 3N-6 dimensions, one for each independent oscillator. This is what we might expect for clusters at very low temperatures. In the liquid form, we might expect the cluster to explore as much of the 6N-dimensional phase space as the constraints allow, and there are just 7 constraints: constancy of energy, angular mamentum and momentum. Hence the dimension of the space explored by the phase point might be expected to be 6N-7. Determination of the dimension on which a point moves is dificult but an algorithm is available to do it[22], at least for small systems. For the cluster of three argon atoms, convergent results have been obtained[23] for two temperatures, both well below 30 K, at which the system develops soft modes. At an energy corresponding to 4 K, the fractal or Hausdorff dimension of the space in which the phase point moves is 3.29 ± 0.01, not at all far from 3N-6. At 19 K, this dimension is 5.71 ± 0.01. At the time this manuscript is being prepared, no converged results were available for higher temperatures. One may conjecture that the melting transition will be characterized by a sharp increase in this dimensonality. The correctness of this conjecture is still entirely open.

3. EXPERIMENTS

No experiments have yet been carried out that are designed to study the phase equilibrium of solid and liquid clusters. Some diffraction experiments have been done which indicate the existence of amorphous clusters[24-26]. Those af Valente and Bartell have been interpreted to imply that the amorphous forms are liquid. Fragmentation experiments[27] have also been interpreted to suggest the presence of both solid and liquid phases. However the only results from experiment that imply coexistence are the interpretations by Eichenauer and Le Roy[28] of infrared studies of SF_6 inside argon clusters[29]. The shift and shape of one broad band was shown by a combination of simulation with experimental results to be consistent with the coexistence of both forms in a molecular beam; the authors are properly cautions.

4. ACKNOWLEDGMENTS

The author wishes to thank the Physical Chemistry Laboratory, Oxford, for kind hospitality and the National Science Foundation for its support of the research of his group in this area.

5. REFERENCES

1. D. J. McGinty, J. Chem. Phys. **58**, 4733 (1973); R. M. Cotterill, W. Damgaard Kristensen, J. W. Martin, L. B. Peterson and E. J. Jensen, Comput. Phys. Comm. **5**, 28 (1973); C. L. Briant and J. J. Burton, Nature **243**, 100 (1973); J. K. Lee, J. F. Barker and F. F. Abraham, J. Chem. Phys. **58**, 3166 (1973). These seem to be the earliest works to find two-phase behavior in simulations of clusters; perhaps an interesting indication of the ripeness of the subject is the appearance of all four of these in 1973. See Ref. 2 for a recent review.

2. R. S. Berry, T. L. Beck, H. L. Davis and J. Jellinek, in *The Evolution of Size Effects in Chemical Dynamics,* Adv. Chem. Phys., edited by I. Prigogine and S. A. Rice (1987; in press) reviews much of the background.

3. C. L. Briant and J. J. Burton, J. Chem. Phys. **63**, 2045 (1975).

4. R. D. Etters and J. B. Kaelberer, Phys. Rev. A **11**, 1068 (1975); J. B. Kaelberer and R. D. Etters, J. Chem. Phys. **66**, 3233 (1977); R. D. Etters and J. B. Kaelberer, *ibid.* **66**, 5112 (1977).

5. J. J. Burton (private communication, ca. 1980).

6. G. Natanson, F. Amar and R. S. Berry, J. Chem. Phys. **78**, 399 (1983).

7. N. Nishioka, Phys. Rev. A **16**, 2143 (1977).

8. K. Yamada and M. Winnewisser, Z. Naturforsch. A **31**, 134 (1976).

9. S. Gartenhaus and C. Schwartz, Phys. Rev. **108**, 432 (1957).

10. R. S. Berry, J. Jellinek and G. Natanson, Chem. Phys. Lett. **107**, 227 (1984); *ibid,* Phys. Rev. A **30**, 919 (1984).

11. F .G. Amar, M. E. Kellman and R. S. Berry, J. Chem. Phys. **70**, 1973 (1979); M. E. Kellman, F. G. Amar and R. S. Berry, J. Chem. Phys. **73**, 2387 (1980); G. S. Ezra and R. S. Berry, J. Chem. Phys. **76**, 3679 (1982).

12. J. Jellinek, T. L. Beck and R. S. Berry, J. Chem. Phys. **84**, 2783 (1986).

13. F. G. Amar and R. S. Berry, J. Chem. Phys. **85**, 5943 (1986).

14. T. L. Beck, J. Jellinek and R. S. Berry, J. Chem. Phys. (in press, 1987).

15. N. Quirke and P. Sheng, Chem. Phys. Lett. **110**, 63 (1984).

16. E. Blaisden-Barojas and D. Levesque, in *The Physics and Chemistry of Small Clusters,* P. Jena, ed. (Plenum, New York, in press, 1987).

17. H. L. Davis, J. Jellinek and R. S. Berry, J. Chem. Phys. (in press, 1987).

18. S. Nosé, Mol. Phys. **52**, 255 (1984); *ibid,* J. Chem. Phys. **81**, 511 (1984).

19. M. S. S. Challa, D. P. Landau and K. Binder, Phys. Rev. B **34**, 1841 (1986).

20. J. Luo, U. Landmann and J. Jortner, in *The Physics and Chemistry of Small Clusters,* P. Jena, ed. (Plenum, New York, in press, 1987); results more explicitly relevant to this problem can be expected for alkali halide clusters from these authors.

21. See, for example, R. S. Berry, T. L. Beck, H. L. Davis and J. Jellinek, in *The Physics and Chemistry of Small Clusters,* P. Jena, ed. (Plenum, New York, in press, 1987); R. S. Berry, in Pro. 1st NEC Symposium on Fundamental Approach to New Material Phases, Hakone, 1986 (Springer, New York, in press); R. S. Berry, Proc. Symposium of the Topical Group on the Few-Body Problem, Washington Meeting of the American Physical Society, April, 1987 (American Physical Society, New York, in press).

22. P. Grassberger and I. Procaccia, Phys. Rev. A **28**, 2591 (1983).

23. T. L. Beck, D. M. Leitner and R. S. Berry (in preparation).

24. J. Farges, M. F. de Faraudy, B. Raoult and G. Torchet, J. Chem. Phys. **59**, 3454 (1973); *ibid,* **78**, 5067 (1983); *ibid,* **84**, 3491 (1986).

25. B. G. DeBoer and G. Stein, Surf. Sci. **106**, 84 (1981).

26. E. J. Valente and L. F. Bartell, J. Chem. Phys. **80**, 1451, 1458 (1984).

27. A. J. Stace, Chem. Phys. Lett. **99**, 470 (1983); A. J. Stace, D. M. Bernard, J. J. Crooks and K. L. Reid, Mol. Phys. **60**, 671 (1987).

28. D. Eichenauer and R. J. Le Roy, Phys. Rev. Lett. **57**, 2920 (1986); J. Chem. Phys. (in press, 1987).

29. T. E. Gough, D. G. Knight and G. Scoles, Chem. Phys. Lett. **97**, 155 (1983).

ELECTRON LOCALIZATION IN WATER CLUSTERS

Uzi Landman, R.N. Barnett and C.L. Cleveland
School of Physics, Georgia Institute of Technology
Atlanta, Georgia 30332, U.S.A.   and
Joshua Jortner, School of Chemistry
Tel-Aviv University, 69978 Tel-Aviv, Israel

ABSTRACT

Electron attachment to water clusters was explored by the quantum path integral molecular dynamics method, demonstrating that the energetically favored localization mode involves a surface state of the excess electron, rather than the precursor of the hydrated electron. The cluster size dependence, the energetics and the charge distribution of these novel electron-cluster surface states are explored.

I.   INTRODUCTION

Isolated clusters provide ways and means for the exploration of the evolution of size effects on energetic, dynamic and chemical phenomena in large, finite systems[1]. Studies of the structure, the level structure and the dynamics of clusters do not solely manifest the "transition" from molecular to condensed matter systems, but also provide a bridge between molecular and surface phenomena, as is evident from experimental studies of microscopic catalysis on metal clusters[2] and from theoretical investigations of surface states for an excess electron on alkali halide clusters[3]. The formation of a cluster-electron surface state is facilitated by the large surface to volume ratio of the cluster, being determined by the cluster structure, its degree of aggregation and the nature of the electron-cluster interactions. We shall demonstrate, using quantum path integral molecular dynamics simulations[3,4] (QUPID), that the energetically stable excess electron states in small water clusters[5-7] involve surface states rather than internally localized states which may be regarded as precursors of the celebrated hydrated electron.[8]

   The existence of the solvated electron was experimentally demonstrated in 1863 for liquid ammonia[9] and in 1962 for water.[8] The localization of an excess electron in the bulk of a polar fluid originates from the combination of long-range and short-range attractive interactions,[10] and is accompanied by a large local molecular reorganization. Nonreactive electron localization in water clusters was

J. Jortner et al. (eds.), Large Finite Systems, 145–152.
© 1987 *by D. Reidel Publishing Company.*

experimentally documented to originate either from electron binding
during the cluster nucleation process,[5,6] or by electron attachment to
preexisting clusters.[7] The occurrence of a weakly bound state in
$(H_2O)_2^-$ (vertical electron binding energy -3 meV for the equilibrium
state[11,12] and -13 to -27 meV for a persistent metastable state),[12]
characterized by a diffuse excess electron charge distribution (radius
of gyration of $\sim 36 a_o$)[11,12] can be understood on the basis of QUPID
calculations, to originate from weak electron-dipole interactions. On
the other hand, the experimental observation of stable $(H_2O)_n^-$ $(n > 11)$
clusters,[5-7] which are characterized[7] by a large vertical electron
binding energy, i.e., -0.7 eV (for $n = 11$) to -1.2 eV (for $n = 20$), poses
a challenging theoretical problem. Quantum mechanical calculations[13,14]
for $(H_2O)_6^-$ and $(H_2O)_8^-$ reveal that the adiabatic electron binding energy
of these, and presumably also larger, water clusters will be positive,
precluding the existence of such stable excess electron clusters, in
contrast with experiment.[5-7] These theoretical studies followed faith-
fully the conventional wisdom in the field of solvated electron theory,[10]
invoking the implicit assumption that the excess electron state in $(H_2O)_n^-$
constitutes an interior localization mode. QUPID calculations are ideally suited to explore alternative localization modes of the excess electron in water clusters.

II   ENERGETICS AND INTERACTIONS

The QUPID method rests on an isomorphism between the quantum problem and
a classical one, wherein the quantum particle is represented by a
necklace of P pseudo-particles ("beads") with nearest-neighbor harmonic
interactions.[3,4] Invoking previous formalism and notation[3] the average

total energy of the system is $E = 3N/2\beta + <V_c> + K + P^{-1} < \sum_{i=1}^{P} V(\vec{r}_i) >$ with

$K = 3/2\beta + K_{int}$, where V_c is the interaction potential between the
classical particles (whose number is N), $V(r_i)$ is the cluster-electron

interaction for the ith pseudoparticle, $K_{int} = (1/2P) \sum_{i=1}^{P} <(\partial V(\vec{r}_i)/\partial \vec{r}_i) \cdot$

$(\vec{r}_i - \vec{r}_p)>$, $\beta = 1/kT$ and $< >$ indicates statistical averaging. The water
molecules in this study were treated classically. The choise of the
number, P, of beads representing the excess electron is temperature
dependent. As a rule of thumb, adequate discretization is achieved
for $PkT \gtrsim e^2/a_o$.
   A key issue in modeling the system is the choice of interaction
potentials. Fortunately, for neutral small water clusters, interaction
potential functions which provide a satisfactory description for a range
of properties are available. We have used the RWK2-M model[15] for the
intra and inter-molecular interactions. For the electron water interaction we have constructed a pseudo-potential (Fig. 1) in the spirit
of the density functional theory, which consists of Coulomb, polarization, exclusion, and exchange contributions:

$$V(\vec{r}_e,\vec{R}_0,\vec{R}_1,\vec{R}_2) = V_{coul} + V_p + V_e + V_x \quad . \tag{1}$$

The position of the oxygen and hydrogen nuclei of the water molecule are given by $(\vec{R}_0,\vec{R}_1,\vec{R}_2)$ and $\vec{r}_e$ is the position of the electron. The Coulomb interaction is

$$V_{coul}(\vec{r}_e,\vec{R}_0,\vec{R}_1,\vec{R}_2) = - \sum_{j=1}^{3} q_j e/\max(|\vec{r}_e - \vec{R}_j|,R_{cc}) \quad , \tag{2}$$

where $\vec{R}_3 = \vec{R}_0 + (\vec{R}_1+\vec{R}_2-2\vec{R}_0)\delta$ is the position of the negative point charge of the RWK2-M model and $R_{cc} = 0.5a_0$. The values $q_1 = q_2 = 0.6e$, $q_3 = -1.2e$, $\delta = 0.22183756$ a_0 were chosen[15] to give a good representation of the dipole and quadrupole moments of H_2O. The polarization interaction is given by

$$V_p(\vec{r},\vec{R}_0) = -0.5\alpha e^2/(|\vec{r}_e-\vec{R}_0|^2+R_p^2)^2 \quad , \tag{3}$$

where $\alpha = 9.7446$ a.u. is the spherical polarizability of the water molecule. The form of V_p and the value of $R_p = 1.6a_0$ were chosen to fit the adiabatic polarization potential as calculated by Douglass et al.[16] The exclusion, V_e, and exchange, V_x, contributions both require the electron density $\rho(\vec{r},\vec{R}_0,\vec{R}_1,\vec{R}_2)$, of the water molecule[17] which, in the regions of importance, is adequately approximated by the simple expression (see Fig. 1)

$$\rho(\vec{r};\vec{R}_0,\vec{R}_1,\vec{R}_2) = 8\,a_o^{-3}\,e^{-3|\vec{r}-\vec{R}_0|/a_0} + a_o^{-3}\sum_{j=1}^{2} e^{-3|\vec{r}-\vec{R}_j|/a_0} \quad . \tag{4}$$

The repulsion, due to the exclusion principle, is modeled as a "local kinetic energy" term,

$$V_e(\vec{r}_e,\vec{R}_0,\vec{R}_1,\vec{R}_2) = 0.5\,e^2a_0\,(3\pi^2\rho)^{2/3} \quad . \tag{5}$$

The exchange interaction is modeled via the local exchange approximation,

$$V_x(\vec{r}_e,\vec{R}_0,\vec{R}_1,\vec{R}_2) = -\alpha_x e^2 a_0(3\pi^2\rho)^{1/3}/\pi \tag{6}$$

The parameter α_x was taken to be $\alpha_x = 0.3$ in order to obtain good agreement between our simulation results and the SCF results of Rao and Kestner[13] for $(H_2O)_8^-$ at a fixed configuration of the water molecules.

III   SURFACE STATES OF EXCESS ELECTRON ON WATER CLUSTERS

Equipped with these potentials we have embarked upon an investigation of the energetics and geometry of $(H_2O)_n^-$ (n = 8–18) clusters. In correspondence with the alternative experimental preparation methods[5-7] we invoked two initial conditions. (i) First condensing the water molecules around a classical negatively charged particle with a radius of $5a_0$, and subsequently replacing the classical particle with the electron necklace. (ii) Placing a compact distribution of beads next to an

equilibrated neutral cluster. For the smaller clusters n ≤ 12 a surface state develops rapidly, regardless of the initial setup of the
calculation, while for n = 18 (i) and (ii) yield an "internal" and
"surface" state, respectively.

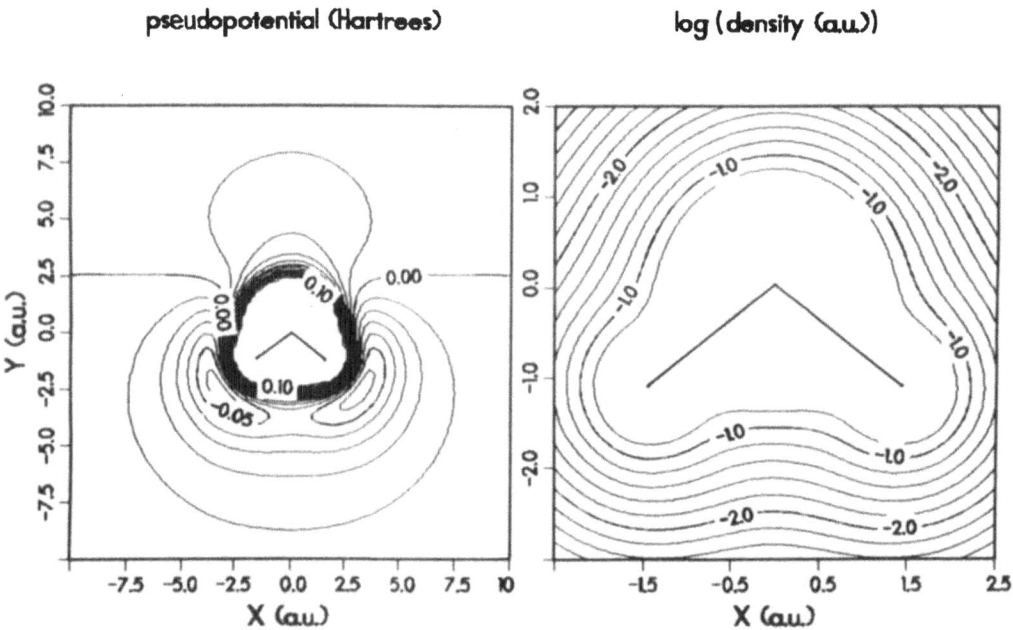

Figure 1: Contours of electron-water interaction, left, and the
electron density ($Log_{10}\rho(r)$), right, (see Eq. 4) in the plane containing
the nuclei. The oxygen is located at the origin.

In table I we summarize the energetic data for the electron vertical binding energy, $EVBE = PE(e-(H_2O)_n) + K_{int}$ where PE is the averaged
interaction potential energy between the electron and the water molecules, the cluster reorganization energy, E_c, and the electron adiabatic
binding energy, $EABE = EVBE + E_c$. The neutral cluster reference states
were obtained by simulated annealing. The lowest energy configuration
for each cluster size was then used to calculate E_c (the difference
between the molecular potential energies of the negatively charged and
neutral clusters). The energetic stability of the negatively charged
cluster with respect to the equilibrium neutral cluster plus free
electron is inferred from the magnitude and sign EABE, (negative values
corresponding to a stable bound state). The bead distributions for the
excess electron (Fig. 2) are characterized by the radius of gyration,
$R_g^2 = \frac{1}{2P^2} \langle \sum_{i,j}(\underline{r}_i-\underline{r}_j)^2 \rangle$, and the degree of localization by the complex

time correlation function[19] $R(t-t') = <|\underset{\sim}{r}(t)-\underset{\sim}{r}(t')|^2>^{1/2}$ for $(t-t')\varepsilon(0,\beta h)$, yielding the correlation length $\underset{\sim}{R}(\beta h/2)$, (Table 1), which for a free particle is denoted by $R_f = \sqrt{3}\lambda_T/2$, where λ_T is the thermal wavelength of the particle. All calculations were at constant temperature with the velocity form of the Verlet integration algorithm.[20,21]

From these results we assert that there is a remarkable quantitative difference between internal and surface states of the excess electron in water clusters. The value of E_c is considerably lower for a surface state than for an internal state insuring relative energetic stability of the former (Table I). As is apparent for $(H_2O)^-_{18}$, $|EVBE|$ is consierably higher (and outside the range of the experimental values) for the interior state, however, the high value of E_c results in EABE = 0.245 eV, precluding a stable internally localized state. On the other hand, for the electron surface state of $(H_2O)^-_{18}$ the value of EABE is close to zero (and the value of EVBE is in the range of measured values) favoring this mode of localization. For $(H_2O)^-_{12}$ (Table I and Fig. 2), only a surface state is found. Finally, for $(H_2O)^-_8$, a very small electron binding energy is found and the state is characterized by a diffuse charge distribution (Fig. 2 and Table I).

From these results we conclude that: (1) The electron localization mode in $(H_2O)^-_n$ clusters involves the formation of a surface state. (2) The onset of electron localization in a tightly bound state in $(H_2O)^-_n$ clusters is exhibited for n > 8, in accord with experiment[5-7] (n > 11). (3) The vertical electron binding energy for the cluster-electron-surface-state in $(H_2O)^-_{12}$ and $(H_2O)^-_{18}$ (Table I) are in adequate agreement with the experimental photo-electron spectroscopic data.[7] Additionally, $|EVBE|$ rises sharply in the range n = 8-12. For the $(H_2O)^-_8$ cluster the diffuse nature of the excess electron distribution could lead to a large collision induced electron detachment cross-section, which may account for the absence of n < 11 clusters from the experimental spectra.[5-7] Considering the complexity of the system, statistical uncertainties and those implicit in the model interaction potentials, we are encouraged by our results which provide a consistent energetic and structural picture of electron localization in small water clusters.

IV   CONCLUSIONS

We have demonstrated the prevalence of the surface localization mode in $(H_2O)^-_n$ clusters, which is qualitatively different from the localization mode of the hydrated electron in the bulk. Consequently, we infer that long-range attractive interactions play an important role in electron localization in the bulk. In this context we conjecture that the striking difference[5] between the lowest coordination number for electron localization in water (n $\gtrsim$ 11) and that in ammonia (n $\gtrsim$ 35) clusters, may originate from a weaker electron-molecule interaction in ammonia, which renders the surface state unstable in $(NH_3)^-_n$ for small n. Consequently localization may require in this case the buildup of long-range attractive interactions thus resulting in large coordination numbers.

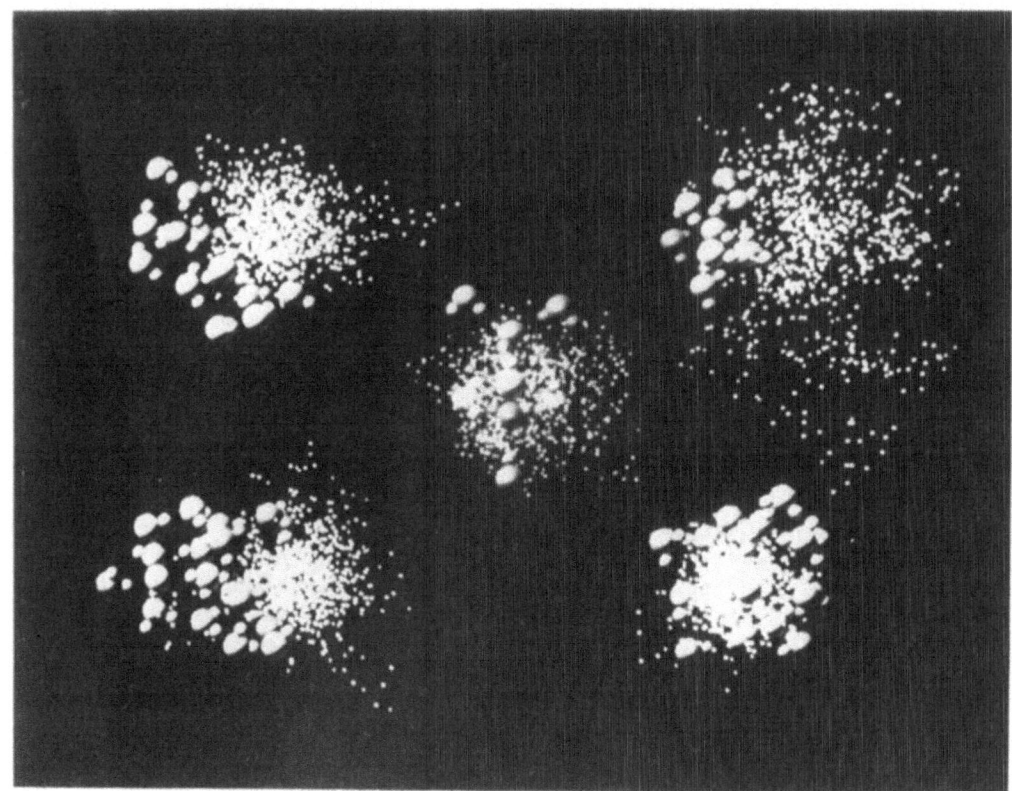

Figure 2: Cluster configurations of $(H_2O)_n^-$, via quantum path-integral molecular dynamics simulations. Balls, large and small, correspond to oxygen and hydrogen, respectively. The dots represent the electron (bead) distributions. Shown at the center is $(H_2O)_8^-$, for a static molecular configuration as in Ref. 13. From top right and going counterclockwise: (i) diffuse surface state of $(H_2O)_8^-$; (ii) surface state of $(H_2O)_{12}^-$; (iii) surface state of $(H_2O)_{18}^-$ and (iv) internal state of $(H_2O)_{18}^-$.

ACKNOWLEDGEMENTS

This research was supported by U.S. DOE Grant No. FG05–86ER45234 (to U.L.) and by the U.S.-Israel Binational Science Foundation (to J.J. and U.L.).

Table I. Energetics and excess electron charge distribution for $(H_2O)_n^-$ clusters. All calculations were at T = 79°K, using P = 4096 beads for the electron necklace. Energies in eV, radius of gyration, R_g', of the electron charge distribution in a_0 units.

CLUSTER	$(H_2O)_8^-$ Diffuse	$(H_2O)_{12}^-$ Surface	$(H_2O)_{18}^-$ Surface	$(H_2O)_{18}^-$ Internal
EVBE	−0.190	−0.97	−1.31	−1.96
E_c	0.136	0.871	1.333	2.204
EABE	−0.054	−0.136	0.023	0.245
R_g	10.6	6.1	5.5	4.1
$\dfrac{R(\beta\hbar/2)}{R_f}$	0.28	0.15	0.14	0.11

REFERENCES

1. Ber. Bunsenges. Physik. Chem. 88 (1984); Surface Sci. 156, 1–1072 (1985).

2. R.L. Whetten, D.M. Cox, D.J. Trevor, A. Kaldor, Phys. Rev. Lett. 65, 1494 (1985); M.E. Geusic, M.D. Morse, R.D. Smalley, J. Chem. Phys. 82, 590 (1985).

3. U. Landman, D. Scharf and J. Jortner, Phys. Rev. Lett. 54, 1860 (1985).

4. D. Chandler and P.G. Wolynes, J. Chem. Phys. 79, 4078 (1981); M. Parrinello and A. Rahman, ibid. 80, 860 (1985).

5. H. Haberland, H.G. Schindler and D.R. Worsnop, Ber. Bunsenges. Chem. 88, 3903 (1984); J. Chem. Phys. 81, 3742 (1984).

6. J.V. Coe, D.R. Worsnop and K.H. Bowen (J. Chem. Phys., 1987).

7. M. Knapp, O. Echt, D. Kreisle and E. Recknagel, J. Chem. Phys. 85, 636 (1986); J. Phys. Chem. (preprint, 1986).

8. E.J. Hart and J.W. Boag, J. Am. Chem. Soc. 84, 4090 (1962).

9. W. Weyl, Ann. Phys. 197, 601 (1863).

10. J. Jortner, J. Chem. Phys. 30, 839 (1959); D.A. Copeland, N.R. Kestner and J. Jortner, J. Chem. Phys. 53, 1189 (1970).

11. A. Wallquist, D. Thirumalai and B.J. Berne, J. Chem. Phys. 85, 1583 (1986).

12. U. Landman, R.N. Barnett, C.L. Cleveland, D. Scharf and J. Jortner, J. Phys. Chem. (1987), in print.

13. B.K. Rao and N.R. Kestner, J. Chem. Phys. 80, 1587 (1984) and references therein.

14. N.R. Kestner and J. Jortner, J. Phys. Chem. 88, 3818 (1984); M.D. Newton, J. Chem. Phys. 58, 5833 (1973).

15. J.R. Reimers, R.O. Watts and M.L. Klein, Chem. Phys. 64, 95 (1982); J.R. Reimers and R.D. Watts, ibid. 85, 83 (1984). This paper as well as the description of the potential in the first one (Eq. 13 and Table 1) contain several ambiguities and typographical errors. When corrected we reproduce their results.

16. C.H. Douglass, Jr., D.A. Weil, P.A. Charlier, R.A. Eades, D.G. Truhlar and D.A. Dixon, Chemical Applications of Atomic and

Molecular Electrostatic Potentials, Eds., P. Politzer and D.G. Truhlar, (Plenum, New York, 1981), p. 173.

17. C.W. Kerr and M. Karplus in Water, F. Franks, ed., (Plenum, New York, 1972), p. 21; M.W. Ribarsky, W.D. Luedtke and U. Landman, Phys. Rev. B32, 1430 (1985).

18. J.N. Bardsley, Case Studies in Atomic Physics 4, 299 (1974); G.G. Kleiman and U. Landman, Phys. Rev. B8, 5484 (1973).

19. A.L. Nichols, D. Chandler, V. Singh and D.M. Richardson, J. Chem. Phys. 81, 5109 (1984).

20. J.R. Fox and H.C. Anderson, J. Phys. Chem. 88, 4019 (1984).

21. In the QUPID method the averaged results do not depend on the dynamic masses used to generate the classical trajectories. We used a mass of 1 amu for the classical particles and 0.025 amu for the beads. The integration time step was 2.625×10^{-16} sec. Prior to averaging the systems evolved till no discernable trend was observed. Averaging was then performed typically over 2×10^{4} time steps.

RARE-GAS SOLVENT CLUSTERS: ORDER – DISORDER TRANSITIONS AND SOLVENT SHELL EFFECTS

Samuel Leutwyler and Jürg Bösiger
Institut für Anorganische, Analytische und Physikalische Chemie
Freiestr.3
CH-3000 Bern 9
Switzerland

ABSTRACT. Electronic spectra of carbazole·Ar_n clusters in supersonic beams exhibit specific size-dependent spectral features, showing only sharp intermolecular vibrational bands for n = 1 to 3, both broad and narrow spectral bands for n = 4 to 6, and only broad bands for $n \geq 7$. Monte Carlo simulations indicate that the change from sharp to broad structure is due to surface melting transitions of the solvent clusters relative to the substrate, and predict a finite width solidlike-liquidlike coexistence range in agreement with the experimental observations. For clusters between n=8 to 20, spectral shifts and bandwidths are influenced by single or multiple atom transfers to the second solvation layer or the second, unsolvated side of the molecule. Closure of the first solvent shell takes place at n=28.

1. INTRODUCTION

The study of melting transitions in clusters and small particles is important for a deeper microscopic understanding of the solid $\leftrightarrow$ liquid phase transition. Two main points are of interest, (1) the size dependence of the melting transition temperature, and (2) the width and shape of the transition in finite systems. The melting points of small particles have long been predicted to decrease with decreasing particle size [1]. Several molecular-dynamics (MD) and Monte-Carlo (MC) simulation studies on Ar_n clusters (n=3 to 100) have found size-dependent and sharp melting transitions [2-5]. However, Berry et al. have predicted by quantum-statistical model calculations that clusters should show unequal freezing and melting temperatures T_f and T_m, and a coexistence range (T_f,Tm) over which both solidlike and liquidlike forms should coexist [6,7]. This prediction is supported by an MD study on Ar_{13} [8].
  Experimental information is still sparse: melting-point measurements of supported Pb, Sn, Bi, In, and Au particles agree well with the Pawlow equation [1] down to sizes of 10 Å [9-12]. Infrared photodepletion spectroscopic measurements of SF_6 – Ar_n clusters performed by Gough et al.[14] have been interpreted by Eichenauer and

153

J. Jortner et al. (eds.), Large Finite Systems, 153–164.
© *1987 by D. Reidel Publishing Company.*

LeRoy [15] in terms of cluster structures involving "surface" and "surrounded" SF_6 chromophores. Their MC simulations suggest coexistence of these structural forms at fixed temperature. Unfortunately, in the work of Gough et al., neither the average cluster size nor the higher moments of the cluster distribution, nor the cluster temperatures were known.

We have measured the electronic spectra of carbazole•Ar_n rare-gas solvent clusters with n=1 to 35, by resonant two-photon ionization (R2PI) spectroscopy using mass spectrometric detection, at the near-UV $S_1 \leftarrow S_0$ ($^1A_1 \leftarrow {}^1A_1$) transition of the carbazole substrate molecule. Since cluster fragmentation can be shown to be negligible under carefully controlled conditions [16,17], cluster sizes are unequivocally determined, and species-specific electronic spectra are obtained.

2. EXPERIMENTAL

The molecular-beam apparatus and time-of-flight mass spectrometer have been previously described [16,17]. The rare-gas solvent clusters were formed by adiabatically expanding carbazole at a vapor pressure of 0.1-0.3 mbar, seeded in Ar (25 atom%)/ Ne (75 atom%) carrier gas through a pulsed circular 0.5 mm diameter   nozzle heated to 395 K; stagnation pressures were between 1.2 to 2.0 bar. R2PI was performed with the tunable UV light from a pulsed Nd:YAG pumped, frequency-doubled dye laser, using pulse energies of ≈ 100 μJ and peak intensities of $2 \cdot 10^5$ Wcm^{-2}.

3. RESULTS AND DISCUSSION

3.1 Resonant two-photon-ionization spectra

The R2PI spectra of the carbazole•Ar_n clusters with n= 1 to 16 are highly individual, see Fig. 1. A number of the salient features of these spectra will be discussed and very briefly interpreted. A fuller discussion of the structural, dynamic, and energetic aspects of rare-gas solvent clusters will then be given in the subsequent sections.

(1) Clusters with n= 1 to 6 exhibit sharp electronic origins followed, towards higher frequencies, by vibrational bands which correspond to intermolecular carbazole-Ar_n vibrations in the S_1 state. These observations imply the existence of (a) single predominant cluster structures and (b) well-defined solvent cluster vibrations.

(2) The evolution of the red shifts $\delta\nu$ as a function of n in the size range n = 1 to 6 indicates that the solvent cluster growth starts and continues on the same side of the carbazole molecule, yielding solvent clusters of binding topology (n + 0). Alternative hypotheses for cluster growth, e.g., adsorption on alternating sides of the substrate, are inconsistent with the observed red shift pattern.

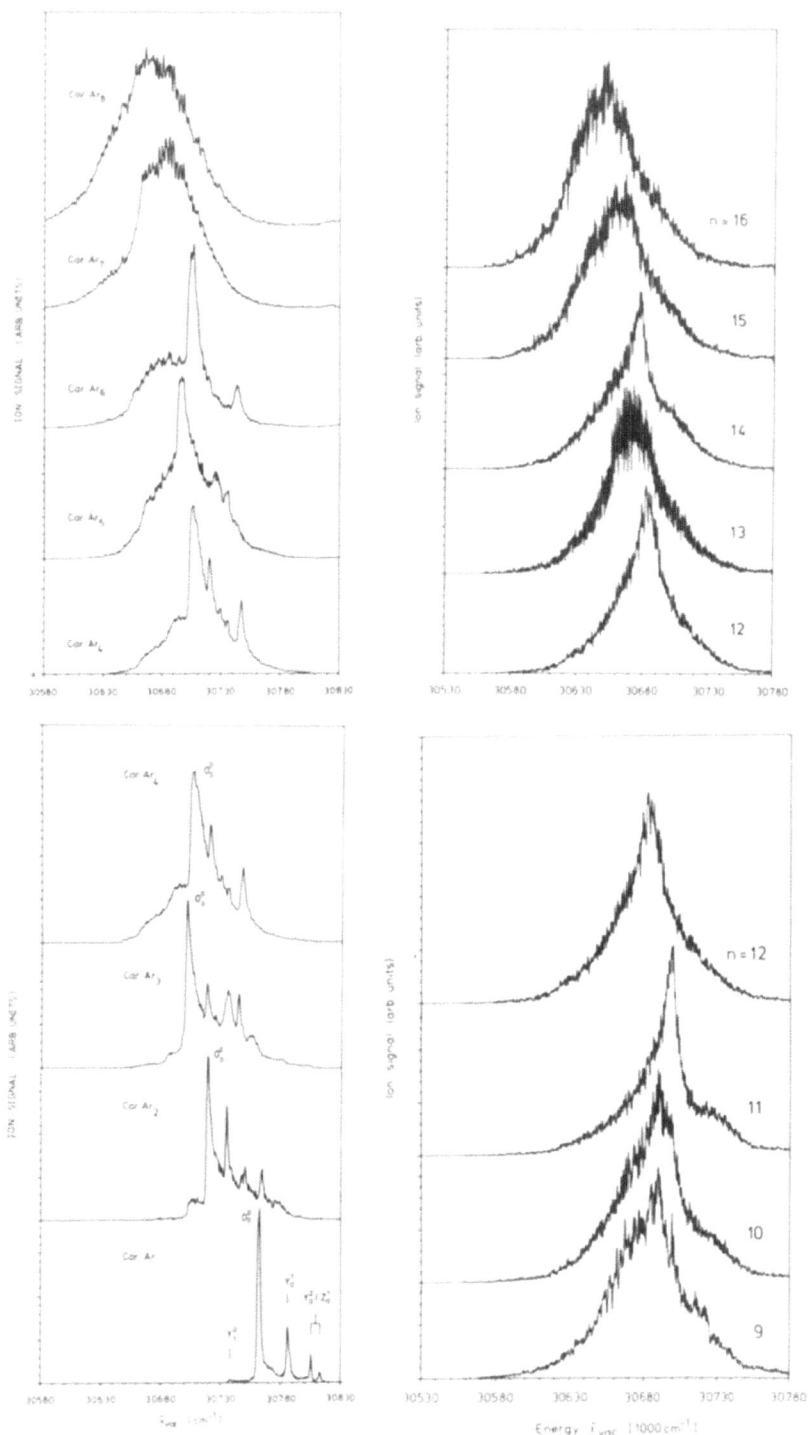

Fig. 1: Resonant-two-photon-ionization spectra of carbazole Ar_n with n = 1 - 16.

(3) The rapid saturation of the red shift after n=3 to n=6, together with the calculations of cluster geometry described below indicates that for atoms adsorbed over the central part of the substrate (the carbon ring framework) there is a differential red shift of 40 -45 cm^{-1} per added carbon atom, but essentially zero red shifts or even small blue shifts arise from atoms adsorbed over the peripheral hydrogen atoms.

(4) At n=4 an additional broad, red-shifted, smooth band appears about 60 cm^{-1} to the red of the 0_0^0 transition, extending into the region of sharp spectral bands. The MC simulations presented below show that the solvent clusters with n ≥ 4 exhibit a *surface melting transition* at temperatures above 14 to 18 K; the resulting frequency spectrum of the intermolecular motions is expected to be quasi-continuous, in agreement with the observed band shape.

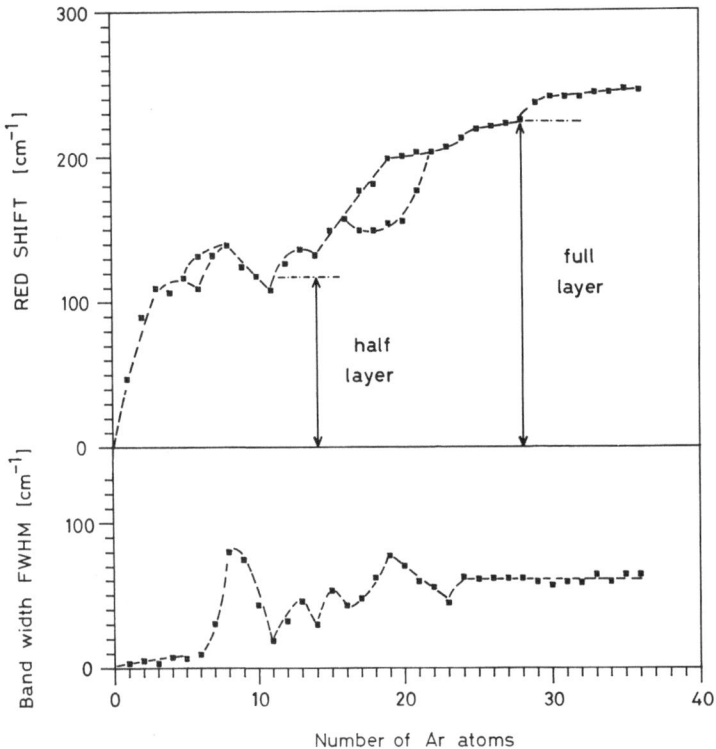

Fig.2: Red shifts and bandwidths (FWHM) of the electronic origin bands of carbazole•Ar_n clusters with n = 1 - 36.

(5) Clusters with n = 4 to 6 show discrete peaks and broad, diffuse bands simultaneously, suggesting that localized and disordered, fluxional structures can coexist in this size range.

(6) For n=4 and 5, the average frequencies of both the narrow and broad bands coincide, implying that for both the localized and the fluxional solvent clusters the cluster topology is same-sided, the spectral differences reflecting the distribution function of the atoms over the molecular surface. For n=6 on the other hand, there is a *bifurcation of spectral structure*: the broad band is shifted about 20 cm^{-1} further to the red than the narrow band. This implies that the topology for the disordered/fluxional cluster corresponds to partial or part-time atom-transfer to the second side of the molecule.

(7) For n=7 and larger, the sharp bands disappear altogether, and only broad structures remain (Fig. 1(b)), implying that the larger clusters have undergone complete surface melting.

(8) From n=6 to 8, both the red shifts and the half-widths of the electronic origins first increase, and then decrease from n=8 to 11; this pattern is repeated a second time from n=11 to 14. The structure calculations indicate that for medium-sized clusters in the size range n=8 to 22, *double-layer single-sided* clusters are energetically competitive with *single-layer contiguous* clusters. In the single-layer, "wetting", cluster geometries, side-transfer of Ar atoms is energetically favorable, and the complexation of the previously "free" second substrate side leads to differential red shifts; on the other hand, in the double-layer, "non-wetting", cluster geometries the contact area between the carbazole substrate and the Ar_n solvent cluster is smaller - approximately the same as for n=7 - leading to differential blue shifts. Thus the differential shift and shape (or moments) of the electronic origin band depend sensitively on the balance between these two types of geometries.

(9) The red shifts $\delta\nu(n)$ for n=3 to 14 vary within a comparatively narrow range of 110 cm^{-1}< $\delta\nu$ < 140 cm^{-1}; this implies that the number of Ar atoms that reach the central area of the other substrate side is, roughly speaking, between 0 and 0.75 Ar atoms. Starting with n=14 a fairly continuous increase of the red shift $\delta\nu$ from $\delta\nu(14)$ = 115 cm^{-1} to $\delta\nu(28)$ = 230 cm^{-1} is observed. This allows a rough separation of the solvation process into two halfs: solvation of one side of the substrate molecule is complete at n≈14 with a red shift level of $\delta\nu$ = 115 ± 15 cm^{-1}; solvation of the second side of the carbazole substrate is complete at n = 28 = 2•14, with a "terminal" red shift level of $\delta\nu$ = 230 cm^{-1} = 2•115 cm^{-1}.

3.2. Numerical simulations of cluster structures and dynamics

Both MD and MC calculations were performed on the carbazole•Ar$_n$ clusters. These indicate that the trends in spectral shifts and bandshapes described above may be only poorly understood in terms of cluster structures, and that a detailed study of the finite-temperature behavior of the solvent clusters is necessary.

Intermolecular potentials: The intermolecular potential energy was modeled using an atom-atom potential which has previously been shown to yield near-quantitative results for structural and energetic properties of aromatic-molecule/rare-gas van-der-Waals complexes [18-21]. The Lennard-Jones 12-6 potential parameters were ε_{ArC}=61.923 K, ε_{ArH}=42.768 K, ε_{Ar-Ar}=142.1 K, and σ_{ArC}=3.3854 Å, σ_{ArH}=3.2072 Å, σ_{Ar-Ar}=3.36 Å.

Structure calculations: As a starting-point for the MC calculations, minimum-energy solvent cluster structures were determined for n=1 to 9 and 11 by slow cooling in the molecular-dynamics simulation. Varying the time steps from $1 \cdot 10^{-14}$ to $5 \cdot 10^{-14}$ s had no influence on the resulting structures. Up to n=5, the minimum-energy structures found by the MD simulations are all "same sided", with the solvent cluster forming a contiguous two-dimensionally close-packed structure covering

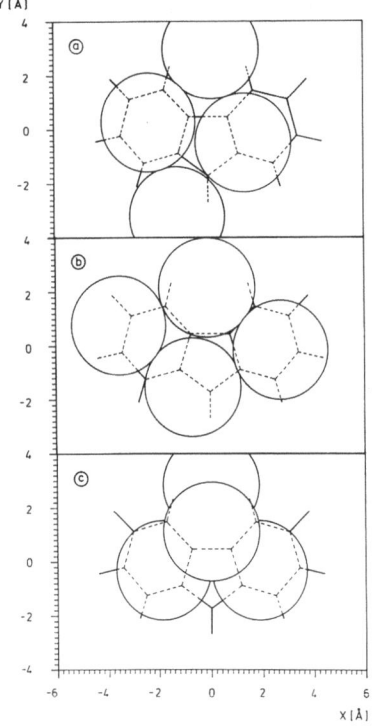

Fig.3: Car•Ar$_4$ cluster structures:
(a)   minimum-energy-structure with total binding energy V_4=-2032 cm^{-1};
(b)   lowest-lying single-layer solvatomer, V_4=-2020 cm^{-1};
(c) lowest-lying double-layer isomer, V_4=-1885 cm^{-1}.

the central part of one surface of the substrate. A single energetically low-lying structure was found for n = 1 to 3. The larger
clusters exhibit many distinct isomers: the number of isomers within 50
cm^{-1} of the absolute energy minimum (not counting enantiomers) is 2, 7,
7, and 1 for n=4, 5, 6, and 7 respectively. Fig.3 shows some
representative isomers for the carbazole•Ar_4 cluster. The appearance of
energetically low-lying structural isomers with increasing cluster size
is suggestive of a heterogeneous spectral broadening mechanism;
however, the 0 K structure calculations predict heterogeneous broadening to be moderate for n=4, important for n=5 and 6, but unimportant
for n=7, which is in disagreement with the experimental observations.

3.4. Order-disorder and phase transitions

Cluster order-disorder transitions were investigated using the
Metropolis MC simulation technique with the same potentials as for the
MD calculations. Initial equilibrations were performed for 100'000 to
300'000 steps and results were recorded for 500'000 to 2'000'000
steps. Since our interest is mainly in the structural transitions of
the solvent clusters, we have evaluated (a) the spatial distribution
functions of the Ar atoms relative to the carbazole substrate frame
$\rho_i=\rho(\{\vec{r}_i\};T)$, i=1...n; (b) the standard deviation of the atomic
position vector of the i-th Ar adatom, σ_i, relative to the carbazole
center of mass, which characterizes the *rigidity of the Ar cluster
relative to the substrate frame*; and (c) the normalized root-mean-
square Ar-Ar bond length fluctuation

$$\delta \ (d_{ij}) = \{<d^2_{ij}> - <d_{ij}>^2\}^{1/2} \ / \ <d_{ij}>$$

between Ar atoms i and j, which characterizes the *internal rigidity of
the solvent cluster*.
Examination of these structural order parameters shows an astonishing
series of order-disorder transitions of the Ar_n solvent clusters. We
have so far found *six* qualitatively different order-disorder
transitions, four of which are important for the understanding of the
various spectral features.
In order of rising temperature, these are: (1) surface isomeriza-
tion/racemization transitions, (2) cluster rigid-fluxional transitions,
(3) surface melting transitions, (4) second-layer atom promotion
("wetting-nonwetting-transition"), (5) atom side-crossing transitions,
and (6) cluster melting transitions.
   At the isomerization/racemization transitions, interconversion
takes place between different isomers (for n=6) or enantiomers (for
n=4, 5, 6 and 7). Isomerization is characterized by steplike increases
of several σ_i values. Since the transition interconverts mirror-
symmetric forms, the spectroscopic changes accompanying the transition
should be barely observable, with a possible exception at n=6. an
illustration of an isomerization transition in carbazole•Ar_4 is
provided by Fig.4(a,b), which show the transition from a non-symmetric
and chiral cluster structure (a) to the C_s symmetric form (b). The

transition takes place at T_{rac}=1.5 K !.

At the rigid-fluxional transition temperatures, the solvent clusters undergo structural changes which involve some, but not all, intracluster Ar-Ar bonds, accompanied by steplike increases in the corresponding δ_{ij} values.

Surface melting transitions are fairly gradual, taking place over several degrees and are characterized by increases in *all* of the σ_i values from $\sigma_i \leq 0.6$ Å at the lower end (freezing temperature) to $\sigma_i \geq 1.8$ Å at the upper end of the transition range (melting temperature). The transition ranges are 14-19 K (n=4), 16-19 K (n=5), and 18-21 K (n=6,7). Fig 4(b,c) exemplifies the effects of a combined rigid-nonrigid and surface melting transitions in carbazole•Ar_4.

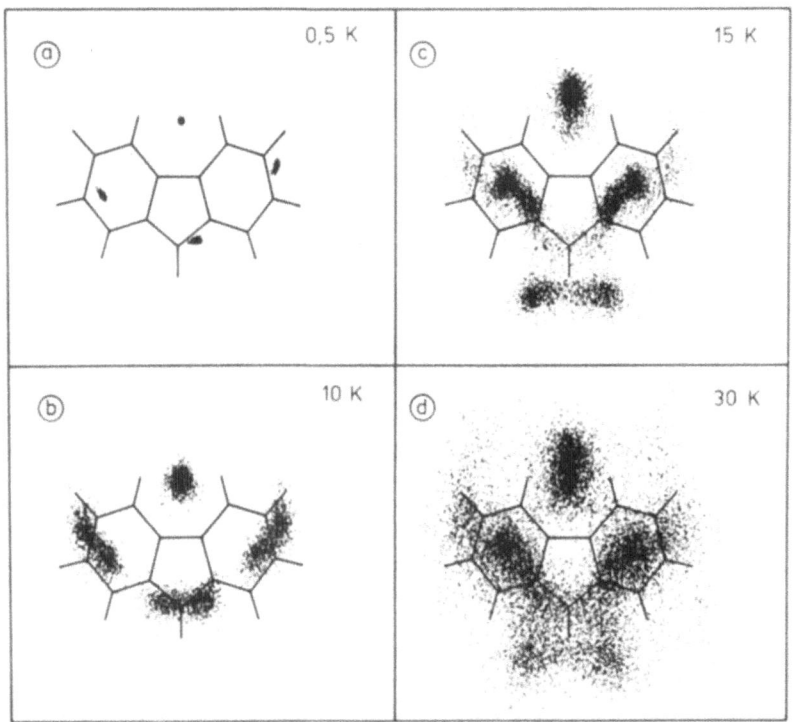

Fig.4: Solvent cluster probability density distributions for carbazole•Ar_4 at (a) 0.5 K,(b) 10 K, (c) 15 K, and (d) 30 K.

For large fractions of the Markov chain (up to $5\cdot10^5$ steps) the solvent clusters exhibits either a localized or a fluxional structure with respect to the substrate,which interconvert by a series of surface rotational/translational hops. This behavior is typical for all solvent

clusters with n = 4 to 7 over the surface melting transition range.
Although some intracluster structural rearrangements occur, the solvent
cluster itself is *not* liquidlike, since the large majority of the intracluster bonds remain stiff, with $\delta_{ij} < 10\%$, up to 25-30 K.

Second-layer promotion of an atom was not observed in the MC
simulations up to n=4. For larger clusters, however, this type of
structural transition becomes rapidly favorable: for n≥5 the process
occurs at 33 K, and drops to the vicinity of 25 K for larger clusters.

The crossing-over of an atom to the other substrate range is also
energetically unfavorable for small clusters, being only observed at T
> 37 K for n=4 and T > 35 K for n=5. However, the barrier to side-
crossing rapidly decreases with increasing cluster size, occurring at T
= 30.5 K for n = 6, T = 27.5 K for n = 7, and T = 20.5 K for n = 8.
Preliminary calculations on n = 8 and 11 solvent clusters indicate that
second-layer-promotion and side-crossing-transitions are competitive
and that the relative importance of these two transitions may depend
sensitively on n in this size range.

The intrinsic melting of the solvent cluster itself is
characterized by an increase of all δ_{ij} values above 10-15 %; this is
in excellent agreement with the Sutherland-Lindemann criterion for bulk
melting. However, unlike the bulk melting process, the transition takes
place over a temperature range of 4 to 5 K. As in the surface melting
transitions, the simulations reveal the coexistence of solidlike and
liquidlike cluster phases at the same temperature; thus the δ_{ij}'s show
hysteresis-type loops as a function of temperature over the melting
range. Fig. 4(c,d) allows direct visualization of the cluster melting
process, where (c) shows the system within the surface melting
temperature range, but below the cluster intrinsic melting range range.
The latter is between 23 to 30 K for this cluster.

3.5 Spectroscopic effects of order-disorder and phase transitions

Because the electronic transition is localized in the carbazole
molecule, the main factor influencing the width and shape of the elec-
tronic-vibrational bands is the adatom positional distribution <u>relative</u>
<u>to the substrate molecule</u>. The surface melting transitions dramatically
expand the region of configuration space accessible to the solvent
clusters, resulting in a smearing of the positional distributions of
the adatoms and a broadening of the electronic transition. In addition,
new low-frequency vibrational modes appear, which may couple to the
electronic transition.

The internal temperatures of the n=1 to 3 clusters can be determined by hot bands appearing on the low-frequency side of the 0_0^0 transitions: approximate vibrational temperatures are 11 ± 2 K (n=1), 13 ± 2
K (n=2) and 15 ± 3 K (n=3). Extrapolating linearly to higher n we estimate internal temperatures in the range T_{vib} = 16 to 22 K for n=4 to
7. The observed and extrapolated vibrational temperatures together with
the calculated ranges for the surface melting transitions for n=1 to 7
are plotted in Fig.5.

For n=2 and 3, the internal temperatures are below the surface
melting temperatures, which are sharp for these two clusters. However,

for n=4 to 6 the internal temperatures are estimated to lie within, and for n=7 , slightly above the surface melting temperature range. These trends parallel the qualitative behavior exhibited in Fig. 1(a),(b): solvent clusters that are below the surface freezing point show sharp bands, clusters with internal temperatures falling inside the surface melting temperature range exhibit both broad and sharp bands simultaneously, and the clusters with internal temperatures above the surface melting point show only a broad band contour. Thus, the calculations provide strong support for the interpretation of the spectral bandshapes in terms of solvent cluster surface melting transitions with coexistence ranges of finite width.

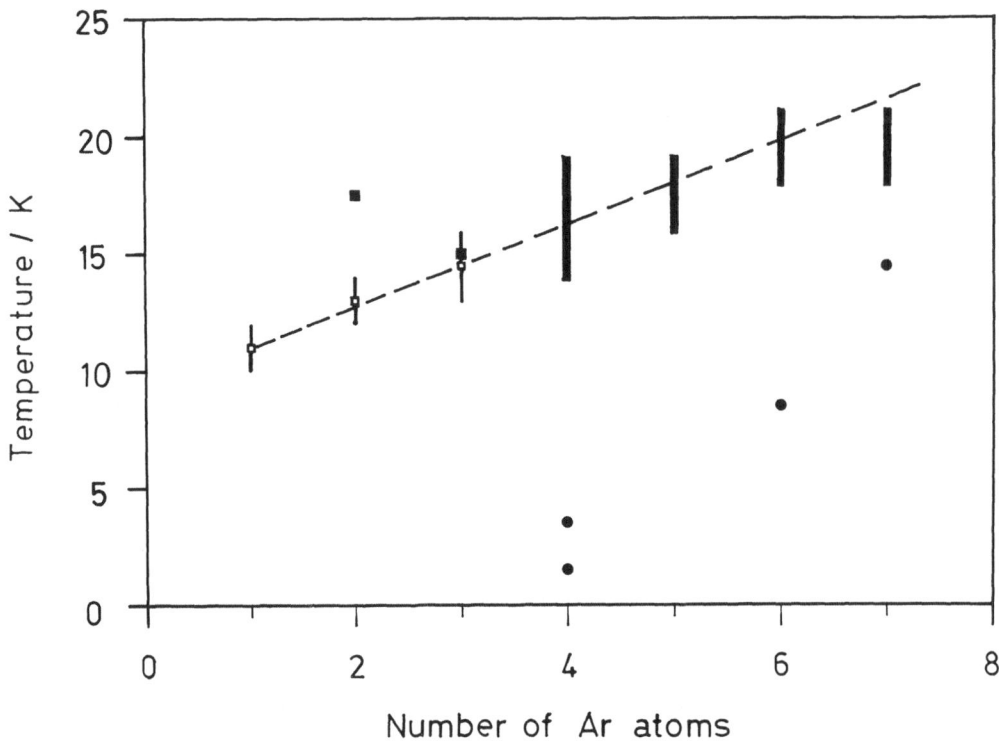

Fig.5: Comparison of calculated surface melting temperatures or melting temperature ranges (■ or vertical bars), with experimental vibrational temperatures (□) for carbazole·Ar$_n$, (n = 1 to 3), extrapolated to n=7 (---). Also included are the calculated isomerization temperatures (●).

The semicyclic blue shift/red shift patterns observed for n=8 to 11 and 11 to 14 can be interpreted on the basis of calculations performed

on the n=8 and 11 clusters. These show that for n=8 and 9, a large
number of different single-layer and double-layer clusters exist in
close energetic proximity. This static disorder by itself leads to
inhomogeneous broadening, since in the double-layer clusters some of
the atoms are not in direct contact with the carbazole substrate, and
therefore contribute little to the red shift. in addition, the single-
layer clusters exhibit side-crossing transitions in the vicinity of T=
20 K; the time-averaged transfer of one or more atoms to the previously
uncomplexed side of the molecule leads to increased red shifts, with
the amount of shift depending on the average fraction of atoms
transferred. Thus both static and dynamic disorder contribute to the
large width of the band.

The n=11 cluster, on the other hand, exhibits a very stable
double-layer single-sided cluster geometry with seven first-layer and
four second-layer atoms, which is ~ 150 cm-1 stabler than any single-
layer structure found so far. Since the side-crossing transition
temperature for this geometry is comparatively high (27 - 30 K), the
spectroscopic behavior is very probably dominated by this single
structure of high differential stability. The expected spectroscopic
effects are (a) a relatively blue-shifted electronic origin, due to the
reduced direct contact with the substrate; (b) spectral intensity
contributions from disordered/molten clusters appear on the red side of
the origin, since the single-layer and double-sided geometries are more
probable at higher energies.

We speculate that the same line of reasoning is appropriate to the
cluster series n=11 to 14.

4. CONCLUSIONS

The electronic band structures of a cold, isolated aromatic molecule
which is partially solvated by rare-gas atoms exhibit a great variety
of shifts, widths and shapes, which reflect both static and dynamic
aspects of the behavior of the microsolvent shell adsorbed on or
surrounding the planar chromophore. The spectral phenomena discussed in
section 3.1. can be rationalized by a series of characteristic and
size-dependent aspects of the solvent cluster finite-temperature
behavior:
(1) geometrically localized, single isomer structures for n = 1 to 3.
(2) cluster growth which proceeds on one side of the substrate
molecule, up to $n \approx 6$ to 8.
(3) rigid-fluxional and surface melting transitions, which are
spectroscopically manifested for n ≥ 4.
(4) coexistence of solidlike and liquidlike phases for n = 4 to 6.
(5) complete surface melting for 7 < n < 10.
(6) side-crossing transitions for n ≥ 7.
(7) partial "soldification" due to cluster structures of special
differential stability for n = 11, 14, and 20.

(8) solvent shell closure at n = 28.

Further work on direct simulation of the spectroscopic band structure, using the classical Franck-Condon principle coupled with the MC simulation method is presently under way, and is expected to yield further insight into the details of the experimental results.

Experimentally, it will be interesting to further investigate the question of homogeneous vs. inhomogeneous contributions to the broad and diffuse bands by hole-burning and laser fluorescence line narrowing techniques.

ACKNOWLEDGEMENTS

We thank Dr.R.Bombach for performing the MD structure minimizations. This work was supported by the Schweizerische Nationalfonds under grants No.2.668-0.85 and 2.065-0.86.

REFERENCES

[1] P.Pawlow, Z.Phys.Chem. **65**, 1 (1909)
[2] R.D.Etters and J.B.Kaelberer, Phys.Rev. A **11**, 1068 (1975) and J.Chem.Phys. **66**, 5112 (1977).
[3] J.B.Kaelberer and R.D.Etters, J.Chem.Phys. **66**, 3233 (1977).
[4] C.L.Briant and J.J.Burton, J.Chem.Phys **63**, 2045 (1975).
[5] N.Quirke and P.Sheng, Chem.Phys.Lett. **110**, 63 (1984).
[6] R.S.Berry, J.Jellinek, and G.Natanson, Chem.Phys.Lett. **107**, 227 (1984)
[7] R.S.Berry, J.Jellinek, and G.Natanson, Phys.Rev. A **30**, 919 (1984).
[8] J.Jellinek, T.L.Beck, and R.S.Berry, J.Chem.Phys. **84**, 2783 (1986).
[9] M.Takagi, J.Phys.Soc.Japan **9**, 359 (1954).
[10] C.J.Coombes, J.Phys. F **2**, 441 (1972).
[11] Gl.S.Zhdanov, Soviet Phys.-Cryst. **21**, 706 (1976).
[12] P.Buffat and J.-P.Borel, Phys.Rev. A **13**, 2287 (1976).
[13] J.-P. Borel, Surf.Sci. **106**, 1 (1981).
[14] T.E.Gough, D.G.Knight, and G.Scoles, Chem.Phys.Lett. **97**, 155 (1983).
[15] D.Eichenauer and R.J.LeRoy, Phys.Rev.Lett. **57**, 2920 (1986).
[16] E.Honegger, R.Bombach, and S.Leutwyler, J.Chem.Phys. **85**, 1234 (1986).
[17] O. Cheshnovsky and S.Leutwyler, Chem.Phys.Lett. **121**, 1 (1985).
[18] M.J.Ondrechen, Z.Berkovitch-Yellin, and J.Jortner, J.Am.Chem.Soc. **103**, 6586 (1981).
[19] U.Even, A.Amirav, S.Leutwyler, M.J.Ondrechen, Z.Berkovitch-Yellin, and J.Jortner, Farad.Discuss.Chem.Soc. **73**, 153 (1982).
[20] S.Leutwyler, A.Schmelzer, and R.Meyer, J.Chem.Phys. **79**, 4385 (1983).
[21] S.Leutwyler, J.Chem.Phys. **81**, 5480 (1984).
[22] J.Bösiger and S.Leutwyler, submitted to J.Chem.Phys.

Spectra, Structure and Dynamics of SF_6-$(Ar)_n$ Clusters

Robert J. Le Roy, John C. Shelley and Dieter Eichenauer
Guelph-Waterloo Centre for Graduate Work in Chemistry
University of Waterloo
Waterloo, Ontario N2L 3G1,
Canada

Monte Carlo and molecular dynamics simulations have been used to investigate the structure, dynamics and infrared spectra of a realistic model for mixed clusters of the form SF_6-$(Ar)_n$. In earlier work based on Monte Carlo studies alone [D. Eichenauer and R.J. Le Roy, *Phys. Rev. Lett.* **57**, 2920 (1986)], it was concluded that features of the simulations pointed out the coexistence of two distinct classes of cluster structures which were associated there with different phases. That concusion *now appears to have been somewhat premature.* However, a re-examination of those results, together with preliminary molecular dynamics simulations, yields a new interpretation which suggests that the inert diluent tends to form a two-dimensional liquid film on the surface of the SF_6 chromophore. Evidence for and possible observable manifestations of this behaviour are presented herein.

1. INTRODUCTION

The past decade has seen a growing interest in the properties of micro-samples of condensed phases ranging in size from a few to several thousands of atoms or molecules. Many of the earlier studies of such systems consisted of computer simulations of the properties of extremely small samples of "Lennard-Jonesium" $(LJ)_n$, a condensed phase of spherical particles interacting through pairwise Lennard-Jones (12,6) interaction potentials.[1-11] In particular, following Hoare and Pal's determination of the minimum potential energy structures for small $(LJ)_n$ clusters,[1] subsequent studies discovered the existence and examined the properties of the melting transitions for cluster sizes ranging from $n = 3$ to 429.[2-11] On the experimental side, electron diffraction studies have been performed on a number of systems in order to determine whether the clusters in question had crystalline, amorphous solid, or liquid-like structures,[12-16] while other inferences about cluster properties and behaviour have been made on the basis of chemical reaction rate measurements[17] and the infrared spectra[18-21] of mixed clusters.

A particularly intriguing result of the simulations mentioned above was the apparent observation of solid-liquid phase coexistence, a situation in which the melting temperature for a solid cluster was somewhat higher than the freezing temperature of the corresponding liquid. In the early work, this behaviour appeared to manifest itself as a kind of hysteresis or loop in the caloric curve obtained from the molecular dynamics simulations.[2] That observation is now believed to have arisen from incomplete convergence of the simulations.[7,9] However, evidence leading to essentially the same conclusion, that an equilibrium between solid- and liquid-like phases exists over a finite temperature range, does appear to be provided by the properties of the short-time averaged kinetic energies in the transition region.[7,9-11] Moreover, this behaviour is predicted by a simple quantum statistical mechanical model developed by Berry and co-workers.[22,23] Unfortunately, it is not clear how the kinds of properties examined in these simulations of $(LJ)_n$ clusters would be reflected in experimental observables,

165

J. Jortner et al. (eds.), Large Finite Systems, 165–172.
© *1987 by D. Reidel Publishing Company.*

and experimental studies of pure clusters have as yet provided no direct evidence for the kinds of phase transition behaviour predicted by the simulations.

In view of the above, the spectrum of an infrared active molecule clustered with an inert diluent offers considerable promise as a technique for possibly observing these phenomena. Perturbations of the vibrational levels of the chromophore should be quite sensitive to the packing and dynamical behaviour of the surrounding "solvent" particles. Work in this laboratory[20,21] has focussed attention on the infrared spectra of clusters consisting of an SF_6 molecule bound to a numbers of Ar atoms. This is a particularly appropriate system for study, both because of the existence of reliable experimental data,[1,24] and because the availability of accurate intermolecular pair potentials[25,26] allows simulations of these species to be quite realistic.

Our previous work on this topic consisted of the development of a simple yet accurate method for predicting the shift of the v_3 vibrational band of an SF_6 molecule perturbed by Ar atoms, and the use of Monte Carlo simulations to predict the infrared spectrum and other properties of SF_6-$(Ar)_n$ clusters.[20,21] In a preliminary communication,[20] it was stated that those results identified a distinctive spectroscopic signature for phase coexistence in such clusters. It now appears that this claim[20] was somewhat rash. The present communication surveys the results of our Monte Carlo simulations and describes the results of our initial molecular dynamics studies of this system.[27] Combining these results with an examination of the spectral properties of simple model cluster structures leads to an improved understanding of the nature of these mixed clusters and of the type of information contained in their infrared spectra.

2. METHODOLOGY

Our method for predicting the shift in the vibrational frequency of an SF_6 chromophore due to perturbation by the other atoms in the cluster is described in detail in Ref. 21. It is based on the observation that the dominant contribution to such shifts is provided by the interaction between the instantaneous dipole arising in an SF_6 molecule as it distorts during its vibrational cycle, and the dipoles it induces in neighbouring Ar atoms. The increasing magnitude of such distortions with the degree of vibrational excitation gives rise to a net red shift relative to the free molecule transition frequency. Of course, other types of interaction terms will also contribute to these band shifts. However, an examination of a generalized version of the best potential energy surface available for this system[25] has shown that their contributions are small, compared to uncertainties associated with the potential energy surface itself, and hence they may reasonably be ignored.[21]

Using the approach described above, the shift of the v_3 band of SF_6 may be readily calculated for any given cluster configuration. Monte Carlo or molecular dynamics simulation procedures may then be used to average over the ensemble of accessible cluster configurations to determine the resulting averaged spectrum for the specified temperature or total energy, respectively. The interaction potential used in these simulations was a pairwise additive function based on the accurate Ar-Ar potential of Aziz and Slaman[26] and the full anisotropic Ar-SF_6 potential of Pack et al.[25]

The Monte Carlo simulations were performed using the force bias modification[28-30] of the original Metropolis et al.[31] algorithm. The Ar atoms were initially randomly distributed on the surface of a sphere of radius 6Å centred on the SF_6, and to allow for equilibration, the results of the first 20,000 accepted moves were discarded. Results were then collected as the cluster configuration evolved through 1×10^6 or 3×10^6 accepted moves. For more details regarding this procedure, see Ref. 21.

The molecular dynamics simulations, in which the classical motion of a system is followed by numerical integration of Newton's equations of motion,[32] were performed using a program adapted for one kindly provided to us by F.G. Amar.[33] These simulations used the same potential energy functions mentioned above,[25,26] with the SF_6 again being treated as a rigid molecule. The initial cluster configuration was typically taken as the final configuration yielded by a 1×10^6-configuration Monte Carlo run, with initial random velocities being chosen from a Gaussian distribution appropriate to the desired simulation temperature.[33] The simulation was then run for a time interval sufficient to allow the average kinetic energy to become a reliable measure of the effective cluster temperature. All of the kinetic energies would then be scaled so as to bring this effective temperature closer to the desired value, and this process repeated until this caloric temperature had converged. Following this initialization phase, the actual simulation run would be allowed to proceed for, typically, 10^5 time steps of 5.0×10^{-15} sec. In such runs, energy was conserved to better than one part in 10^4, and linear and angular momentum were conserved to approximately this same accuracy.

3. THE SYNTHETIC SPECTRA AND THEIR INITIAL (MIS)INTERPRETATION

By analogy with the results of experiments on pure Ar clusters,[16] the effective internal temperature associated with the mixed SF$_6$-(Ar)$_n$ clusters studied experimentally[18,24] is believed to lie in the range 30-50K. Monte Carlo simulations for cluster sizes ranging from $n = 1$ to 1000 were therefore performed at $T = 30$ and $50 K$. The synthetic spectra thus obtained for $n \leq 18$ are shown in Fig. 1.

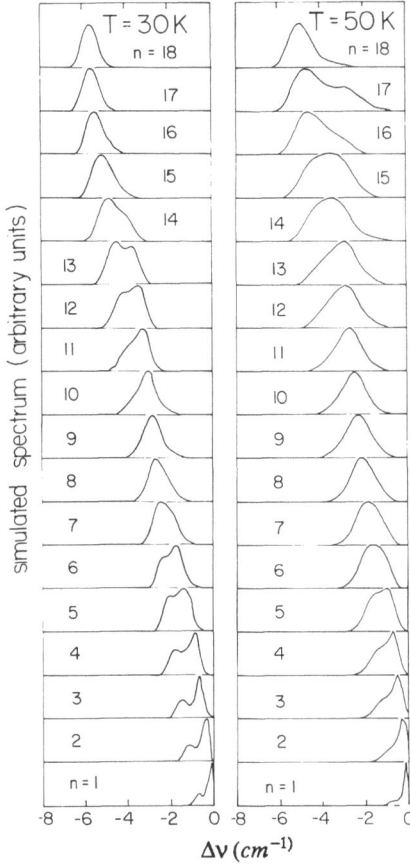

Figure 1. Simulated spectrum of the ν_3 band of SF$_6$ in SF$_6$-(Ar)$_n$ for cluster sizes $n = 1$-20 at $T = 30$ and $50 K$; frequencies are expressed as displacements from the free monomer transition energy of $948 \, cm^{-1}$.

Aside from the bimodal structure for cluster sizes $n \leq 6$ and (at $30 K$) $n = 11$-14, structure which might be expected to be washed out on averaging over the distribution of cluster sizes contributing to the experiment, the nature of the simulated spectra and their dependence on cluster size are qualitatively very similar to the experimental results.[18] The agreement improves further upon averaging the calculated spectra over a realistic assumed cluster size distribution.[21] This agreement gives us confidence in both the validity of our model for the frequency shifts and the reliability of our simulations.

As was pointed out in Ref. 20, for a single perturber atom, the ν_3 band is split into two components, a doubly degenerate perpendicular band plus a singly degenerate parallel band, with the red shift of the latter being four times as large as that for the former.[21] This is the source of the splitting in the $n = 1$ spectra seen in Fig. 1. This splitting due to the net perturber anisotropy persists for $n = 2$-6, but it is increasingly washed out by the thermal averaging, particularly at the higher temperature, and disappears by $n = 7$. However, it is clearly purely spectroscopic in origin, so it might be expected to provide little information about the dynamical properties of the cluster.

In the analysis of these Monte Carlo simulations presented in Ref. 20, the bimodal structure of the synthetic spectra obtained for $n = 11$-14 at $T = 30 K$ and for slightly larger n values at $50 K$, were identified as the key features of the Monte Carlo simulations. It was argued there that the two peaks are due to two different classes of cluster structures, or two different phases, and that their simultaneous presence in a single spectrum, which gives rise to a narrow maximum in a plot of the overall bandwidth as a function of cluster size, was evidence for phase coexistence.[20] It is shown below that that interpretation is wrong. However, it is also shown that such spectra may indeed contain a wealth of information about the structure and properties of these clusters, and that they promise to be an incisive experimental tool for discerning the effects of phase transitions in such systems.

4. RESULTS OF THE MOLECULAR DYNAMICS SIMULATIONS

Our initial molecular dynamics simulations of SF_6-$(Ar)_n$ clusters focussed attention on $n = 13$ clusters at $T = 30 K$, a case for which the (incorrect) initial interpretation[20] of the synthetic spectrum had led us to expect to find phase coexistence. Following Jellinek et al.,[9] we first calculated the mean square displacement of all particles in the cluster as a function of time, averaged over n_t separate run segments,

$$\langle R^2(t) \rangle = \frac{1}{N n_t} \sum_{j=1}^{n_t} \sum_{i=1}^{N} [\mathbf{R}_i(t_{0j} + t) - \mathbf{R}_i(t_{0j})]^2 \tag{1}$$

where N is the number of particles (here, $N = 14$). The slope of the long-time part of a plot of this quantity is the corresponding diffusion coefficient:[9]

$$D(E_{tot}) = \frac{1}{6} \frac{d\langle R^2(t) \rangle}{dt} \tag{2}$$

The $\langle R^2(t) \rangle$ functions for SF_6-$(Ar)_{13}$ generated by molecular dynamics runs at total energies E_{tot} corresponding to temperatures of 15 and $30 K$ are shown in Fig. 2. They clearly indicate that these clusters are liquid at $30 K$ but solid at $15 K$.

In order to locate the phase transition region more precisely, we then calculated values of the other key diagnostic property,[6,7,9–11] the relative root mean square bond length fluctuation δ, defined as:

$$\delta = \frac{2}{N(N-1)} \sum_{i<j=1}^{N} \frac{(\langle R_{ij}^2 \rangle_t - \langle R_{ij} \rangle_t^2)^{1/2}}{\langle R_{ij} \rangle_t} \tag{3}$$

where $\langle \ \rangle_t$ is the time average calculated along the entire trajectory, and $R_{ij} = |\mathbf{R}_i - \mathbf{R}_j|$ the distance between particles i and j. For the same $n = 13$ clusters considered above, these displacements are plotted as functions of temperature in Fig. 3. The near linear behaviour at low temperatures followed by a steep rise and a flattening off near a value of $\delta = 0.3$ is exactly the type of behaviour associated with the phase transitions studied in Refs. 6, 7 and 9-11.

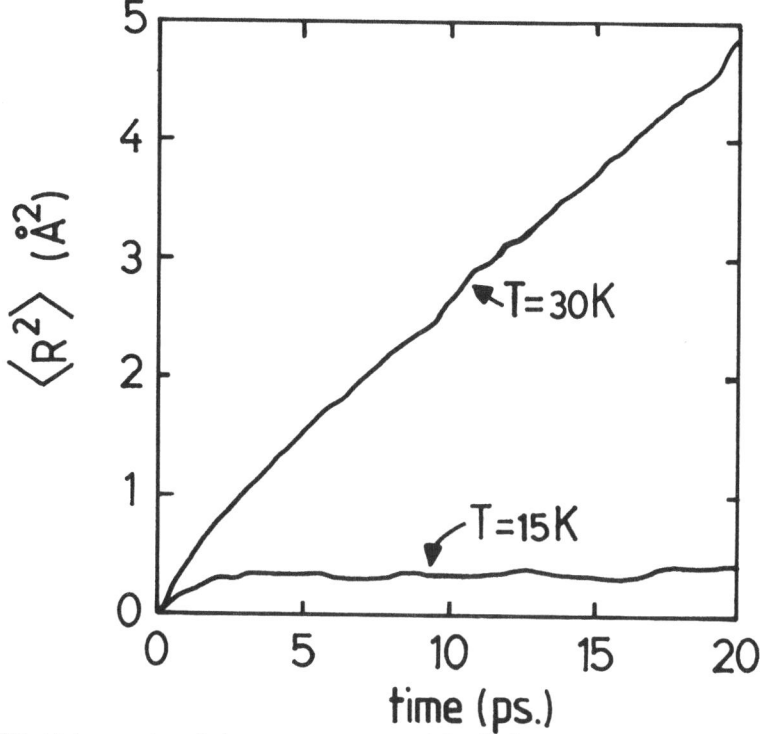

Figure 2. For SF_6-$(Ar)_{13}$, a plot of the mean square particle displacement vs. time at energies E_{tot} corresponding to temperatures of ca. $15 K$ and $30 K$.

From the results shown in Figs. 2 and 3, it seems clear that SF_6-$(Ar)_{13}$ clusters are liquid-like at $T = 30 K$ and solid-like at $T = 15 K$. Hence, the bimodal structure of the spectrum for this case (see Fig. 1) *can not* be associated with phase coexistence or a phase transition. However, it is shown below that this structure *does* provide very clear information about the structure and aspects of the dynamical behaviour of these clusters.

5. A REINTERPRETATION OF THE SYNTHETIC SPECTRA, AND ITS IMPLICATIONS

Since the molecular dynamics calculations described above demonstrate the absence of solid-like behaviour for SF_6-$(Ar)_{13}$ at $30 K$, it seemed appropriate to look for a more mechanical explaination of the double peak in its predicted spectrum. The ν_3 band of SF_6 is a triply degenerate transition which is generally split into three components when the molecule is perturbed by anisotropic surroundings. However, within an approximate model for this perturbation, the splitting was found to be *exactly zero* for a spherical chromophore exactly half buried in the plane surface of semi-infinite continuum of perturbers.[21] This suggests that the disappearance by $n = 9$ of the splitting found in the the $30 K$ synthetic spectra for smaller n values merely indicates that at this temperature, the nine Ar atoms form an approximately hemispherical "cap" on the SF_6 molecule. The structure observed for both $n \leq 6$ and (at $30 K$) $n = 11$-14 may then simply be attributed to the splitting of the parallel and perpendicular components of the ν_3 band for an SF_6 molecule in an anisotropic environment.

As a test of the above hypothesis, frequency shifts were calculated for simplified model systems consisting of close-packed hard sphere Ar atoms of radius $1.9 A$ fixed to the surface of a spherical SF_6 chromophore of radius $2.6 A$ (corresponding to an Ar-SF_6 distance of $4.5 A$). The representative

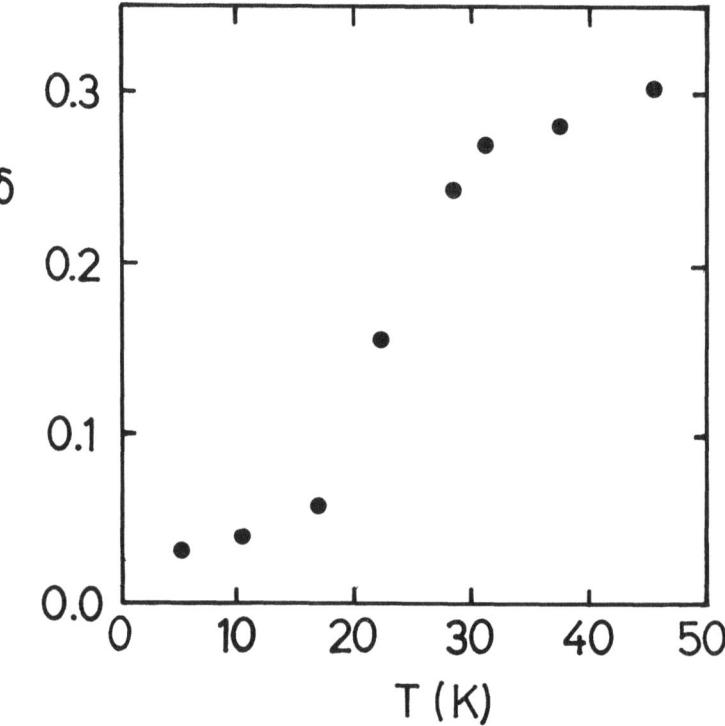

Figure 3. For SF_6-$(Ar)_{13}$, values of the relative root mean square bond length fluctuation δ at various temperatures.

SF_6-$(Ar)_n$ configuration used for each value of n was determined by starting from and holding fixed the characteristic configuration determined for $(n-1)$ Ar's, and determining the position of the additional atom by requiring that the pairwise sum of the Ar-Ar interaction potentials[26] be minimized. The resulting spectral shifts are plotted vs. n in Fig. 4. Note that the assignment of the perpendicular (round points) and parallel (triangular points) components of the band was based on the pattern of the shift values and their dependence on n.

From consideration of Figs. 1 and 4, it seems evident that the above hypothesis is valid, and that the bimodal structure of the synthetic spectrum for $n = 11$-14 merely represents the splitting of the parallel and perpendicular components of the band when the surface of the SF_6 chromophore is *more* than half covered by a unimolecular layer of Ar atoms. While this assignment is quite satisfying from a spectroscopic point of view, it also has very intriguing dynamical implications. In particular, we note that the splitting of the peaks in the synthetic spectra for $n = 11$-13 at $T = 30K$ is largely washed out at $50K$. Recalling (see Figs. 2 and 3) that these clusters are already liquid-like at $30K$, this suggests that the thermal motion of the Ar atoms at the higher temperature effectively removes the persistent perturber anisotropy associated with our model of a close packed $(Ar)_n$ film on the surface of the SF_6. This in turn implies that at $30K$ the Ar atoms in these clusters are indeed coalesced as a two-dimensional liquid, while at $50K$ they behave more like a two-dimensional gas or three-dimensional liquid. This then leads to the conclusion that the phase transition associated with the steep rise in the δ values for SF_6-$(Ar)_{13}$ seen in Fig. 3 corresponds to a solid to liquid transition for a *two-dimensional* film of Ar atoms wrapped around the SF_6 chromophore.

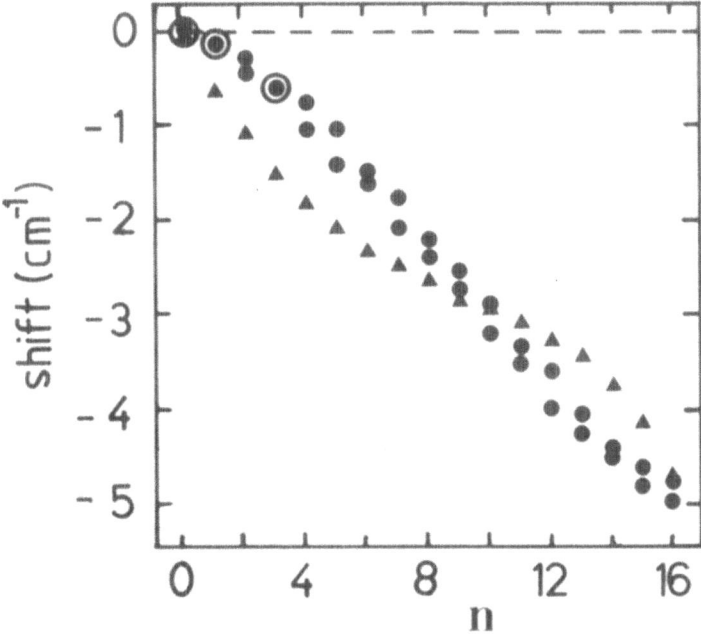

Figure 4. Frequency shifts predicted for the three components of the v_3 band of SF_6 when a close packed unimolecular film of rigid Ar atoms is placed on the surface of a spherical model for the chromophore. Round points identify the perpendicular components and triangular points the parallel component of the band.

6. CONCLUDING REMARKS

In view of the preliminary nature of the present study, the "dynamical" and phase change implications outlined above must be regarded as somewhat speculative. For example, the pronounced anisotropy of the real Ar-SF_6 potential surface[25] might raise some doubts regarding the validity of conclusions based on the simple spherical particle model used to generate Fig.4. However, the "spectroscopic" reassignment of the low temperature splitting in the synthetic spectra for $n = 12$-14 clusters does provide a very convincing overall explaination of the properties of the synthetic spectra of Fig.1.

It clearly would be very exciting if an examination of *experimental* band shapes analogous to those shown in Fig.1 could be used to determine the degree of coverage and phase behaviour of a unimolecular film on the surface of a bulky chromophore. For neutral species, such effects may remain difficult to resolve because of the effects of the finite cluster size distribution in the beam.[21] Thus, ionic cluster beams, from which a single cluster size can readily be selected for study, may provide the most fruitful area for finding these effects.

Acknowledgements

We are very pleased to acknowledge helpful discussions with and the healthy skepticism of Professors T.E. Gough and G. Scoles, and we would like to thank Professor F.G. Amar for providing us with a copy of his molecular dynamics program. This research was supported by the Natural Sciences and Engineering Research Council of Canada.

References

1. M.R. Hoare and P. Pal, *Adv. Phys.* **20**, 161 (1971).

2. D.J. McGinty, *J. Chem. Phys.* **58**, 4733 (1973).

3. W.D. Kristensen, E.J. Jensen and R.M.J. Cotterill, *J. Chem. Phys.* **60**, 4161 (1974).

4. C.L. Briant and J.J. Burton, *J. Chem. Phys.* **63**, 2045 (1975).

5. R.D. Etters and J.B. Kaelberer, *Phys. Rev.* **A11**, 1068 (1975).

6. J.B. Kaelberer and R.D. Etters, *J. Chem. Phys.* **66**, 3233 (1977).

7. R.D. Etters and J.B. Kaelberer, *J. Chem. Phys.* **66**, 5112 (1977).

8. N. Quirke and P. Sheng, *Chem. Phys. Lett.* **110**, 63 (1984).

9. J. Jellinek, T.L. Beck and R.S. Berry, *J. Chem. Phys.* **84**, 2783 (1986).

10. F.G. Amar and R.S. Berry, *J. Chem. Phys.* **85**, 5943 (1986).

11. T.L. Beck, J. Jellinek and R.S. Berry, *J. Chem. Phys.* **87** (1987, in press).

12. R.K. Heenan, E.J. Valente and L.S. Bartell, *J. Chem. Phys.* **78**, 243 (1983).

13. R.K. Heenan and L.S. Bartell, *J. Chem. Phys.* **78**, 1265 (1983).

14. E.J. Valente and L.S. Bartell, *J. Chem. Phys.* **80**, 1451 (1984).

15. E.J. Valente and L.S. Bartell, *J. Chem. Phys.* **80**, 1458 (1984).

16. J. Farges, M.F. de Feraudy, B. Raoult and G. Torchet, *J. Chem. Phys.* **78**, 5067 (1983).

17. A.J. Stace, *Chem. Phys. Lett.* **99**, 470 (1983).

18. T.E. Gough, D.G. Knight and G. Scoles, *Chem. Phys. Lett.* **97**, 155 (1983).

19. T.E. Gough, M. Mengel, P.A. Rowntree and G. Scoles, *J. Chem. Phys.* **83**, 4958 (1985).

20. D. Eichenauer and R.J. Le Roy, *Phys. Rev. Lett.* **57**, 2920 (1986).

21. D. Eichenauer and R.J. Le Roy, University of Waterloo Chemical Physics Research Report CP-312 (1987).

22. G. Natanson, F. Amar and R.S. Berry, *J. Chem. Phys.* **78**, 399 (1983).

23. R.S. Berry, J. Jellinek and G. Natanson, *Chem. Phys. Lett.* **107**, 227 (1984); *ibid, Phys. Rev.* **A30**, 919 (1984).

24. B. Zhang, X.J. Gu, N.R. Isenor and G. Scoles (1987, unpublished work).

25. R.T. Pack, J.J. Valentini and J.B. Cross, *J. Chem. Phys.* **77**, 5486 (1982).

26. R.A. Aziz and M.J. Slaman, *Mol. Phys.* **58**, 679 (1986).

27. J.C. Shelley, R.J. Le Roy and F. Amar, unpublished work (1987).

28. C. Pangali, M. Rao and B.J. Berne, *Chem. Phys. Lett.* **55**, 413 (1978).

29. M. Rao and B.J. Berne, *J. Chem. Phys.* **71**, 129 (1979).

30. M. Rao, C. Pangali and B.J. Berne, *Mol. Phys.* **37**, 1773 (1979).

31. N. Metropolis, A.W. Rosenbluth, M.N. Rosenbluth, A.H. Teller and E. Teller, *J. Chem. Phys.* **21**, 1087 (1953).

32. B.J. Alder and T.E. Wainwright, *J. Chem. Phys.* **31**, 459 (1959); *ibid* **33**, 1439 (1960); *ibid, Phys. Rev.* **127**, 359 (1962).

33. F.G. Amar, private communication (1986).

LASING AND STIMULATED RAMAN PROCESSES IN SPHERICAL AND SPHEROIDAL DROPLETS

G. Kurizki
Chemical Physics Department
Weizmann Institute of Science
Rehovot, Israel 76100

and A. Nitzan
School of Chemistry
Tel-Aviv University
Ramat-Aviv, Israel 69978

Abstract. A theory is presented which explains the main spectral and spatial features observed experimentally by Chang and coworkers in lasing and stimulated Raman emission from micrometer-size droplets. Instead of the usual plane-wave expansion this theory involves the expansion of the emitted field inside the droplet in spherical or spheroidal modes, taking account of its strong reflection at the surface whenever its frequency is near a Mie resonance. The amplification condition and spectral shift from Mie resonance are obtained for each such mode, whether in lasing or in stimulated Raman scattering, by taking the expectation value of the appropriate susceptibility in that mode.

A series of pioneering experiments conducted in Yale have demonstrated the occurrence of lasing[1,2] and stimulated Raman scattering (SRS)[3,4] in micrometer-size pure or dyed spherical liquid droplets. The salient features revealed in these experiments are:
a) Lasing occurs at a number of frequencies within the emission band that are red-shifted relative to the morphology-dependent resonances (MDR's), whose positions are obtainable from the Mie theory of light scattering by the droplet[5].
b) Strong SRS can occur at multiorder (combination) Stokes lines when there is an MDR near each of the combination lines.
c) The pumping thresholds for both processes are significantly lower in such a droplet than in bulk.
d) Although the pumping region in the droplet is confined to the narrow entry and exit (focal) spots of the incident laser beam, the stimulated emission from the droplet is nearly isotropic.

The Yale group pointed out that the occurrence of lasing and multiorder SRS can be attributed qualitatively to the high internal reflection by the spherical dielectric boundary of amplified emission at MDR's, causing the buildup of large internal field intensity. They not-

173

J. Jortner et al. (eds.), Large Finite Systems, 173–177.

ed that calculations pertaining to lasing from such systems must combine laser equations with Mie theory and that "the normal plane-wave growth equations for the SRS and four-wave mixing processes need to be modified into multipass spherical-wave nonlinear equations"...

In this Paper we outline the required reformulation of the conventional semiclassical theory of stimulated linear and nonlinear processes so as to allow for the spherical or spheroidal boundary of the active medium and the nonisotropic character of the pumping. The results pertaining to the aforementioned experimentally observed features are emphasized.

Fluorescence and Raman emission (non-stimulated) in a spherical particle have been described previously[6] as linear response to the pump field with a constant polarizability. A description of the self-consistent field-dipole interaction (which is the essence of stimulated processes) in a sphere has been given thus far only for the initiation regime of superfluorescence[7,8], for a homogeneous emitter distribution in the sphere and under the assumption of full population inversion. One of these treatments[7] represented the problem by a scalar wave equation for the field. However, it used boundary conditions which were not justified for this representation, and lead to MDR positions in disagreement with previous treatments[5,6]. The other treatment[8] does not include a dielectric boundary at all (the sphere being a homogeneous collection of two-level atoms), thus making difficult the connection to Mie theory. (In principle, one can calculate the dielectric response of the system of two-level atoms, then use it within the standard Mie theory. This will lead to MDR's governed by the resonant atomic transition).

The basis of our treatment are the two coupled Maxwell-Bloch equations for the field $\vec{E}$ and polarization $\vec{P}$ at each point in space and time. On making the rotating-wave approximation and assuming that the population inversion $N^{(0)}(\vec{r})$ established by pumping and relaxation is time-independent (steady-state, small signal regime[9]), these equation reduce to the following from for the Fourier components $\vec{E}_\lambda(\vec{r})$, $\vec{P}_\lambda(\vec{r})$ associated with frequency Ω_λ (Ω_λ is taken to be positive):

$$[\vec{\nabla}\times\vec{\nabla}\times - k_\lambda^2(1+i/Q_\lambda)]\vec{E}_\lambda = 4\pi k_{0\lambda}^2 \vec{P}_\lambda \qquad (1)$$

$$\vec{P}_\lambda = \chi_\lambda^{(1)}(\vec{r})\vec{E}_\lambda = \frac{\mu_\omega^2 N^{(0)}(\vec{r})}{\hbar(\Omega_\lambda-\omega-i\gamma)} \vec{E}_\lambda \qquad (2)$$

Here $k_\lambda=\Omega_\lambda \sqrt{\varepsilon}/c$ and $k_{0\lambda}=\Omega_\lambda/c$, ε being the dielectric constant of the optically-unexcited medium (solvent). The absorptive part of the wavevector ik_λ/Q_λ accounts for the finite lifetime of the mode, $1/Q_\lambda$ being the ratio of the MDR width to the spacing between adjacent MDR's of the solvent. The frequency, linewidth and dipole moment of the fluorescing molecular transition are denoted by ω, γ and μ_ω, respectively.

A major complicating circumstance is the strong, anisotropic spatial dependence of the susceptibility $\chi_\lambda^{(1)}$, because popoulation inversion is confined to the focal region of the pumping field. We have allowed for this spatial dependence by using a perturbative expansion in field-dipole coupling, in which $\chi_\lambda^{(1)}$ is taken to be a small parameter (compared to the solvent dieletric index ε). This means that either the pumping power or the concentration of active molecules is not too large, an assumption consistent with the small-signal (near-threshold) lasing regime.

To zeroth order, we neglect both $1/Q_\lambda$ and $\chi_\lambda^{(1)}$. Then (1) yields the following field eigenmodes within the sphere[5,6].

$$
|\ell m\rangle_M = \frac{j_\ell(k_\lambda r)}{\sqrt{\ell(\ell+1)}} \ \vec{L}\ Y_{\ell m} \ ; \ |\ell m\rangle_E = \vec{\nabla}\times |\ell m\rangle_M \qquad (3)
$$

where M(E) refer to magnetic (electric) type modes and $\vec{L}$ is the angular momentum operator. A signal field of an outgoing spherical wave $a_\lambda^{(0)}h_\ell^{(1)}(k_\lambda r) \ \vec{L}\ Y_{\ell m}e^{-i\Omega_\lambda^{(0)}t}$ will produce a response, via boundary reflection, in the form $b_\lambda^{(0)}|\ell m\rangle \ e^{-i\Omega_\lambda^{(0)}t}$. The amplitude ratio $b_\lambda^{(0)}/a_\lambda^{(0)}$ is given by the standard Mie denominator[5], whose poles correspond to the solvent MDR's $\Omega_\lambda^{(0)}$.

The first-order terms in (1) determine whether a given field mode lases, i.e. whether its steady-state oscillation at Ω_λ is not damped. This can be ascertained on substituting (2) into (1), then transforming it into the following scalar form by applying the angular-momentum operator $\vec{L}$

$$
\{\nabla^2 + k_\lambda^2[1 + i/Q_\lambda + \chi_\lambda^{(1)}] + \frac{(\vec{L}\chi_\lambda^{(1)})}{\vec{L}\cdot\vec{E}_\lambda^{(0)}} \cdot \vec{E}_\lambda^{(0)}]\}\vec{L}\cdot\vec{E}_\lambda^{(1)} = 0 \qquad (4)
$$

Here $\vec{E}_\lambda^{(0)}$ or $\vec{E}_\lambda^{(1)}$ are, to zeroth or first iteration, the sum of the signal and response (reflected) fields, associated with an MDR λ.

The eigenvalues of (4), which must be real at steady state, are given to first order by the expectation value of the terms inside the square brackets (with respect to $|\ell m\rangle$). Hence the steady-state condition implied by (6) near a magnetic-type MDR is:

$$
\frac{1}{Q_\lambda} = \mathrm{Im}\langle \ell m|\tilde{\chi}_\lambda^{(1)}|\ell m\rangle \equiv \mathrm{Im}\langle \ell m|[\chi_\lambda^{(1)} + \frac{(\vec{L}\chi_\lambda^{(1)})\cdot j_\ell(k_\lambda r)\vec{L}\ Y_{\ell m}}{j_\ell(k_\lambda r)Y_{\ell m}\ \sqrt{\ell(\ell+1)}}]|\ell m\rangle \qquad (5)
$$

A narrow symmetric distribution of $\tilde{\chi}^{(1)}$ about $\theta=0$ (the focal point of pumping gives rise to nearly isotropic threshold in (5), (weakly dependent on ℓ in the range $\ell_{min}<\ell<\ell_{max}$ where ℓ_{max} and ℓ_{min} differ typically[1,2] by 10-20) thus explaining the observed[1] isotropic lasing.

The real part of $\tilde{\chi}^{(1)}_{\lambda,\ell m}$ determines the frequency pulling, i.e. the observed lasing red shift

$$\Omega_\lambda^{(1)} \simeq \Omega_\lambda^{(0)}(1-\text{Re}<\ell m|\tilde{\chi}_\lambda^{(1)}|\ell m>) \qquad (6)$$

Third-order expansion in field-dipole coupling[10] yields a polarization $\vec{P}^{(3)}$ proportional to the third order of the field. This polarization is responsible for lasing at combination frequencies, four wave mixing and SRS. For the latter process the Stokes-mode polarization $\vec{P}^{(3)}_{\lambda s}$ is related to the pump field $(\vec{E}_{in})_\lambda^{(0)}$ by the Bloembergen expression[10]

$$\vec{P}_{\lambda s}^{(3)}(\omega_s \simeq \Omega_{\lambda s}) = \vec{E}_{\lambda s}|(\vec{E}_{in})_\lambda^{(0)}|^2 \frac{N^{(0)}|M_R|^2}{\hbar\,(\Omega_\lambda-\Omega_{\lambda s}-\omega_v+i\gamma_v)} \equiv \chi_s^{(3)}\vec{E}_{\lambda s}|(E_{in})_\lambda^{(0)}|^2 \quad . \qquad (7)$$

Here M_R is the Raman-scattering matrix element, $\omega_s=\omega-\omega_v$ is the Raman-shifted frequency, μ_ω the corresponding dipole moment and γ_v the linewidth of the lower state of the Raman transition.

Following the same procedure as above we obtain from the scalar equation for $\vec{L}\cdot\vec{E}_{\lambda s}$ a steady-state condition which has the same form as (5), on making the substitution

$$\chi_\lambda^{(1)} \rightarrow \chi_s^{(3)}|E_{in}^{(0)}|^2_\lambda \qquad (8)$$

If the steady-state condition is satisfied, then the field

$$(\vec{E}_{in})_{\lambda s} = \vec{E}_{\lambda s} + (\vec{E}_{ref})_{\lambda s} \qquad (9)$$

where $(\vec{E}_{ref})_{\lambda s}$ is the boundary-reflection response to $\vec{E}_{\lambda s}$, can be the pump for SRS in the next Stokes order. This is seen from (7) on setting

$$\Omega_\lambda \rightarrow \Omega_{\lambda s};\Omega_{\lambda s}\rightarrow\Omega_{\lambda s} \quad -\omega_v \simeq \omega-2\omega_v \quad . \qquad (10)$$

The relative intensities of successive Stokes orders at steady state are strongly dependent on those of the corresponding $(\vec{E}_{ref})_{\lambda S}$, i.e. on their Mie denominators.

The entire above analysis can be repeated for a spheroidal particle, such as were used in one of the experiments[2]. The angular momentum states in a prolate spheroid are defined relative to the axis on which the foci $z=\pm a/2$ are located. The only difference in formal results is that spheroidal outgoing waves[11]

$$e^{im\phi}S_{m\ell}(h,\eta)he_{m\ell}(h,z) \qquad\qquad (11)$$

replace their spherical counterparts $Y_{\ell m}h_{\ell}^{(1)}(k_{\lambda}r)$, where $h=(1/2)\,ak_{\lambda}$, $\eta=\cos\theta$. Likewise, the boundary-reflected waves are now

$$e^{im\phi}S_{m\ell}(h,\eta)je_{m\ell}(h,z) \qquad\qquad (12)$$

instead of $Y_{\ell m}j_{\ell\lambda}(kr)$. When $h>1$, spherical Mie resonances are split into m-labelled components and the field becomes anisotropic, acquiring a definite polarization.

References

1. H.M. Tzeng, K.F. Wall, M.B. Long and R.K. Chang, Opt.Lett. 9, 499 (1984).

2. S.X. Qian, J.B. Snow, H.M. Tzeng and R.K. Chang, Science 231, 486 (1986).

3. J.B. Snow, S.X. Qian and R.K. Chang, Opt.Lett. 10, 37 (1985).

4. S.X. Qian and R.K. Chang, Phys.Rev.Lett. 56, 926 (1986).

5. M. Kerker, The Scattering of Light and Other Electromagnetic Radiation (Academic Press, New York, 1969).

6. H. Chew, P..J. McNulty and M. Kerker, Phys.Rev. A13, 396 (1976).

7. J. Mostowsky and B. Sobolevska, Phys.Rev.A 28, 2943 (1983).

8. S. Prasad and R. Glauber, Phys.Rev.A 31, 1583 (1985).

9. H. Haken, Laser Theory (Springer, Berlin, 1983).

10. Y.R. Shen, The Principles of Nonlinear Optics (Wiley, New York, 1984).

ELECTRON ATTACHMENT TO CLUSTERS

O. Echt, M. Knapp[*], C. Schwarz[+], and E. Recknagel
Department of Physics
University of Konstanz
P.O. Box 5560
D-7750 Konstanz, West Germany

ABSTRACT. Weakly bound clusters are formed by adiabatic expansion of CO_2, N_2O, H_2O, SF_6, Xe, and some halocarbons. Electron attachment to these clusters is investigated in the low-pressure environment of the collimated cluster beam. The energy of the electrons in this study ranges from ~0 to ~10 eV. Size distributions of the resulting cluster anions are compared to cluster cation mass spectra, obtained by electron impact. Resonances in the yield of anions are increasingly redshifted with increasing cluster size, due to the electronic solvation shift. New resonances, in addition to the shape resonances known from electron attachment to monomers, are observed at low electron energy in some, but not in all, cases.

1. INTRODUCTION

Mass spectrometry of clusters, grown by homogenous nucleation, commonly exploits electron impact ionization or multiphoton ionization; positively charged clusters being detected in both cases. Negatively charged clusters may be formed if electrons are present in a supersaturated gas, e.g. if electrons are injected into a supersonic jet /1/, or if a laser is used to generate a plasma in the gas prior to expansion /2/. These techniques yield intense beams of anions, suitable for experiments like field detachment /3/, photodetachment /4/, photoelectron spectroscopy /5, 6/, or post-ionization /7/.

A different method for formation of cluster anions utilizes electron attachment to neutral clusters in a low-pressure environment: a collimated beam of clusters is crossed by a beam of free electrons of low energy /8 - 10/, or by a beam of atoms with a low ionization potential, such as alkali atoms /11/ or rare gas atoms, excited into a high Rydberg state /12/. This technique yields relatively low ion intensities, but a study of the production process itself may give

179

J. Jortner et al. (eds.), Large Finite Systems, 179–193.

valuable information on the properties of the clusters. This is especially true if the ion yield of the clusters is determined as a function of the kinetic energy of the particles: The yield of anions,
formed by attachment to clusters, crucially depends on the kinetic
energy of the incoming electrons, because a transition is made from a
free to a bound state. The corresponding resonances and appearance
potentials may reflect electron affinities and, in case of dissociative
attachment, energy thresholds. The dependence of these quantities on
cluster size reveals their convergence towards bulk values which are,
unfortunately, unknown in most cases.

     In this contribution, we present results obtained from electron
attachment to a beam of clusters, grown by adiabatic expansion of CO_2,
N_2O, H_2O, SF_6, Xe, and some halocarbons (C_2F_6, $C_2F_4Cl_2$, C_2F_3Cl). The
size n of these clusters ranges from one to several hundred in some
cases. We discuss the composition of the resulting anions, the energy
dependence of their intensities, the efficient quenching of autodetachment or (intramolecular) fragmentation due to the large size of the
systems and due to the presence of decay channels with low activation
energy, and the appearance of an additional resonance for attachment of
low-energy electrons.

2. EXPERIMENTAL

The salient features of the experimental setup are evident from Fig.1:
The gas in question thermalizes in a copper cylinder (temperature
variable from 150 K to 400 K) and expands through a nozzle (diameter
100 μm, length ~ diameter) into vacuum, neutral clusters being formed
by homogeneous nucleation in the supersonic beam downstream of the
nozzle. The walls of the expansion chamber and of the subsequent differential pumping chamber are cooled to 77 K in order to cryo-pump the
gas which bounces off the heated collimators. The background downstream
of the second collimator is ~10^{-7} torr during operation of the cluster
source.

     At a distance of ~10 cm from the nozzle, the collimated cluster
beam is crossed by a pulsed electron beam (pulse duration ~1μs). The
repetition rate of the electron beam (~1 kHz to ~30 kHz) is chosen to
be larger than the inverse time-of-flight of the largest cluster ions
detectable in the mass spectrum. The electron energy can be varied
between ~0 eV and a few hundred eV, although the range of interest in
this study is ~0 to ~10 eV. The electrons are emitted from a directly
heated filament, resulting in an energy spread of ~1 eV (full-width at
half-maximum). The ion extraction field is kept at zero as long as the
electron beam is on. Furthermore, a magnetic field helps to align the
electrons. Their current is recorded in a Faraday cup and stabilized by

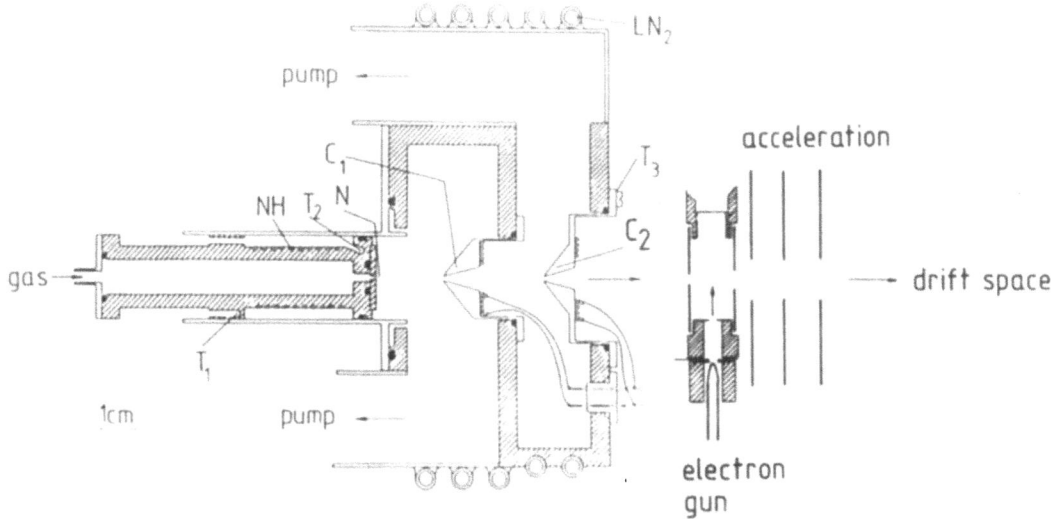

Figure 1. Source for clusters, generated by adiabatic expansion of gases, and ionizer for the time-of-flight mass spectrometer.

automatic readjustment of the power supply for the filament.

The anions being produced by electron attachment are accelerated into the grounded drift tube by a potential drop of 2 kV. They are detected at the end of the drift tube (length ~1.1 m) by the usual single-ion counting techniques. A home-built time-to-digital converter in combination with a multichannel analyzer is used to accumulate time-of-flight mass spectra of the ions. More details of the experimental setup are to be found in two recent publications from this laboratory /13, 14/.

3. SIZE DISTRIBUTIONS OF CLUSTER ANIONS AND CATIONS

If weakly bound clusters are formed in a jet under mild expansion conditions, the size distribution of their cations, formed by electron impact ionization, is generally found to be a quasi-exponentially decaying function /15/. Superimposed on this distribution one often observes local intensity anomalies ("magic numbers"), the appearance of them being hardly affected by the expansion conditions or by the energy of the ionizing electrons /16/. Similar distributions are obtained even if different methods of ionization are employed, such as Penning ionization /17/ or nonresonant multiphoton ionization /18/. There is agreement among most researchers in this field that these "magic numbers" reflect the enhanced stability of positively charged clusters of a

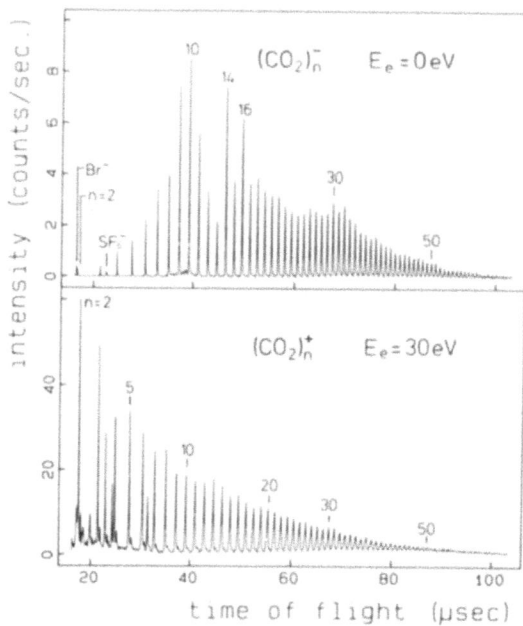

Figure 2.   Time-of-flight mass spectra of carbon dioxide clusters:
$(CO_2)_n^-$, obtained by electron attachment at ~0 eV (top), and $(CO_2)_n^+$,
obtained by electron impact ionization at 30 eV (bottom). Mass
peaks are labeled by cluster size n.

particular size, because rapid fragmentation following vertical ionization and subsequent relaxation in the cluster enriches the most stable
cluster sizes /16 - 19/.

Size distributions of cluster <u>anions</u>, formed by electron attachment, are completely different from cation spectra in most cases being
investigated so far. Fig. 2, e.g., presents a comparison of a mass
spectrum of $(CO_2)_n^-$, formed by attachment of electrons at ~0 eV, and of
$(CO_2)_n^+$, formed by electron impact ionization at 30 eV. The expansion
conditions for the jet and hence the size distributions of the neutral
clusters were identical in both cases. Two qualitative differences are
obvious from Fig. 2: i) The ion intensity of the anions is extremely
small for n = 1 and 2, it increases with increasing size until it
reaches a maximum, the position of which will in general depend on the
expansion conditions and, to a small extent, on the electron energy.
ii) The pronounced magic numbers in the distribution of cluster anions
(n = 10, 14, 16 in case of $(CO_2)_n^-$) are independent of electron energy,
and they do not occur in the distributions of cluster cations. Qualitatively similar observations have been made for clusters of nitrous
oxide and of water /13, 20/.

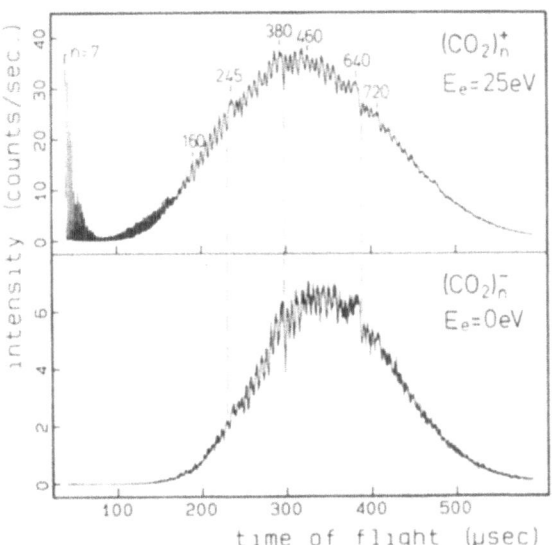

Figure 3. Mass spectra of $(CO_2)_n^+$, obtained by electron impact ionization (top spectrum), and of $(CO_2)_n^-$, from electron attachment at ~0 eV. Numbers above the spectra indicate the cluster size n.

The fact that the same magic numbers in $(CO_2)_n^-$ spectra are observed at around 3 eV (which constitutes the first shape resonance of small CO_2 clusters, see below), and that they also occur if CO_2 vapor condenses at electrons in a jet /21/, strongly suggests that the magic numbers in mass spectra of cluster <u>anions</u> also arise from fragmentation: the energy released in the relaxation after electron attachment, even at ~0 eV, cannot be accommodated in the cluster without intermolecular dissociation. This may not be true, though, for very large clusters, or if the cluster is extremely cold prior to attachment.

In some cases, however, the distributions of cluster anions and cations are remarkably similar. Sulfur hexafluoride constitutes such a case: Local intensity maxima occur at a cluster size n = 13, 59, 78, 91 in both cases, if the composition of the cluster ions is denoted as $(SF_6)_n^-$ and $(SF_6)_{n-1} \cdot SF_5^+$, respectively /20/. (Intramolecular fragmentation of SF_6 is quenched very efficiently for electron attachment, but not for electron impact ionization, alos see below).

It is also remarkable that there is a general trend for anion and cation spectra to be similar if the cluster size n becomes large: Spectra of carbon dioxide clusters (Fig. 3) reveal a series of characteristic steps in the distribution of clusters beyond a size n ~ 150. Recently, we have been able to record $(CO_2)_n^+$ spectra in this size range at high mass resolution. We find that <u>groups</u> of several neighbouring mass peaks are more abundant than others. This effect probably has the same origin as in Ar_n^+ distributions, as discussed recently by Northby and coworkers /22/: They are more abundant if geometric sub-shells (faces on top of an ordered cluster core) are occupied or, at least, nearly filled. The packing of neutral molecules in large clus-ters is obviously rather independent of the exact geometry of the innner, charged cluster core. This is also supported by size distribu-tions of $Xe_n \cdot O^-$ which display intensity drops beyond cluster size n = 55 and 71 /20/, the same magic numbers being observed in mass spectra of Xe_n^+ /16/. The ions $Xe_n \cdot O^-$ arise from dissociative attachment to xenon clusters containing N_2O impurities. Cation mass spectra indicate that most of the clusters are actually free of impurities, but we have not yet been able to detect pure Xe_n^-. This may be due to an extremely small cross section for electron attachment, or to the smallness of their adiabatic electron affinity. We estimate that the anions would rapidly autodetach if their electron affinity is less than +0.2 eV. In comparison, the bottom of the conduction band in solid bulk xenon is 0.4 eV below the vacuum level /28/.

4. SIZE DEPENDENCE OF SHAPE RESONANCES, NEW RESONANCES AT LOW ENERGY

Electron attachment to small molecules like CO_2 or N_2O does not result in long-lived anions with the corresponding composition /8 - 10/, because i) their adiabatic electron affinities are very small (N_2O) or even negative (CO_2), whereas ii) their vertical electron affinities are strongly negative. The transient anions formed by resonant electron capture are therefore highly excited (by approximately 3 eV in case of CO_2 and N_2O), and will autodetach within $\sim 10^{-14}$ s. Dissociation into stable product ions like O^- may compete favorably with autodetachment only if the electron exceeds the energy threshold for this process. This condition is fulfilled for N_2O (vertical electron affinity -2.23 eV, the reaction $N_2O + e \longrightarrow N_2 + O^-$ is endothermic by 0.2 eV /23/). CO_2 constitutes a different case: Its vertical electron affinity is -3.8 eV, but the heat of the endothermic reaction $CO_2 + e \longrightarrow CO + O^-$ is even more negative by 0.2 eV /23/. Thus, the only anion being produced by electron attachment to CO_2 is O^-, with a maximum yield at 4.4 eV electron energy. The yield below 3.8 eV is zero although the adiabatic electron affinity is about -0.6 eV, and although CO_2^- in its vibronic ground state would be long-lived with respect to autodetach-ment /23/.

This picture changes drastically if electrons are attached to CO_2 clusters. We observe ions with the composition $(CO_2)_n^-$, with n being as small as 1, because i) rapid intermolecular fragmentation may quench autodetachment, ii) the electron energy need not exceed the vertical electron affinity of the parent cluster because, in contrast to O^- production, no highly endothermic reaction occurs, and

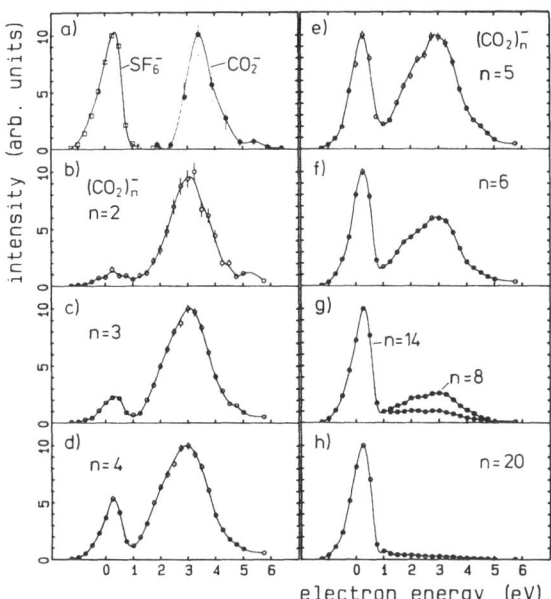

Figure 4. Yield of $(CO_2)_n^-$, with n ranging from 1 to 20, versus electron energy. The SF_6^- yield (top left trace) indicates the position of
a "true" 0 eV resonance, with the shape of the resonance being controlled by the energy distribution of the electrons.

iii) the increase in the number of degrees of freedom helps to dissi-
pate the excess energy (difference between the electron energy and the
adiabatic electron affinity). Consequently, CO_2^- is observable if electrons are attached to clusters rather than to monomers /10/. Its yield
(Fig. 4 top, left spectrum) exhibits a resonance at 3.4 eV (corrected
electron energy), well below the vertical electron affinity of the
monomer, because the precursor of CO_2^- is a cluster (whose size, unfortunately, cannot be determined unambiguously).

We are thus witnessing the change of the vertical electron affinity with increasing cluster size or, in other words, the <u>electronic</u>
solvation shift of CO_2^- in a cluster: The energy difference between
$(CO_2)_n$ (having relaxed inter- and intramolecular coordinates) + e
and $(CO_2)_n^-$ (geometry of the neutral precursor) decreases due to the
rapid response of the electrons of the individual, linear CO_2 molecules
to the incoming excess electron. (The information obtained from photoelectron spectroscopy /5/ is, of course, complementary to this quantity, because it reflects the energy difference between the geometrically
relaxed cluster <u>anion</u> and a neutral cluster of the same geometry).
Accordingly, the resonance in the $(CO_2)_n^-$ yield is increasingly red-
shifted with increasing size n. The concurrent increase in the width of
the resonance is probably related to the fact that different precursor

sizes contribute to the same cluster ion size. Similar solvation shifts versus cluster size have been determined for $(CO_2)_{n-1} \cdot O^-$, $(N_2O)_n^-$ and $(N_2O)_{n-1} \cdot O^-$ /8, 10, 20, 24, 25/, and for ions produced by attachment to clusters of O_2 /9/, SO_2 /26/, and halocarbons /27/ (also see below).

The most remarkable feature in Fig. 4, however, is the appearance of an additional resonance for electrons of low energy. The existence of this new resonance, not known from experiments at CO_2 monomers, was established by Märk and coworkers for the trimer and larger cluster anions /25/. Fig. 4 suggests that this resonance already occurs in the dimer ion. Anyhow, this new feature becomes dominant in the yield curves of clusters containing more than ~5 molecules. Absolute cross sections cannot be determined, because the partial densities of the neutral clusters in the beam are, of course, unknown. The mass spectra in Fig. 2, though, indicate that the yield of, say, $(CO_2)_{10}^-$ from electron attachment at ~0 eV is approximately as large as that for $(CO_2)_{10}^+$, generated by electron impact at 30 eV. In these measurements, the properties of the neutral cluster beam were identical, while the electron current at 30 eV is certainly larger than at ~0 eV. Thus, attachment of low-energy electrons to large CO_2 clusters is an extremely efficient process.

The exact position of this low-energy resonance has to be determined by comparison with the yield curve of other well-known anions. The energy scale in Fig. 4 displays the potential difference between the tip of the filament and the ionizer, corrected for contact potentials, thermal energies of the electrons, etc., by a calibration procedure. The true, mean electron energy will, however, necessarily deviate from this "corrected electron energy" if it falls below the width of the energy distribution, which is ~1 eV. The energy distribution will then become skewed and broadened even further. The top left spectrum in Fig. 4 displays the yield curve of SF_6^-, obtained from attachment to SF_6 under identical conditions concerning the electron gun and the ionizer. This molecule is known to have an extremely large cross section for attachment of electrons at energies ≤ 20 meV /29/. Thus, a comparison of the yield curves in Fig. 4 reveals that the $(CO_2)_n^-$ resonance occurs very close to zero eV, with an experimental uncertainty of 0.1 eV.

The observation of $(CO_2)_n^-$, $n \geq 2$, at ~0 eV electron energy implies that the adiabatic electron affinities of neutral CO_2 clusters as small as the dimer is non-negative, although that of the monomer is -0.6 eV /23/. This conclusion is in agreement with a calculation of the energetics of $(CO_2)_2^-$ /30/, but at variance with a more recent theoretical investigation /31/.

What is the nature of this resonance at low energy? The strongly negative vertical electron affinity of CO_2 makes it unlikely that the incoming electron directly couples with an individual molecule in the cluster. Antoniewicz et al have pointed out that a polarizable cluster

will always provide a long range attractive potential for an electron
outside of the cluster. The binding energy for these surface electrons
will depend on the size and the dielectric constant of the cluster.
Even in the extreme case of helium, a bound surface state for electrons
is known to exist at 0.7 meV below the vacuum level, at least for
planar He films, corresponding to infinite cluster size /33/. The
transfer of the excess energy of the incoming electron to the cluster,
which is required for electron attachment, may be accomplished by
excitation of vibrational van der Waals modes, in analogy to the elec-
tron-ripplon coupling in helium films.

If these ideas are correct, efficient electron capture at low
energies should be a rather general phenomenon for large clusters. One
has to bear in mind, though, that the lifetime of these cluster anions
and hence their observability may depend on the availability of other,
deeper traps for electrons in the cluster. Also, other processes may be
more efficient for electron attachment in some cases. SF_6, e.g., features an adiabatic electron affinity of +1.0 eV and a high cross section for attachment of thermal electrons /29/. It is therefore not
surprising to find large yields of $(SF_6)_n^-$ at low energies, the electron probably being localized at an individual molecule in the cluster.
In halocarbons containing chlorine, on the other hand, dissociative
attachment will easily yield Cl^- being solvated in the cluster, even
for low electron energy, because the dissociation energy of the C-Cl
bond is typically 3.5 eV in these molecules, i.e. approximately equal
to the electron affinity of Cl.

We have determined the yield of anions, produced by electron
attachment to clusters of $C_2F_4Cl_2$, C_2F_6 and C_2F_3Cl /27/. $(C_2F_4Cl_2)_n^-$
does exhibit a resonance close to 0 eV, but this may be related to the
first shape resonance of the monomer: the yield of Cl^- peaks at 0.3 eV.
$(C_2F_6)_n^-$ is observed only if the electron energy is about 3 to 4 eV;
this resonance is obviously related to the resonance in the F^- yield.
C_2F_3Cl is the only molecule for which an additional resonance in the
yield of cluster anions is observed, see Fig. 5: The Cl^- yield from
attachment to the monomer peaks at 1.3 eV (also see /34/), whereas
$(C_2F_3Cl)_n^-$, $n \geq 2$, features a resonance at ~0 eV in addition to the
(slightly redshifted) resonance which mirrors the Cl^- yield. The surprising feature in Fig. 5, though, is the relatively high intensity of
the new resonance at ~0 eV already for the dimer ion.

N_2O behaves similar: Attachment to the monomer yields O^- with a
maximum cross section at 2.25 eV /10/. Attachment to nitrous oxide
clusters yields $(N_2O)_n^-$, $n \geq 1$, with a corresponding resonance, being
increasingly redshifted with increasing n. An additional resonance
occurs for low, but non-zero electron energy; it is as intense as the
shape resonance at ~2 eV for cluster ions as small as the dimer /24/.

Another interesting system featuring a resonance for attachment of
electrons at ~0 eV is water. The behaviour of low-energy excess elec-

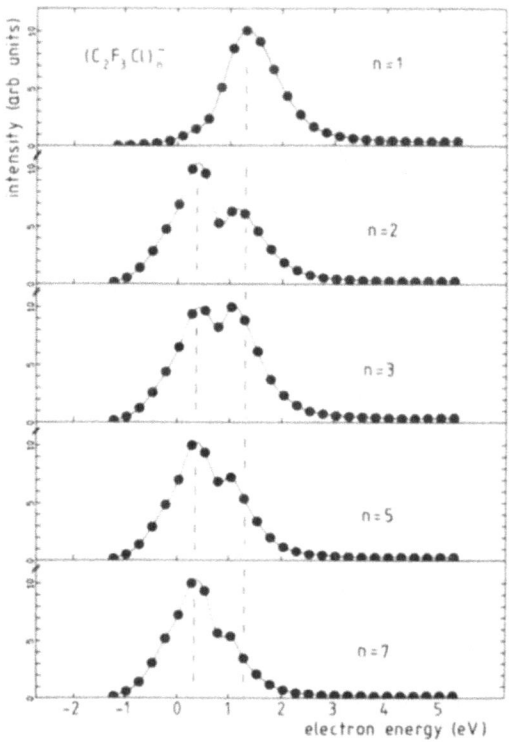

Figure 5. Yield of $(C_2F_3Cl)_n^-$ vs. electron energy. The position of the additional resonance for $n \geq 2$ agrees with a SF_6^- resonance.

trons in liquid water has been subject to numerous studies (see, e.g., references cited in /13/). The adiabatic electron affinity of H_2O is extremely close to zero, whereas the energy of a solvated electron in liquid bulk water is approximately 1.5 eV below the vacuum level. This implies that sufficiently large water clusters should strongly bind an excess electron. In spite of this, several attempts to generate $(H_2O)_n^-$ by attachment of electrons to free water clusters failed /35/. Haberland and coworkers finally succeeded in producing $(H_2O)_n^-$, $n \geq 11$, and $(D_2O)_n^-$, $n \geq 12$, by injecting electrons into the high-density regime of adiabatically expanding water vapor /1, 36/. Theses findings suggested that hot, thermally disordered clusters may temporarily capture electrons, with subsequent relaxation, stabilization and cluster growth being guaranteed by frequent collisions, while relatively cold, structurally relaxed clusters would repel electrons (i.e. their vertical electron affinity being strongly negative, considerable molecular reorientation being necessary to accommodate the excess electron). These ideas were supported by theoretical studies of neutral and charged water clusters /37/. Also, they were in line with estimates according to which the concentration of preexisting deep traps for excess elec-

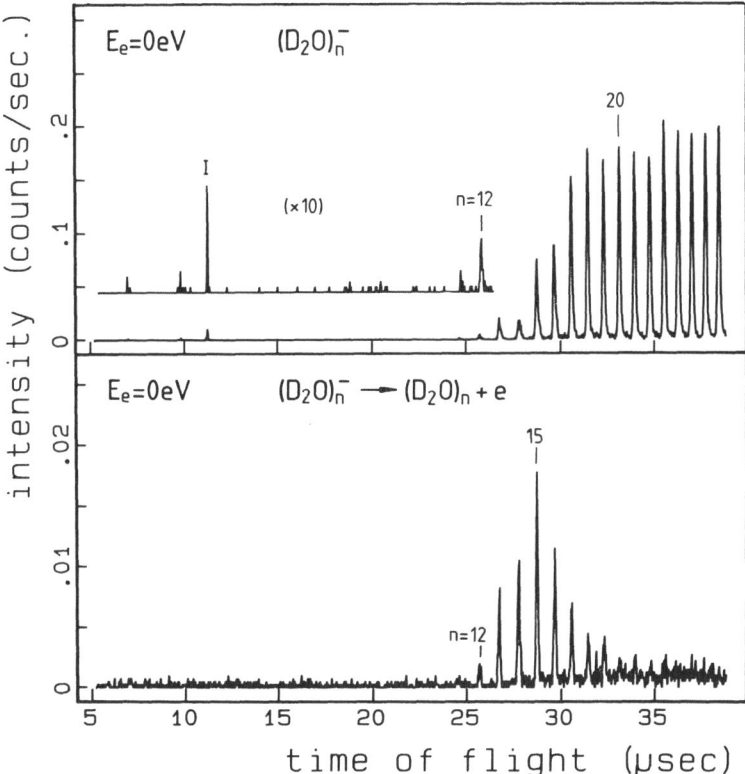

Figure 6. Top: Mass spectrum of $(D_2O)_n^-$ from attachment of electrons to
deuterated water clusters at ~0 eV. Bottom: Intensity of neutral water
clusters from autodetachment reactions in the drift tube.

trons in polar liquids is exceedingly small /38/.

       Recent experiments in our laboratory, however, revealed that
$(H_2O)_n^-$, $n \geq 11$, can be formed very efficiently by electron attachment
to free clusters in a molecular beam if and only if the energy of the
electrons is close to zero /13, 39/. A typical mass spectrum is displayed in Fig. 6 (top) for deuterated water. It is void of small water
cluster ions; but the intensity of $(D_2O)_n^-$ rises steeply above $n = 11$
and levels off at $n \sim 20$, depending slightly on the expansion conditions. (A detailed account of our findings and of their implications
has been published in /13/). These distributions are essentially identical to those obtained by injection of electrons into a water jet, if
neat water is expanded in both experiments.

       Thus, the efficient capture of electrons at free water clusters
may indeed be related to the above-mentioned surface states /32/, they
may at least provide an initial, weakly bound state until accommodation
of the electron into the cluster is made possible by orientational
fluctuations of the individual molecules. We would like to point out,

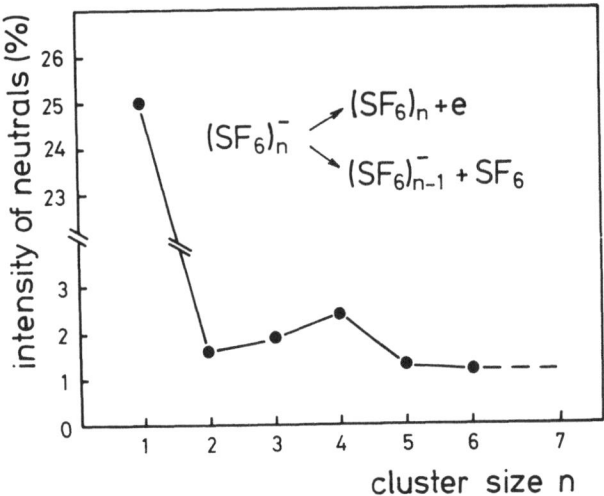

Figure 7. Probability for autodetachment from $(SF_6)_n^-$ in the drift tube

however, that very recent experimental and theoretical studies indicate
that, contrary to previous belief, the thermal disorder in liquid water
at room temperature is sufficiently large to provide a large concentration of deep preexisting traps for excess electrons /40, 41/.

The bottom spectrum in Fig. 6 displays the intensity of neutrals,
being produced by unimolecular reactions of $(D_2O)_n^-$ in the drift tube
of the time-of-flight mass spectrometer. These neutrals are mainly
produced by autodetachment and, to a lesser extent, by "evaporation" of
neutral monomers /13/. A comparison of their intensity with that of
their precursor ions, being injected into the drift tube (see top
spectrum) reveals that the smallest ions $12 \leq n \leq 15$ have a large
tendency to undergo autodetachment, while clusters of size $n \geq 20$ are
essentially stable against autodetachment in the drift tube.

The high probability for autodetachment from a system as large as,
say, $(H_2O)_{15}^-$, may seem surprising: In contrast, ions $(SF_6)_n^-$ formed by
electron attachment to SF_6 clusters are not likely to undergo autodetachment whenever $n \geq 2$, see Fig. 7. Only the monomer ions, SF_6^-, are
prone to autodetach, because they mainly derive from attachment to
monomers which have no other decay channel for stabilization. In general, cluster ions may stabilize by evaporation of neutrals, if the
corresponding activation energy is less than the electron affinity.
This condition is clearly fulfilled for SF_6 clusters, because SF_6
features a high electron affinity of 1.0 eV /29/. The electron affinity
of the water dimer, however, is of the order of 0.01 eV, it appears to
increase rather slowly with size n as long as the electron resides in a
diffuse surface state /3, 42/. The internal energy in the neutral
clusters, formed in a neat expansion, is such that a balance exists
between frequent condensation and re-evaporation of the monomers at

clusters in the jet, further evaporations in the collimated beam will cool the clusters. The heat of sublimation for bulk water is 0.47 eV. Therefore the relaxation energy following attachment of low-energy electrons is negligible; and we may tentatively identify the cluster size n = 15 as defining the region where the electron affinity of water clusters gradually becomes larger than their heat of sublimation.

ACKNOWLEDGEMENT

The help of D. Kreisle is gratefully acknowledged. This work has been financially supported by the Deutsche Forschungsgemeinschaft.

REFERENCES

* Present address: Robert Bosch GmbH, D-7410 Reutlingen
\+ Present address: Hahn-Meitner Institut, D-1000 Berlin 39

1. H. Haberland, H.G. Schindler and D.R. Worsnop,
    Ber.Bunsenges.Phys.Chem. **88** (1984) 270
2. L.A. Bloomfield, M.E. Geusic, R.R. Freeman and W.L. Brown,
    Chem.Phys.Lett. **121** (1985) 33
3. H. Haberland, C. Ludewigt, H.G. Schindler and D.R. Worsnop,
    Phys.Rev.A, in print
4. Y. Liu, Q.L. Zhang, F.K. Tittel, R.F. Curl and R.E. Smalley,
    J.Chem.Phys. **85** (1986) 7434
5. J.V. Coe, J.T. Snodgrass, C.B. Freidhoff, K.M. McHugh and K.H.
    Bowen, Chem.Phys.Lett. **124** (1986) 274
6. D.G. Leopold, J. Hoe and W.C.Lineberger, J.Chem.Phys.**86** (1987) 1715;
    L.A.Posey, M.J.Deluca and M.A.Johnson, Chem.Phys.Lett.**131** (1986) 170
7. R.J. Beuhler, Phys.Rev.Lett. **58** (1987) 13
8. C.E. Klots and R.N. Compton, J.Chem.Phys. **69** (1978) 1636
9. T.D. Märk, K. Leiter, W. Ritter and A. Stamatovic,
    Phys.Rev.Lett. **55** (1985) 2559
10. M. Knapp, O. Echt, D. Kreisle, T.D. Märk and E. Recknagel,
    Chem.Phys.Lett. **126** (1986) 225
11. K.H. Bowen, G.W. Liesegang, R.A. Sanders and D.R. Herschbach,
    J.Phys.Chem. **87** (1983) 557
12. T. Kondow, J.Phys.Chem. **91** (1987) 1307; K. Mitsuke,
    T. Kondow and K. Kuchitsu, J.Phys.Chem. **90** (1986) 1552
13. M. Knapp, O. Echt, D. Kreisle and E. Recknagel,
    J.Phys.Chem., in print
14. O. Echt, Hyperfine Interactions, in print
15. O. Echt, M. Knapp and E. Recknagel, Z.Phys. **B53** (1983) 71

16. D. Kreisle, O. Echt, M. Knapp and E. Recknagel, Phys.Rev.A 33
    (1986) 768; O. Echt, A. Reyes Flotte, M. Knapp, K. Sattler and E.
    Recknagel, Ber.Bunsenges.Phys.Chem. 86 (1982) 860

17. H.P. Birkhofer, H. Haberland, M. Winterer and D.R. Worsnop,
    Ber.Bunsenges.Phys.Chem. 88 (1984) 207

18. O. Echt, M.C. Cook and A.W. Castleman Jr.,
    Chem.Phys.Lett 135 (1987) 229

19. J.M. Soler, J.J. Saenz, N. Garcia and O. Echt, Chem.Phys.Lett. 109
    (1984) 71; H. Haberland, Surf.Sci. 156 (1985) 305

20. M. Knapp, Ph.D. Thesis, University of Konstanz, 1986

21. M.L. Alexander, M.A. Johnson, N.E. Levinger and W.C. Lineberger,
    Phys.Rev.Lett. 57 (1986) 976

22. I.A. Harris, K.A. Normann, R.V. Mulkern and J.A. Northby,
    Chem.Phys.Lett. 130 (1986) 316

23. References to thermodynamic data of CO_2, N_2O and their anions are
    to be found in ref. 10.

24. M. Knapp, O. Echt, D. Kreisle, T.D. Märk and E. Recknagel, in
    "The Physics and Chemistry of Small Clusters", NATO ASI Series,
    P. Jena et al, eds., Plenum Press New York, in print

25. A. Stamatovic, K. Leiter, W. Ritter, K. Stephan and T.D. Märk,
    J.Chem.Phys. 83 (1985) 2942

26. T.D. Märk, P. Scheier and A. Stamatovic, submitted to Z.Phys.D

27. C. Schwarz, Diploma Thesis, University of Konstanz, 1986

28. N. Schwentner, E.-E. Koch and J. Jortner, in "Electronic Excita-
    tions in Condensed Rare Gases" (Springer Tracts in Modern Physics,
    vol. 107), 1985, p. 75

29. E.P. Grimsrud, S. Chowdhury and P. Kebarle, J.Chem.Phys. 83 (1985)
    1059; and references therein

30. Y. Yoshioka and K.D. Jordan, J.Am.Chem.Soc. 102 (1980) 2621

31. S.H. Fleischman and K.D. Jordan, J.Phys.Chem. 91 (1987) 1300

32. P.R. Antoniewicz, G.T. Bennett and J.C. Thompson, J.Chem.Phys. 77
    (1982) 4573

33. A.J. Dahm and W.F. Vinen, Physics Today, February 1987, p. 43;
    and references therein

34. E. Illenberger and H. Baumgärtel, J.Electron Spectr.&Rel.Phenom. 33
    (1984) 123

35. C.E. Klots, Radiat.Phys.Chem. 20 (1982) 51; R.N. Compton, in
    "Electronic and Atomic Collisions", edited by N. Oda and K. Taka-
    yanagi (North-Holland, Amsterdam, 1980), p. 251; D.R. Herschbach,
    in a private communication to J. Jortner

36. H. Haberland, H. Langosch, H.-G. Schindler and D.R. Worsnop,
    J.Phys.Chem. 88 (1984) 3903

37. N.R. Kestner and J. Jortner, J.Phys.Chem. 88 (1984) 3818;
    B.K. Rao and N.R. Kestner, J.Chem.Phys. 80 (1984) 1587

38. M. Tachiya and A. Mozumder, J.Chem.Phys. 61 (1974) 3890

39. M. Knapp, O. Echt, D. Kreisle and E. Recknagel,
    J.Chem.Phys. **85** (1986) 636
40. A. Migus, Y.Gauduel and A.Antonetti, Phys.Rev.Lett. **58** (1987) 1559
41. J. Schnitker and P.J. Rossky, J.Chem.Phys. **86** (1987) 3471;
    J. Schnitker, P.J. Rossky and G.A. Kenney-Wallace, J.Chem.Phys. **85**
    (1986) 2986; M. Hilczer, W.M. Bartczak and M. Sopek, J.Chem.Phys.
    85 (1986) 6813
42. A. Wallqvist, D. Thirumalai and B.J. Berne, J.Chem.Phys. **83** (1986)
    1583; U. Landman, J. Jortner and D. Scharf, presented at the Int.
    Symp. Phys. Chem. of Small Clusters, Richmond, Oc 30 – Nov 3, 1986

BEAMS OF ELECTRONS SOLVATED IN CLUSTERS

Hellmut Haberland, Christoph Ludewigt, Hans-Georg
Schindler and Douglas R. Worsnop*
Fakultät für Physik, Universität Freiburg, Germany

An ion source, based on electron injection into a supersonic
expansion, is described in detail, and data obtained for negatively charged clusters of H_2O, NH_3, CO_2, HCl, $C_2H_4(OH)_2$ and Hg
are discussed.

INTRODUCTION

The negative cluster ions $(H_2O)_n^-$ and $(NH_3)_n^-$ are believed to
provide models of excess electrons in liquid ammonia and water.
However, experimental production of these species has been
problematic for a long time. In the first successful attempt, a
low energy β-emitter was used as an electron source.[1] It was
placed in the high pressure stagnation chamber behind a supersonic expansion. Very weak signals were observed (1 to 100
counts/sec) which were very sensitive to contamination by minute
impurities with high electron affinity such as Cl^-. Previously
unobserved positively charged cluster ions such as $(H_2O)_n$ ($n \geq 3$)
have also been observed with this configuration.[2]

After much development the impurity problems have been eliminated and the intensity increased. Electrons from several sources
(radioactive emission, photoelectrons from CW and pulsed UV
sources, thermal emission, and gas discharges) have been injected
into different parts of a supersonic expansion. In the final
configuration described here, ion beam intensities up to 10^{10}/sec
have been observed at one cluster mass. Results obtained with
this technique have been published elsewhere.[3]

The so-called solvated electron was first observed in 1863
from the blue color characteristic of solutions of sodium metal
dissolved in liquid ammonia. This broad absorption band was later
attributed to excess electrons in the liquid, that is electrons
which were not associated with any negative ion but trapped in
pure solvent. Similar phenomena have been observed upon injection
of electrons into a variety of dielectric vapors, liquids, and
solids.[4-10] Clusters of $(H_2O)_n$ (hydrated electrons) in liquid
water have been used to explain DNA decomposition after interaction with ionizing radiation or high power UV photons.[11]

195

J. Jortner et al. (eds.), Large Finite Systems, 195–207.

Theoretical models describing these electron states have evolved from continuum pictures over detailed molecular calculations on clusters, such as $(H_2O)_n^{-}$ [7,10] to Feynman path integral calculations [12,13] on $(H_2O)_2^{-}$. The motivation for the present work stemmed from the inability to test these detailed models with condensed phase results.

The monomer molecules of the clusters discussed here do not have a bound negative ion state. However, all show stable excess electron states in condensed phases. Therefore the binding energy of the excess electron must change from negative to positive values as a function of cluster size. The mass spectra presented here place an upper limit on this critical size. This limit is probably the experimentally most easily accessible size effect of cluster physics.

THE SOURCE

The key to this ion source is the injection of electrons into the high collision zone of a supersonic expansion, where there are enough collisions both for ion formation and cluster growth. At the same time, supersonic cooling achieves low temperatures so that weakly bound ions and clusters can be produced. Finally, the supersonic jet expands into a collisionless zone from which the cluster ions can easily be extracted.

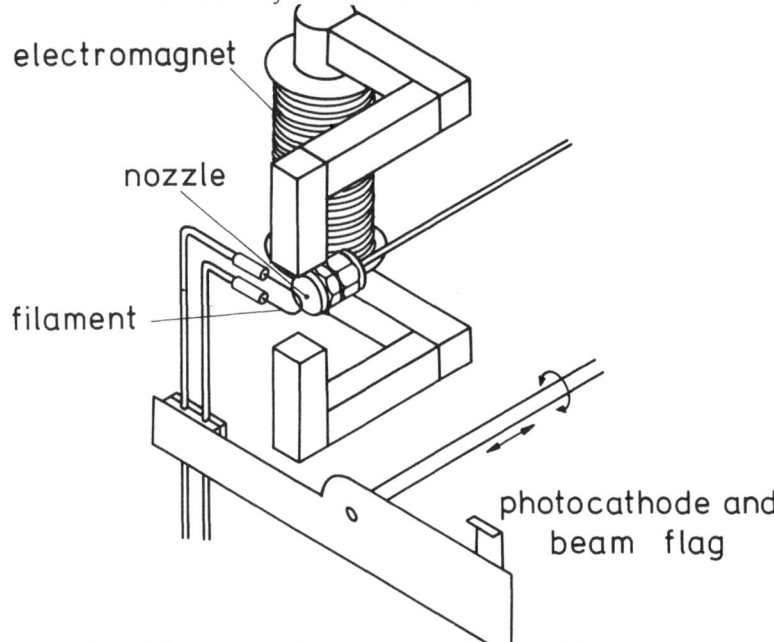

Figure 1. Schematic of the ion source. Electrons are injected into an expanding supersonic jet. This design produces high intensity ion beams of cold large cluster ions some of which have never been synthesized before. Three different operating modes are described in the text. The source can run CW or pulsed.

Figure 1 shows a diagram of the source in its present form. The stagnation gas is fed into a 3 mm stainless steel tube closed with a Swagelok end cap. This unscrewable cap contains the super-sonic nozzle, a central bore of 0,02 to 0,15 mm diameter. No gasket is used. This set-up permits rapid and easy interchange of nozzles. The gas inlet system is bakeable to 200 C. A heatable filament is positioned about 1 mm from the nozzle. Electrons emitted from it are confined by the field of an electromagnet (0.05 Tesla). The skimmer is electrically isolated and coated with graphite. Ions are focused with a simple lens system into a quadrupole mass spectrometer or a time of flight spectrometer. The source can be operated in several different modes:

(1) A sustained glow discharge is struck between the filament and its grounded surrounding. The discharge current is typically 4 mA at 25 V. Highest ion beam intensities are obtained in this mode, but the mass spectra can include undesired cluster ions. For example, in a neat water expansion $(H_2O)_{10}^-$ is the smallest sol-vated electron cluster detected; however, $(H_2O)_n OH^-$ is observed, particularly for masses below $10 \times 18 = 180$ amu. OH^- is a stable negative ion that can be formed by dissociative electron attach-ment of high energy electrons in the discharge. This is a two step process, having a 6.3 eV threshold for a water molecule.[14] Its simplified reaction can be written as:

$$e^- + H_2O \longrightarrow OH^- + H$$

Under the supersonic expansion conditions here, the ion-molecule reaction occurs not only with water molecules but also in water clusters. The overall reaction can be written:

$$e^- + (H_2O)_n \longrightarrow OH^-(H_2O)_m + (n-m-1)H_2O + H$$

(2) The hot filament is kept at the same potential as the nozzle and skimmer. In this mode the energy of the electrons is limited to their thermal kinetic energy plus the voltage drop across the filament (~ 5 Volt). This discriminates against higher energy processes, giving for example much cleaner $(H_2O)_n^-$ mass spectra. The $(H_2O)_n^-$ are formed by attachment of low energy electrons to growing neutral water clusters. However, the ion beam intensity in this more controlled mode is about 100 times smaller than with the discharge (mode 1).

(3) The cleanest $(H_2O)_n^-$ spectra are obtained by injecting photoelectrons into the expansion. The filament is rotated away and a
stainless steel plate is positioned near the nozzle and irra-
diated with light from a mercury discharge lamp. If no potential
is applied between the nozzle and the plate, the energy of the
photoelectrons is ~ 1 eV. Using a 100 Watt lamp, focused with a
quartz lens, the intensity is about 10^5 weaker than in the discharge mode.

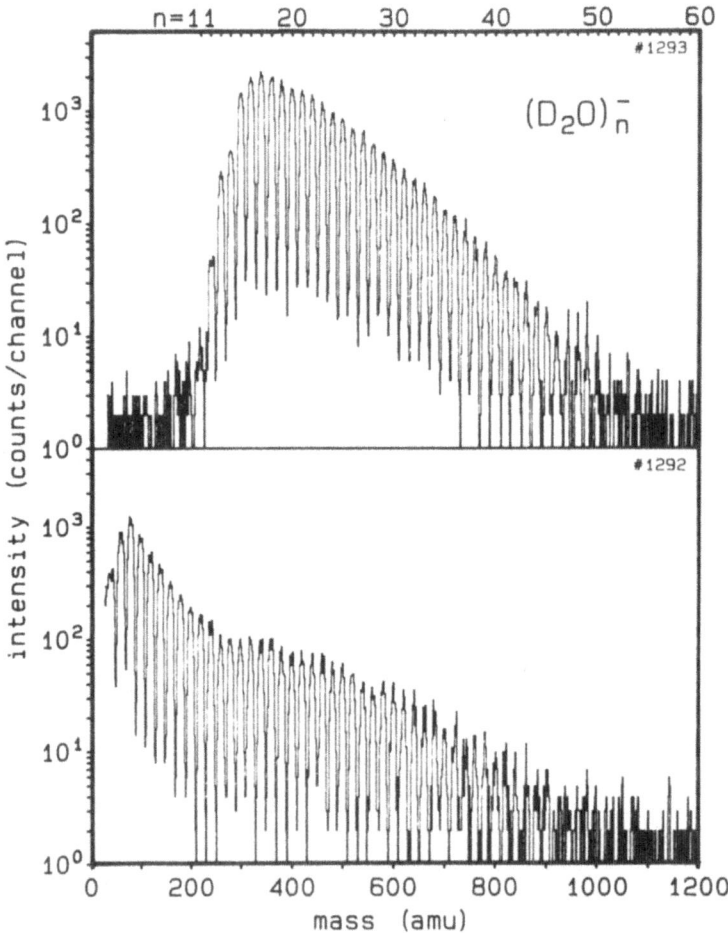

Figure 2. Negative ion mass spectra from a neat heavy water
expansion. (a) Photoelectrons with kinetic energy ~ 1 eV were
used. The onset of the $(D_2O)_n^-$ distribution occurs at n = 11.
(b) The photoelectron source was biased to 12 Volts. The mass
peaks appearing below 200 amu are $OD^-(D_2O)_n$ ions formed from
ion–molecule reaction in neutral clusters. Mass identification
was made under high mass resolution.

Figure 2 shows water cluster negative ion mass spectra using photoelectrons as the electron source. The results are from an expansion of neat heavy water. In Figure 2a, the logarithmic scale shows a minimum $n = 11$ for $(D_2O)_n^-$ under these conditions. In Figure 2b, where the photoelectron source is biased by 12 volts smaller clusters appear due to the appearance of $OD^-(D_2O)_n$. The mass of all the peaks in Figure 2 was established under high mass resolution conditions.

(4) The source can be operated in a pulsed configuration. Either the glow discharge, the electron energy (i.e. the filament bias), or the UV photon source (e.g. an excimer laser) is pulsed. In the case of the second mode, the electron energy cannot be near zero for efficient pulsing. In this pulsed configuration the ion source is particularly useful for coupling to a time-of-flight mass spectrometer.[15] In this mode the source is similar to that described by Johnson et al.[16] In their design, the ion beam is accelerated perpendicularly to the neutral beam. We prefer a collinear arrangement, which leads to less mass discrimination at large masses in our apparatus.

Electrons are introduced directly into the supersonic expansion in all the operating modes discussed above. This eliminates the impurity problem seen in earlier designs in which electrons were introduced in the high pressure stagnation chamber before the expansion.[1,2,17] Any ions formed there undergo many collisions so that ion/molecule chemistry can be dominated by impurities. In the case of negative ions, the charge generally finds the species with the highest electron affinity. Sensitivities to halogen compounds < 1 ppm have been observed.[17] In this mode the source is excellent for producing beams such as $Cl^-(H_2O)_n$ which comes from an expansion of water containing a trace of CCl_4.

RESULTS

a) $(H_2O)_n^-$

As mentioned above, only $(H_2O)_n^-$ clusters with $n > 10$ are observed when electrons are injected into neat water expansions. Smaller clusters can be produced by seeding a small amount of water into a rare gas expansion. Figures 3 and 4 summarize results using Ar. At water mole fractions ~ 0.5% only $(H_2O)_2^-$ and its clusters with Ar are observed (Figure 3a).

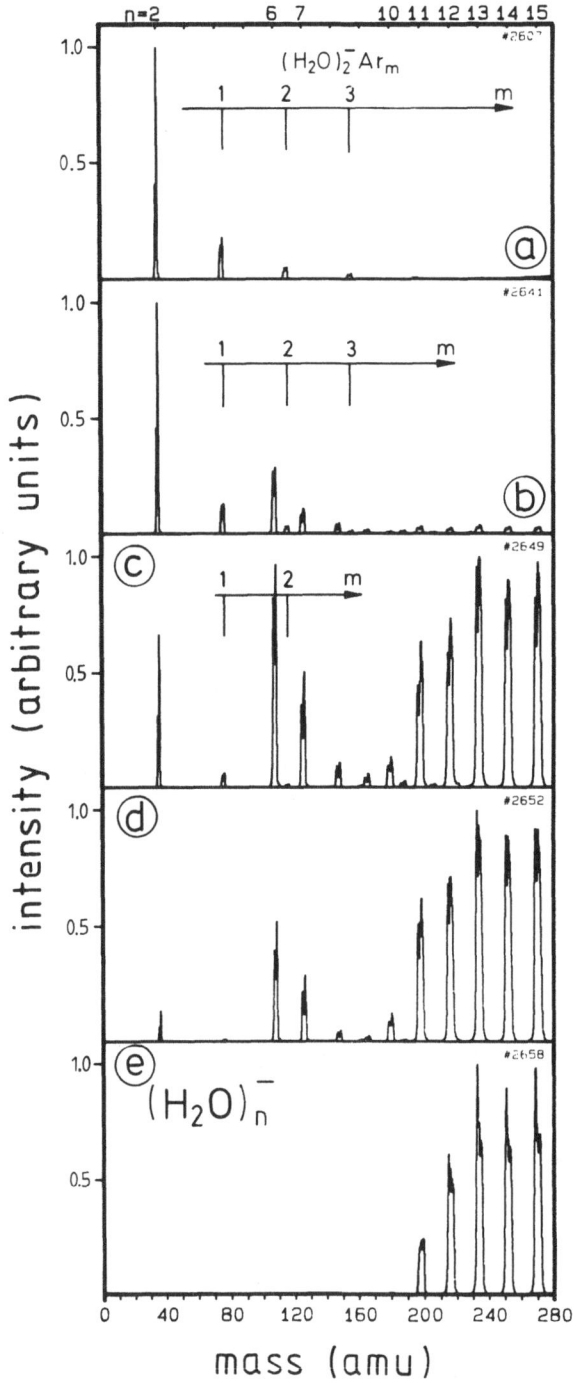

Figure 3. Mass spectra of negativly charged clusters from an expansion of water seeded in Ar (pressure · nozzle diameter = 30 Torr · cm). As the water mole fraction increases from 0.1 to 3.9%, $(H_2O)_n^-$ with $n = 2$ and $n = 6\text{–}9$ successively appear and disappear; above 4% only $n > 10$ is observed, as in neat water expansions. Below 2%, Ar atoms Van der Waals bonded to $(H_2O)_n^-$ are also observed, an indication of how cold the expansion is. The water partial pressures are roughly: a) 0.1%, b) 0.9%, c) 1.6% d) 2.2%, and e) 3.9%

As the H_2O mole fraction increases, larger $(H_2O)_n^-$ clusters appear, with strong intensity at n = 6 and n = 7 (Figure 3b – 3c) As the water mole fraction increases, the n = 2 and then n = 6,7 intensities decrease until at about 4% only n > 10 is observed as in the neat water expansion. (Figure 3d – 3e).

Weak signals have also been observed at n = 8,9. In very strong expansions, n = 3 is observed with an intensity two or more orders of magnitude less than n = 2. The n = 5 cluster has been seen only with seeding in strong Kr and Xe expansions. No signal has been observed at n = 4.

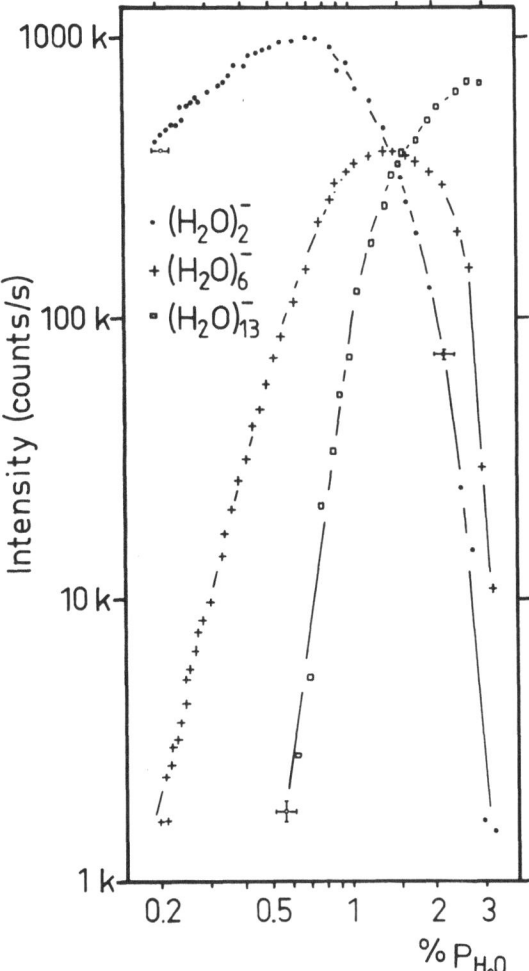

Figure 4. Water mole fraction dependence of $(H_2O)_n^-$ cluster ions (n = 2, 6, 13) under conditions as in Figure 3.

Figure 4 summarizes the water mole fraction dependence, showing how the n = 2 and n = 6 clusters first appear and then disappear until n > 10 dominate. Under these conditions, positively charged clusters show the steep, monotonic decrease in intensity with increasing size that is typical in cluster mass spectrometry.

The pressure dependence of Figure 4 probably is a crude reflection of the relative stability of these $(H_2O)_n^-$. The primary effect of increasing water mole fraction is to warm the expansion, due to the heat of condensation from increased cluster growth. Thus, the more weakly bound small clusters disappear at higher water pressures, in contrast to positively charged clusters. In fact, the electron on $(H_2O)_2^-$ has been detached in an electric field of 31 kV/cm, showing that it is indeed very weakly bound.[3d + f] Recently obtained photoelectron spectra for n > 11 show an electron energy appearance threshold that corresponds to an electron binding energy > 0.5 eV, which is consistent with the apparent enhanced stability of the n > 10 clusters in our mass spectra.[18] Presumably n = 6,7 are intermediate between these cases. The details of the intensity distribution of Figure 3 remain to be analyzed.

Electrons can also be attached to cold clusters far down-stream after the expansion, as has been observed in this laboratory[15] and elsewhere.[19] However, the ion beam intensity is much lower and only clusters with n > 10 are observed.

No clear cut signal was ever observed at a charge to mass ratio of 18 amu. This implies that the H_2O^- ion, which has recently been detected by de Kronning and Nibbering[20] cannot be synthesyzed under our conditions.

b) $(NH_3)_n^-$

Figure 5 shows the very different results observed when electrons are attached to ammonia clusters. In a neat ammonia expansion, the smallest $(NH_3)_n^-$ is n = 36. Seeding in Ar shifts this threshold to n = 35. Also, the onset in intensity is extremely steep (note the logarithmic scale in Figure 5) as compared with the results for $(H_2O)_n^-$. The difference between electron binding in ammonia and water-clusters is not understood, but it suggests the electron states of the smaller $(H_2O)_n^-$ are unbound for ammonia. In the condensed phase large differences are known to exits.[4,6,21] The ammonia dimer has recently been shown to have an unexpected structure compatibel with these results.[22] A very recent calculation by Bennemann and Stampfli was able to predict the n = 35 threshold.[23]

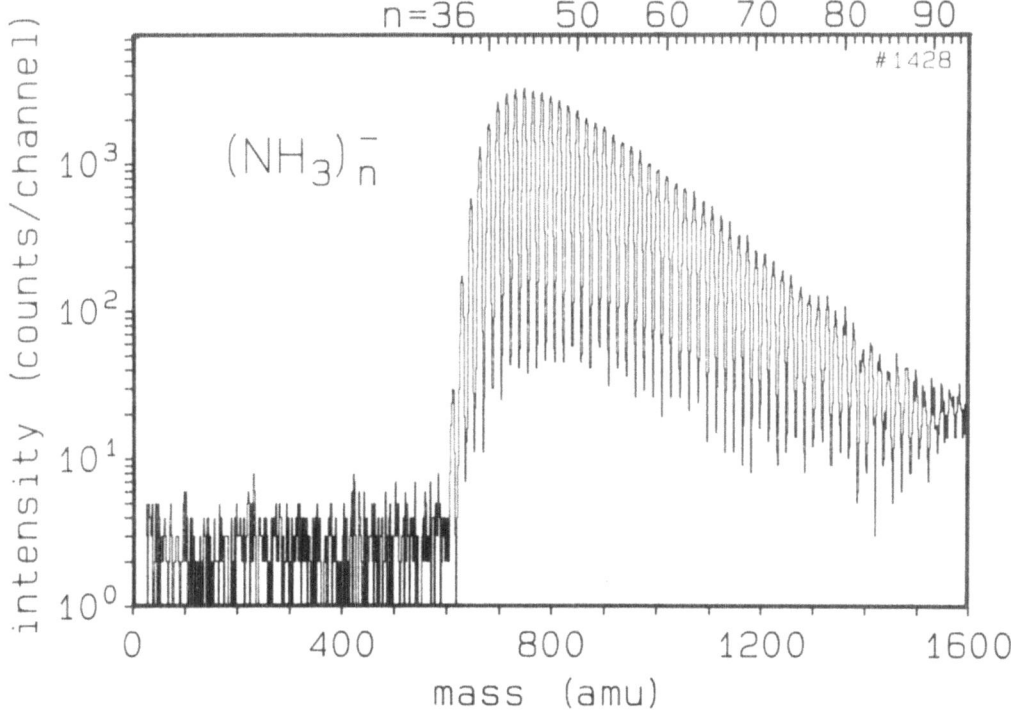

Figure 5. Negative cluster ions from a neat ammonia expansion. Peaks identified as $(NH_3)_n^-$ under high mass resolution. Note the steep onset above n = 36, which has recently been calculated by Stampfli and Bennemann.

c) $(CO_2)_n^-$

The source has been used to produce a variety of clustered ions. Figure 6 shows a $(CO_2)_n^-$ mass spectrum, showing signals peaking at n = 10, 14 and 16. Mass selected ion beam intensities up to 10^8/sec have been obtained. The metastable monomer ion CO_2^- has been detected at a count rate of 10^5/sec. The mass distribution of Figure 6 is independent of expansion conditions, though, as with all these clusters, one can grow larger clusters by seeding in rare gases (for a fixed source chamber pumping capacity). Because CO_2 is a convenient permanent gas with which to work, the $(CO_2)_n^-$ system is routinely used for set-up and testing.

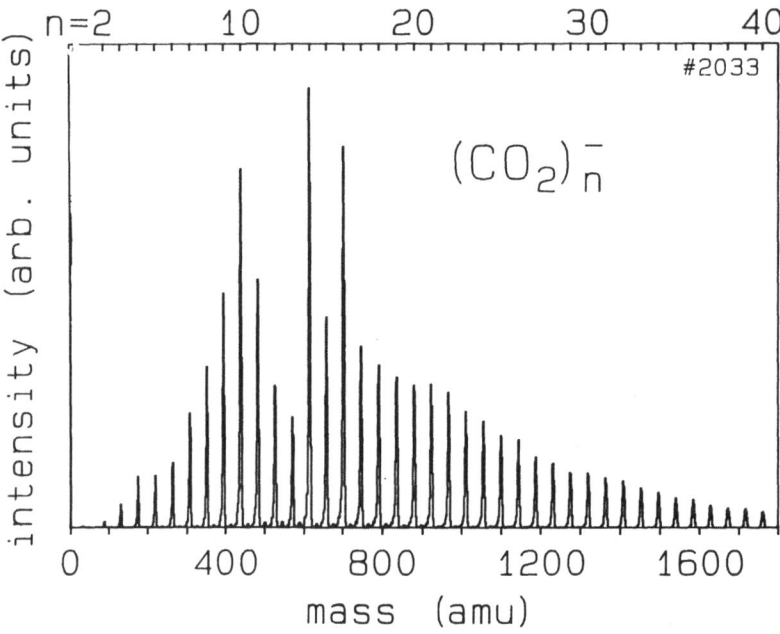

Figure 6. $(CO_2)_n^-$ cluster ions from a neat CO_2 expansion

d) $(C_2H_4(OH)_2)_n^-$

Excess electrons in condensed ethylene glycol show an electronic absorption spectrum shifted to much higher energies than when in water or other alcohols.[24,25] This shift is thought to be due to chelation, since each $(C_2H_4(OH)_2)_n^-$ molecule has two polar OH groups with which to stabilize an electron. The mass spectrum shown in Figure 7 is consistent with this picture. The appearance threshold is at n = 2; under stronger expansion conditions than those used in Figure 7, an unstructured intensity distribution up to n = 30 has been observed. These data indicate that the extra electrons are more strongly bound in $(C_2H_4(OH)_2)_n^-$ than $(H_2O)_n^-$ or $(NH_3)_n^-$, as predicted from the liquid phase results.
In the case of ethylene glycol, seeding in rare gases is necessary because of its low vapor pressure. Attempts to increase[26] the vapor pressure by heating decomposed the ethylene glycol. Also, the distribution of Van der Waals bounded Ar atoms on $(C_2H_4(OH)_2)_n^-$ in Figure 7 is indicative of the cooling effect of rare gas seeding.

e) Other species

Very recently Hg_n^-, n ≥ 3 and $(HCl)_n^-$, n ≥ 2 have been observed.[27]
These results will be published elsewhere.

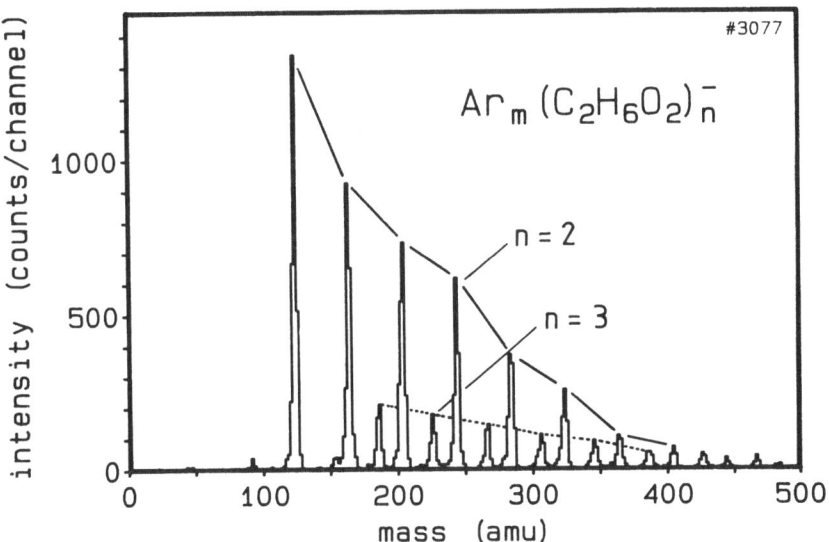

Figure 7. Mass spectrum from ethylene glycol seeded in Ar. The dimer and trimer $(C_2H_4(OH)_2)_n^-$ anion with up to 10 added argon atoms are observed

f) Positively Charged Clusters

Intense beams of positively charged clusters have also been produced with this source, e.g. a discharge in a pure Ar expansion produces Ar_n^+. The intensity is several orders of magnitude larger than in conventional electron impact sources.

SUMMARY

A versatile cluster ion source, based on injection of electrons into a supersonic expansion, has been developed. Intense beams of positively and negatively charged clusters have been produced from gaseous and liquid samples. The source has two advantages: (1) Supersonic cooling permits the synthesis of novel cluster ions. (2) Ion beam intensities can be several orders of magnitude larger than designs based on ionization of or electron attachment to neutral clusters down-stream from the supersonic nozzle.

Acknowledgement: DRW acknowledges the support of the Alexander von Humboldt Foundation in early stages of this work.

REFERENCES

*current address: Aerodyne Research, Inc., Billerica,
Massachusetts 01821, USA

1) M. Armbruster, H. Haberland, and H.G. Schindler,
    Phys. Rev. Lett. 47, 323 (1981)
2) H. Langosch, H. Haberland, Zeitschrift für Physik D,
    Atoms, Molecules and Clusters, 2, 243 (1986)
3)a) H. Haberland, H.G. Schindler, D.R. Worsnop,
    Ber. Bunsenges. Phys. Chem. 88, 270–272 (1984).
  b) H. Haberland, H. Langosch, H.G. Schindler, D.R. Worsnop,
    J. Phys. Chem. 88, 3903 (1984).
  c) H. Haberland, C. Ludewigt, H.G. Schindler, D.R. Worsnop,
    J. Phys. Chem. 81, 3742 (1984).
  d) H. Haberland, C. Ludewigt, H.G. Schindler, D.R. Worsnop,
    Z. Phys. A 320, 151 (1985).
  e) H. Haberland, C. Ludewigt, H.G. Schindler, D.R. Worsnop,
    Surface Science 156, 157 (1985)
  f) H. Haberland, C. Ludewigt, H.G. Schindler, D.R. Worsnop,
    Phys. Rev. A, Rapid Communication, Spring 1987
4) R. Olinger, U. Schindewolf, A. Gaathon, J. Jortner
    Ber. Bunsenges. Phys. Chem. 75, 690 (1971)
5) G.A. Kenney-Wallace, Adv. Chem. Phys. 47, 535 (1981)
6) See complete volumes of: J. Phys. Chem. 84 (1980) and
    J. Phys. Chem. 88 (1984), Proceedings of the Colloques
    Weyl V and VI.
7) M. Newton, J. Chem. Phys. 58, 5833 (1973)
8) J. Jortner, Ber. Bunsenges. Phys. Chem. 88, 188 (1984)
9) N.R. Kestner, J. Jortner, J. Phys. Chem. 88, 3818 (1984)
10) M. Sprik, R.W. Impey, M.L. Klein, J. Chem. Phys. 83
    5802 (1985)
11)a)D.A. Angelov, D.N. Nikogosyan, and A.A. Oraevskii,
    Sov. J. Quantum Electron. 10, 1502 (1981)
  b)D.N. Nikogosyan, and D.A. Angelov, Dokl. Akad. Nauk
    SSSR 253, 733 (1980) and references therein.
12) A. Wallqvist, D. Thirumalai, and B. Berne,
    J. Chem. Phys. 85, 1583 (1986)
13) U. Landman, R.N. Barnett, C.L. Cleveland, D. Scharf, and
    J. Jortner priv. communication, and Proc. Int. Symp. on
    Phys. and Chem. of Small Clusters, 1986 Richmond, Virginia,
    (Ed. P. Jena), to be published as NATO ASI
14) C.E. Klots and R.N. Compton, J. Chem. Phys. 69, 1644 (1978)
15) Ph.D. thesis H.G. Schindler, Freiburg 1986, unpublished

16) M.A. Johnson, M.L. Alexander and W.C. Lineberger, Chem. Phys. Lett. 112, 285 (1984)
17) Diplomarbeit H. Langosch, Freiburg 1984, unpublished
18) J.V. Coe, D.R. Worsnop, and K.H. Bowen, J. Chem. Phys. submitted
19) M. Knapp, O. Echt, D. Kreisle, and E. Recknagel, J. Chem. Phys. 85, 636 (86)
20) L.J. de Kronning and N.N.M. Nibbering, J. Am. Chem. Soc. 106, 7971 (1984)
21) U. Schindewolf, University Karlsruhe, private communication
22) D.D. Nelson, G.T. Fraser, and W. Klemperer, J. Chem. Phys. 83, 6201 (1985)
23) P. Stampfli and K.H. Bennemann, FU Berlin, preprint 1987
24) R.R. Hentz and G.A. Kenney-Wallace, J. Phys. Chem. 78, 514 (1974)
25) F.Y. Jou and G.R. Freeman, Can. J. Chem. 57, 591 (1979)
26) Diplomarbeit Ch. Ludewigt, Freiburg 1985, unpublished
27) H. Haberland. H. Langosch, T. Richter, unpublished results.

THEORETICAL STUDIES OF SODIUM WATER CLUSTERS: THEIR IONIZATION AND EXCITATION ENERGIES

Neil R. Kestner and S. Dhar
Chemistry Department,
Louisiana State University
Baton Rouge, LA 70803 USA

ABSTRACT   Large basis set ab initio calculations of $Na(H_2O)_n$ clusters with n=1 to 4 were performed. The sodium-monomeric water water cluster yields an ionization potential of 4.10 eV (SCF) or 4.34eV (MP2) in excellent agreement with the experimental value of 4.379 eV. Energetic and structural details for the larger clusters are also presented.

1.0   INTRODUCTION

Over the past few years there have been several theoretical studies of small clusters of sodium and water[1,2,3]. It is an interesting system from the viewpoint of bonding. Most of the research, both experimental and theoretical, has been to directed toward the study of the structure and bonding in these systems. Recently some new rather different experimental data has been obtained which gives new impetus for such studies from a very different point of view. It has been possible to directly determine the ionization potential of small clusters of water on sodium atoms in a molecular beam[4]. This new technique can lead to a study of the convergence to the ionization potential in liquid water as a function of cluster size for this chemically important case. In another experiment evidence is given for their role in the radiation chemistry of strongly alkaline water solutions[5].

We have done some rather extensive studies on the first few members of this family and determined the structure and the energetics of ionization as a function of cluster size. In a few cases we were also able to estimate the excitation energies which might be expected for such species. Some structural features are sensitive to the basis set but the ionization potential is, fortunately, much less sensitive. There are also well established shifts due to correlation effects. We are able to only determine these for the very simplest basis and only for the beginning members of this series. The trend is readily established, but since we have only done the first four members, we can not place a definite bound on the ionization in an infinite system.

J. Jortner et al. (eds.), Large Finite Systems, 209–215.
© *1987 by D. Reidel Publishing Company.*

2.0 CALCULATIONAL DETAILS

We have performed ab initio molecular orbital quantum chemistry
calculations on the series $Na(H_2O)_n$ where n varies from one to four
using a 6-31G** basis as our minimal set. Our best basis set used was
for Sodium (15,9,1) contracted to [8,7,1], for Oxygen (13,8,1)
contracted to [7,5,1], and for hydrogen (10,1) contracted to [3,1].
These later basis sets were taken from the literature. In the two
simplest cases the geometry was optimized during the Hartree Fock or MP2
calculations using procedures contained within the Gaussian 82
package[6]. In the case of the larger systems the minimizations had to be
performed by manual linear searches. The geometry is sensitive to the
basis set and to correlation effects in the lower members of the series,
at least. We do not have MP2 results for any of the higher members. We
monitored the orbital energies and charge densities as well as the total
energies for these members. We have also calculated the neutral formed
by vertical electron attachment at the stable ion-water geometry. These
species are of interest as they may be involved in the radiation
chemistry and also they give us clues about the bonding in the higher
members of the series, those members relevant to liquid studies. We
will present a limited amount of data on their binding energy. This
requires a great deal more work since this stability requires a lot of
knowledge of the process of formation or decomposition and is a function
of temperature, i.e. involves entropic effects. Such studies will be
undertaken later by quantum path integral methods.

RESULTS

In Tables I and II we present the main energetic results of the
calculation.

TABLE   I

Number of	Ionization Potentials (in eV)	
Waters	6-31G** Basis	Large Basis
None	4.955	4.951
One	4.11	4.099
Two	3.47	5.531
Three	3.04	3.025
Four	2.34	2.445

These ionization potentials are calculated by comparing the energy of
the neutral with the ion at the same geometry. We have also looked at
the Koopman theorem results for the neutral and they lead to ionization
potentials which differ by only .04 to .07 eV, being in all cases larger
than those listed. An MP2 run was made on the sodium water monomer and
the ionization potential was increased to 4.34 eV. Although we can not
do such calculations because of computer program limitations, that

calculation suggests that the values at the SCF level will be about a few tenths of an electron volt low for all members of the series.

These data are simpler to visualize in the following plot (Figure 1).

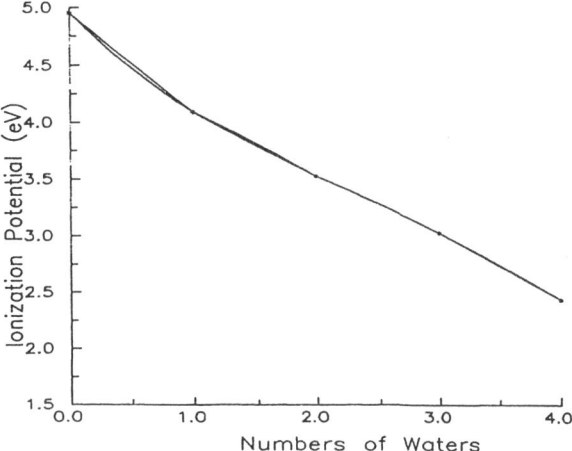

Figure 1. Ionization Potentials

TABLE II

Number of Waters	Dissociation Energy	
	Relative to n Water Molecules	Relative to Performed Optimal Cluster
One	.164 eV	.164 eV
Two	.390	.364
Three	.698	.753
Four	1.331	1.43

In Table II we list the dissociation energies of sodium water clusters calculated in two different ways. In the first approach we simply subtract the energy of n water molecules calculated with the same basis set(all at the SCF level). In the second case we align the waters in the form needed by the sodium and calculate the energy of the n water cluster for that geometry. This is then used to evaluate the difference. The second method corrects for some basis set effects. We did not search for the optimal water cluster and calculate its energy but such data is available form calculations of Clementi and others[7]. The difference are not substantial since the cluster is quite open and there is very little hydrogen bonding present for these cluster orientations.

The detailed geometries of the species are sensitive to the basis set but on the next page we present some of the SCF optimized geometries of the first three members (Figure 2). The better basis set gives the

$Na(H_2O)$

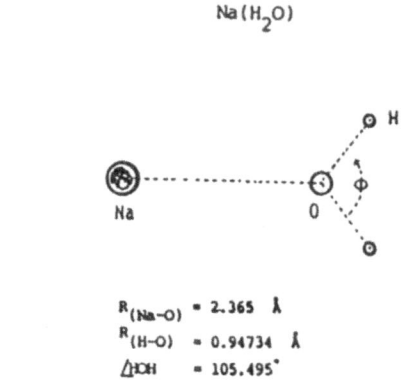

$R_{(Na-O)} = 2.365$ Å
$R_{(H-O)} = 0.94734$ Å
$\angle HOH = 105.495°$

$Na(H_2O)_3$

$Na(H_2O)_2$

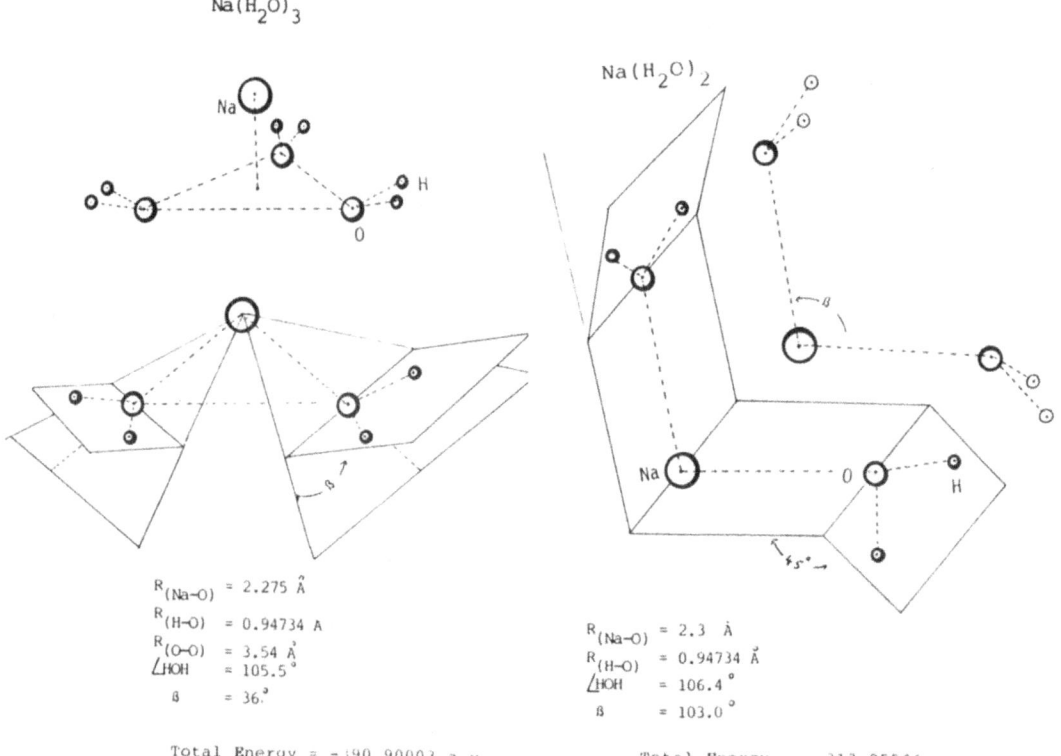

$R_{(Na-O)} = 2.275$ Å
$R_{(H-O)} = 0.94734$ A
$R_{(O-O)} = 3.54$ Å
$\angle HOH = 105.5°$
$\beta = 36.°$

Total Energy = -190.90003 a.u

$R_{(Na-O)} = 2.3$ Å
$R_{(H-O)} = 0.94734$ Å
$\angle HOH = 106.4°$
$\beta = 103.0°$

Total Energy = -313.95566 a.u

Figure 2.   Optimal Hartree Fock Geometries of
'Small Sodium Water Clusters

planar form for sodium-monomeric water but the very recent MP2 results
of Curtiss, Krafa, Gauss, and Cremer[8] suggest that the water is tilted
46 degrees out of plane. For the larger species it appears that the
waters orient in such a way that their lone pairs can participate in
some type of bonding with the sodium. The water orientations are at
least consistent with such an interpretaion.

We have also calculated the positive ion species at their optimum
geometry. The optimum water oxygen distance found by our large basis set
calculation was 2.31 A, in good agreement with molecular dynamics
calculations[9,10]. Surprisingly, the sodium to oxygen distance does not
change much nor do the orbital energies of the occupied orbitals,
suggesting that the species is more like a sodium ion-water cluster plus
electron. In fact when we add an electron to the sodium ion water
cluster at its optimum shape we obtain an energy similar to that for the
fully optimized neutral sodium cluster. The HOMO in the neutral cluster
is quite diffuse, becoming more diffuse as more waters are added (The
orbital energy is .18 in the free atom, dropping to .093 with four
waters in clusters).

The electron density of the sodium water species is also very
interesting. The electron density of the outer electron is very diffuse
and extends over the region occupied by the waters. This is most
clearly seen for the sodium surrounded by four waters. On the other
hand the total electron density is mainly located on the sodium and
oxygen ions, again strongly suggesting an ionic type of bonding.

In Table III we list the calculated ionic dissipation energies
which were obtained during course of our studies.

TABLE III

Dissociation Energy of Ion Hydrates (kcal/mole)[12]

Number of Waters	Our results (Large Basis)	Lybrand and Kollman[11]	Experimental
1	24.0	23.9	24.0
2	(41.1)*	44.7	43.8
3	(61.7)*	61.3	59.6
4	77.8	74.1	73.4
5		86.2	85.7
6		98.0	96.4

*Positive Ion geometry is not completely optimized

The Lybrand and Kollman[11] potential predicts four fold coordination of
the sodium ion while the Jorgenson potential[13] predicts six fold
coordination and an energy of -80.7 for the n=4 case. They find from ab
initio calculations that the four coordinated ion energy is lower than
the six coordinated ion, i.e. the extra two waters prefer to go into the
second coordination shell.

3.0 COMPARISON WITH EXPERIMENTAL DATA

3.1 Beam Data

The most obvious comparison to be made is with the experimental work of
Schultz, Haugstatter, Tittes, and Hertel[4]. They have determined the
ionization potential of the sodium-monomeric water cluster to be 4.379 ±
.002 eV. This value is very close to the predictions in Table I and
even close to our MP2 result mentioned earlier. They also find a great
abundance of the sodium tetrameric water cluster which seems quite
stable in our calculations as well. They suggest that the sodium
trimeric water cluster ionizes around 3.49 eV and that higher clusters
ionize below 3.49 eV. These less rigorous results are also consistent
with our work. We can even predict the ionization potentials of the
higher member of the series; next we plan to use path integral
techniques to study the higher members of this series and approach the
liquid limit. It will obviously require clusters of at least 12-18
water molecules. Since the first coordination layer contains about four
waters, we expect the ionization potential to change more slower when
more waters are added.
      In regard to stability we know that typical hydrogen bonds are
about 5 kcal/mole at the Hartree Fock level. This means that the
observed stability of the four water sodium cluster (30.7 kcal/mole) is
larger than that of the most stable tetramer cluster of pure water
(20.53 kcal/mole by Kistenmacher, et. al.[7] or 26.34 kcal by Vernon, et.
al.[14] using the Watts potential). Similar statements can even be made
about the dimer and trimer, although the stability seems to be
relatively larger for the tetrameric water species. The stability could
increase slightly if the coordination number of sodium can increase a
little more.

3.2 Radiation Chemistry Data

Our studies are also of interest for their application to the work of
Telser and Schindewolf[5]. They have found evidence in highly alkaline
water solutions for an intermediate during the radiolysis of water which
they think involved excited states of neutral hydrated sodium. We have
attempted to calculate the excited state directly but our results are
still incomplete. However, since we find that even four waters decrease
the orbital energy (and the ionization potential) of sodium by over 50%
(2.445/4.951 is .49), we expect a similar reduction in the optical
absorption line. Using that simple idea on the sodium-D line yields a
transition of 1.0 eV for the four water system and probably lower in
liquid water. Telser and Schindewolf find light absorption at 270 nm or
4.59 eV. This would have to be a transition from the sodium to some
state in the water or to a CTTS state (charge transfer to solvent
state). There are many transitions available and since the electron is
so diffuse with a lot of density on the oxygens, one would expect a
substantial oscillator strength also. However the simple shift of the s
to p transition in atomic sodium is not consistent with the band found

in the experiment. Further studies are needed on some of the higher
excited states which are candidates for this transition.

4.0   CONCLUSION

We have presented calculations for sodium atoms surrounded by one to
four water molecules. These results are in good agreement with the
latest beam data as regards ionization potential and stability. Further
work is in progress to obtain their excitation spectra and to study
their role in the radiation chemistry of highly alkaline solutions.
      (This work is supported by DOE GRANT: DOE-FG05-87ER13674)

REFERENCES

  1. J. Bentley, *J. Am. Chem. Soc.* **104**, 2754-2759 (1982).

  2. M. Ternary, H.F. Schaeffer III, and P.A. Kollman, *J. Am. Chem. Soc.*
     99, 3885-3886 (1977).

  3. J. Bentley and I. Carmichael, *J. Phys. Chem.* **85**, 3821-3826 (1981).

  4. C.P. Schulz, R. Haugstaetter, H.U. Tittes, and I.V. Hertel, *Phys.
     Rev. Letters* **57**, 1703-1706 (1986).

  5. Th. Telse and U. Schindewolf, *J. Phys. Chem.* **90**, 5378-5382 (1986).

  6. Gaussian 82 by J.S. Birkley, R.A. Whiteside, K. Raghavachari, R.
     Seeger, D.J. DeFrees, H.B. Schlegel, M.J. Frisch, J.A. Pople and
     L.R. Kahn. Copyright to Carneige Mellon University, USA

  7. H. Kistenmacher, G.C. Lie, H. Popkie, and E. Clementi, *J. Chem.
     Phys.* **61**, 546 (1974).

  8. L.A. Curtiss, E. Kraka,J. Gauss, and D. Cremer, *J. Phys. Chem.* **91**,
     1080-1084 (1987).

  9. E. Clementi, R. Brosotti, J. Fromm, and R.O. Watts, *Theor. Chim.
     Acta* **43**, 101 (1976).
 10. J. Fromm, E. Clementi, and R.O. Watts, *J. Chem. Phys.* **62**, 1388
     (1975).

 11. T.P. Lybrand and P.A. Kollman, *J. Chem. Phys.* **83**, 2923 (1985).

 12. I. Dzidic and P. Kebarle, *J. Phys. Chem.* **74**, 1466 (1970); M.Arshadi,
     R. Yamdagni, and P. Kebarle, *J. Phys. Chem.* **74**, 1475(1970).

 13. W.L. Jorgensen, *J. Am. Chem. Soc.* **103**, 335 (1981).

 14. M.F. Vernon, D.J. Krajnovich, H.S. Kwok, J.M. Lisy, Y.R. Shen, and
     Y.T. Lee, *J. Chem. Phys.* **77**, 47 (1982).

ISOMERIZATION IN ALKALI-HALIDE CLUSTERS INDUCED BY ELECTRON ATTACHMENT

Dafna Scharf[†], Uzi Landman[‡] and Joshua Jortner[†]
[†]School of Chemistry, Tel-Aviv University
69978 Tel-Aviv, Israel
[‡]School of Physics, Georgia Institute of Technology
Atlanta, Georgia 30332, U.S.A.

ABSTRACT. The mechanism of cluster isomerization induced by electron attachment to Na_4Cl_4 clusters was explored by the constant temperature quantum path integral molecular dynamics method over the temperature range 50°K-1200°K. The configurational modifications can be related to the role of an excess electron as a pseudonegative ion and to the effect of partial neutralization of the alkali ion(s) by the excess electron which induce isomerization at moderately low temperatures in comparison with the parent neutral Na_4Cl_4 cluster.

1. INTRODUCTION

The nature of phase transformations in finite systems is a subject of considerable conceptual and practical interest[1-4]. Characteristic to small clusters is the possibility of structural isomerizations between several distinct configurations, while in large clusters melting is exhibited, which is manifested by coexistence between ordered and disordered structures. An alternative approach to the interesting problem of configurational changes in clusters involves cluster isomerization induced by non-reactive electron attachment[5,6]. We have recently explored electron attachment to the small, neutral Na_4Cl_4 cluster over a broad temperature range (50°K-1200°K), which yielded the following information:
(1) Structures, energetics and excess electron charge distribution of isomers induced by electron attachment.
(2) The nature of configurational modifications in $Na_4Cl_4^-$.
(3) A comparison between thermal isomerization and isomerization induced by electron attachment to the same cluster.
(4) Symmetry breaking effects pertaining to the nuclear configuration and to the excess electron charge distribution induced by electron attachment to a cluster at high temperatures.
    Alkali halide clusters (AHC) were chosen for this study because of several reasons. Extensive information is available on the nature of the various interactions in AHCs [7-11]. The interionic interactions are well understood and extensive model calculations were performed for

217

J. Jortner et al. (eds.), Large Finite Systems, 217–229.

neutral and charged AHCs[8]. Reliable information is also available on
an adequate pseudopotential to describe the electron alkali cation interaction[10],[11], which was further studied by us[12]. We have conducted our
study of electron attachment to a Na_4Cl_4 cluster by employing the
constant temperature quantum-path-integral moleculuc-dynamics (QUPID)
method. This approach rests on a discrete version of Feynman's path
integral method[13-15],[11],[5] and provides a powerful method for simulation
studies of excess electron states in clusters.

2. QUANTUM PATH INTEGRAL MOLECULAR DYNAMICS METHOD

In the path-integral formulation of quantum statistical mechanics the
problem of calculating the partition function for the quantum system is
mapped unto an isomorphic classical problem wherein the quantum component (in our case the excess electron) is represented by a closed necklace of P particles (beads). This provides a discretization of the path-
integral formula, which becomes exact for $p \to \infty$.

The excess electron interaction with the AHC of N ions is described
by the Hamiltonian

$$H = - \frac{\hbar^2}{2m} \nabla^2 + V_e(\vec{r}) \tag{2.1}$$

where m is the mass of the electron, and $V_e(r)$ is the electron-AHC
potential. The energy of the system is given by

$$E = - \frac{\partial}{\partial \beta} \ln Z \tag{2.2}$$

where the partition function is given by

$$Z = Tr[\exp(-\beta H)] \tag{2.3}$$

and $\beta = 1/kT$ is the inverse temperature. Invoking a small (β/P) expansion and factorization of the partition function into P discrete
contributions, establishes the isomorphism. The expression for the
partition function is then

$$Z \simeq \left(\frac{Pm}{2\pi\hbar^2\beta}\right)^{3P/2} \int d^3r_1 \ldots d^3r_p \exp[-\beta\, V_{eff}(\vec{r}_1 \ldots \vec{r}_p)], \tag{2.4}$$

with the effective potential, V_{eff} being

$$V_{eff} = \sum_{i=1}^{P} \left[\frac{Pm}{2\hbar^2\beta^2} (\vec{r}_i - \vec{r}_{i+1})^2 + \frac{1}{P} V_e(\vec{r}_i)\right] . \tag{2.5}$$

The effective excess electron potential (2.5) consists of a superposition of nearest neighbor harmonic interactions and the excess electron-
AHC interactions. The electron-AHC interaction consists of a sum of
electron-ion interactions

$$V_e = \sum_{I=1}^{N} \phi_{eI}(r-R_I) \quad , \tag{2.6}$$

which are described by:

(a)   a Coulomb repulsive interaction from a closed shell anion of radius R_A, complemented by the electron induced polarization (α_A) interaction operative at distances larger than R_A and further corrected for the "self energy" of the induced dipole

$$\phi_{eI} = \begin{cases} e^2/r - \dfrac{e^2 \alpha_A}{2r^4} \quad ; \quad r > R_A \\[2mm] e^2/r \quad\quad\quad ; \quad r \leqslant R_A \end{cases} \quad , \tag{2.7}$$

(b)   a local model pseudo-potential for the electron-cation interaction

$$\phi_{eI} = \begin{cases} -e^2/R_c \quad ; \quad r \leqslant R_c \\[2mm] -e^2/r \quad\;\; ; \quad r \geqslant R_c \end{cases} \tag{2.8}$$

where R_c is a cutoff radius, $R_c = 3.24a_0$[9,10,12].

The AHC is treated as a classical system consisting of N ions. The interionic potential within the AHC is taken to be a sum of pair interactions

$$V_{AHC} = \frac{1}{2} \sum_{I \neq J} \sum \phi_{IJ}(R_{IJ}) \tag{2.9}$$

with the interionic potential being given by the Born-Mayer potential as parametrized   by Fumi and Tosi[7], and a Coulomb contribution.

The average energy of the electron-AHC system is now given by the expression

$$QE = \frac{3}{2}NkT + <V_{AHC}> + KE + \frac{1}{P} < \sum_{i=1}^{P} V_e(\vec{r}_i) > \tag{2.10}$$

where $< >$ indicate statistical averages over the Boltzmann distribution as defined in Eq. (2.4). The first two terms on the r.h.s. of Eq. (2.10) correspond to the average kinetic and potential energies of the ionic AHC system. The last term in Eq. (2.10) is the electronic potential energy and the electron kinetic energy, KE, is given by[11]

$$KE = K_{free} + K_{int} = \frac{3}{2}kT + \frac{1}{2P} \sum_{i=1}^{P} <(\partial V(\vec{r}_i)/\partial \vec{r}_i) \cdot (\vec{r}_i - \vec{r}_P) \quad . \tag{2.11}$$

The kinetic energy of the excess electron consists of two contributions: the free particle term and a contribution from the interaction of the excess electron with the ions of the AHC.

The averages over the Boltzmann distribution, Eq. (2.10) and (2.11), can be replaced by averages over phase space trajectories generated by the following Hamiltonian

$$H = \sum_{i=1}^{P} \frac{m^* \dot{\vec{r}}_i^2}{2} + \sum_{I=1}^{N} \frac{M_I \dot{\vec{R}}_I^2}{2} + \sum_{i=1}^{P} [\frac{Pm}{2\hbar^2 \beta^2} (\vec{r}_i - \vec{r}_{i+1})^2 + \frac{1}{P} V_e] + V_{AHC} ,$$

(2.12)

where the mass m^* is arbitrary, M_I are the corresponding masses of the ions and m the electron mass. In our calculations we chose $m^* = 1$ a.m.u. We have utilized a constant temperature version of MD where the particles were subject to stochastic collisions. The equations of motion were integrated using the velocity form of the Verlet algorithm.

In the initial configuration of the system the electron bead particles were randomly distributed over a sphere of radius R_s centered about the center of mass of the AHC. R_s was chosen to be of the order of the thermal wavelength of a free electron at T $\sim$ 1000°K (R_s was about 15 times the crystalline lattice constant). The number of beads P was varied from P = 1000 at 50°K down to P = 520 at 575°K and 750°K and P = 250 at 1000°K.

In a number of independent runs, each corresponding to one particular temperature, the electron beads slowly approached the AHC while being frequently subjected to rescaling of velocities using a fifth-order predictor corrector integration algorithm, and integration step $\Delta t = 0.01$ time unit (t.u.), where 1t.u. = 1.03 x 10^{-15} sec. Subsequently, the cluster was equilibrated to the corresponding temperature applying the constant temperature procedure with Verlet algorithm and typical values of stochastic collision frequencies of 1 x 10^{-3} - 1 x 10^{-4}. A typical equilibration run was about 1 x 10^5 t.u. Averaging was then performed over the subsequent 5 x 10^4 - 1 x 10^5 t.u.

3.   CLUSTER ISOMERIZATION INDUCED BY ELECTRON ATTACHMENT

We have investigated the equilibrium structure of an electron attached to a cluster of Na_4Cl_4 at four temperatures: 50°K, 575°K, 750°K and 1000°K. We have found that different equilibrium configurations are exhibited for $Na_4Cl_4^-$ as a function of the temperature domain [Fig. 1-3]. Three temperature domains can be distinguished:
(a)   The low temperature domain $0° < T < 500°K$. From calculations at 50°K and 375°K we find that the localized state of the excess electron in this low temperature range is exhibited mainly inside the cubic equilibrium configuration of the ions (Fig. 1a). The ionic configuration highly resembles the low-temperature ionic configuration of the neutral Na_4Cl_4 cluster (Fig. 4a), thus no major configurational changes are exhibited in this temperature range.
(b)   The intermediate temperature domain $500°K < T < 750°K$. A configurational change is exhibited. At a temperature of $\sim$575°K a single configuration of the $Na_4Cl_4^-$ cluster prevails which resembles an open umbrella [Fig. 1(b)], with the excess electron bead at one of its corners. The ionic cluster corresponds to a distorted planar configuration.
(c)   The high temperature domain $750°K < T < 1200°K$. Coexistence

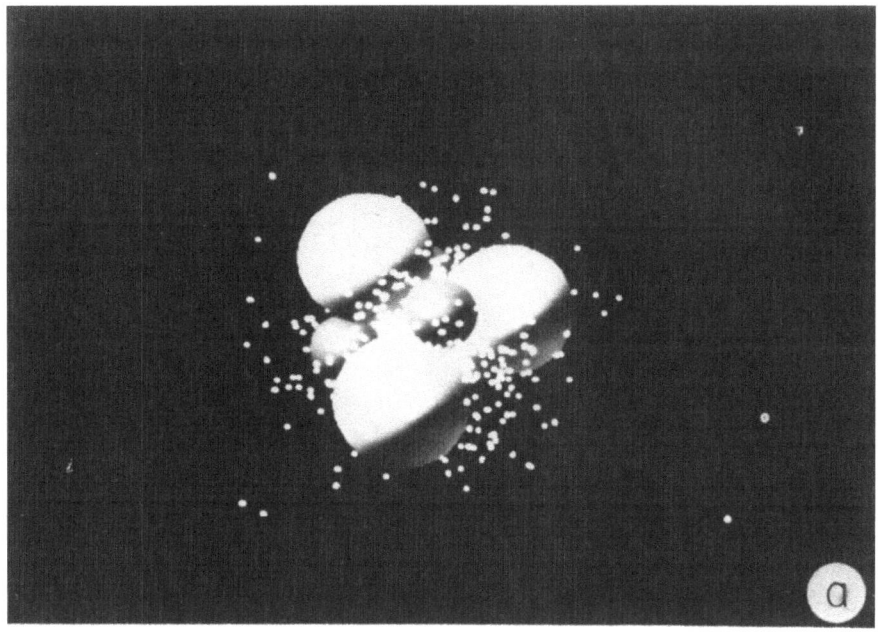

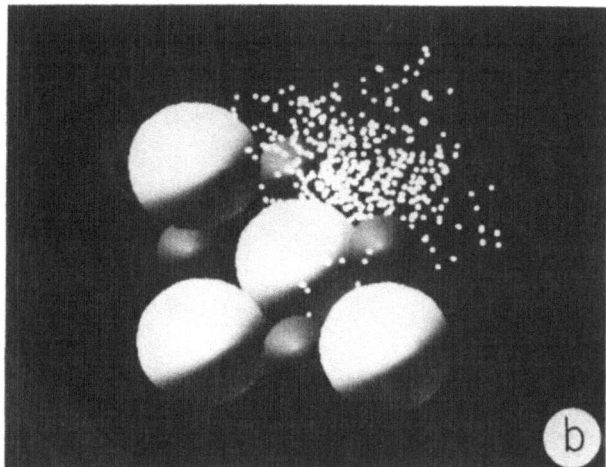

Figure 1: Configurations of $Na_4Cl_4 + e^-$ at 50°K (a) and 575°K (b). The large bright spheres and small dark spheres represent the Cl^- and Na^+ ions (the sizes were scales according to the corresponding ionic radii). The small dots are the locations of the quantum pseudo particles (beads), representing the excess electron charge distribution. Note in the configuration shown in (b) the excess electron occupies the position of a halide anion in the $(Na_4Cl_4)Cl^-$ cluster, (see Fig. 5).

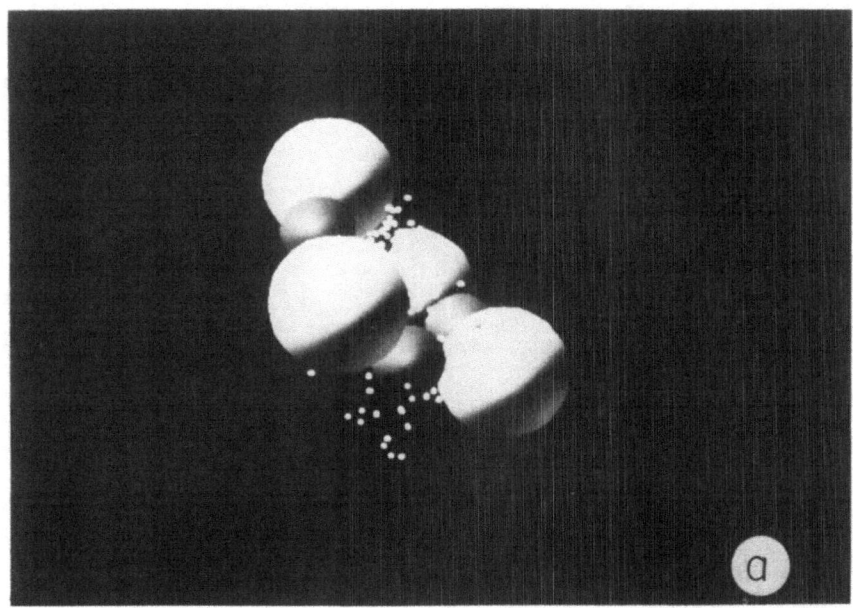

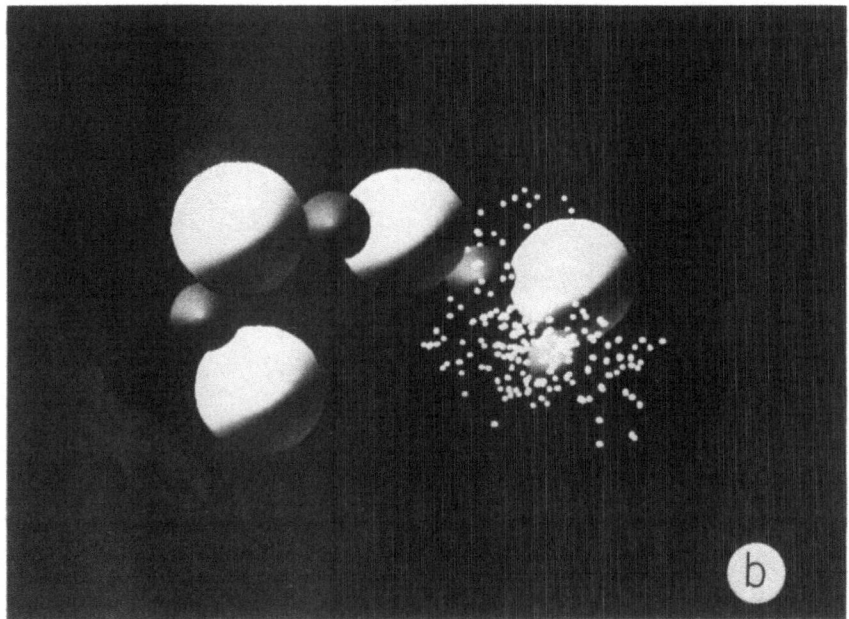

Figure 2: Configurations of $Na_4Cl_4 + e^-$ at 750°K. For details see caption of Figure 1.

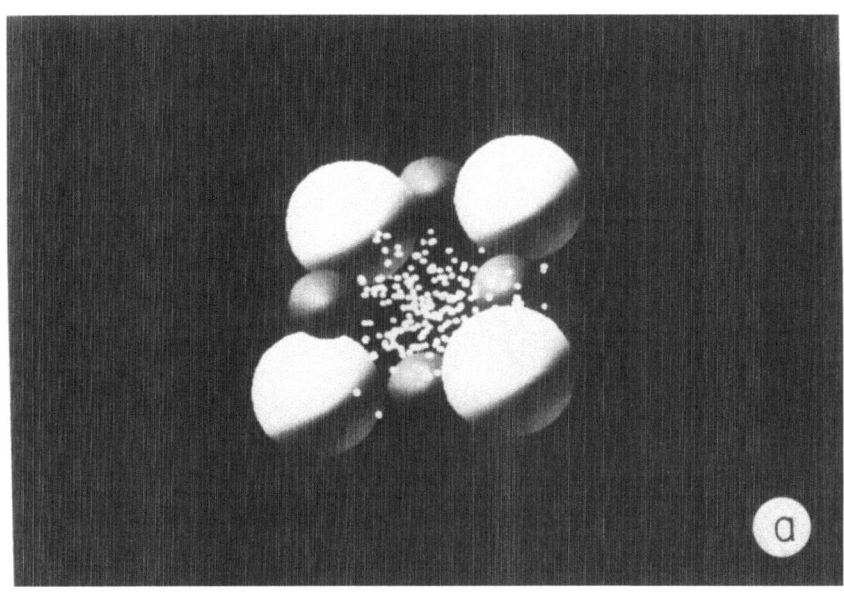

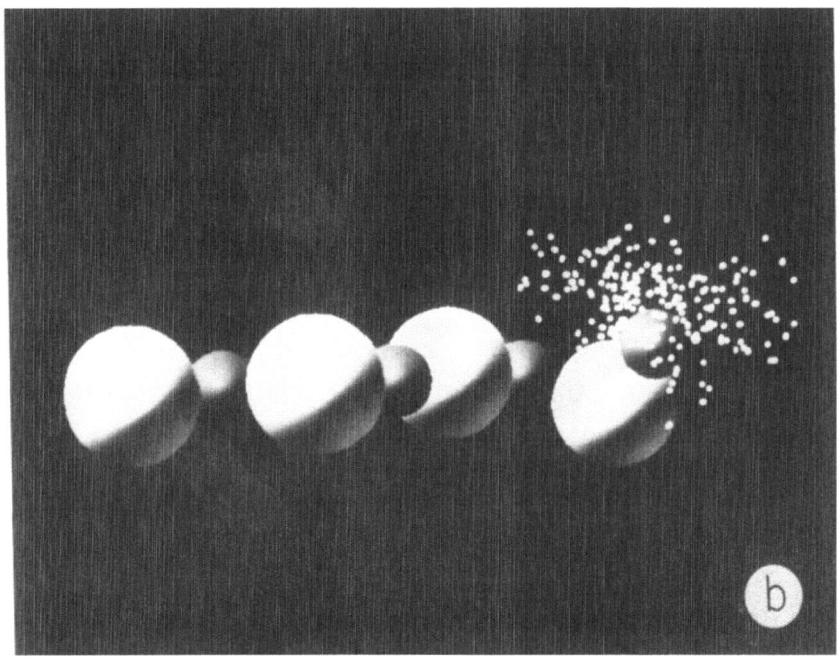

Figure 3: Configurations of $Na_4Cl_4 + e^-$ at 1000°K. For details see caption of Figure 1.

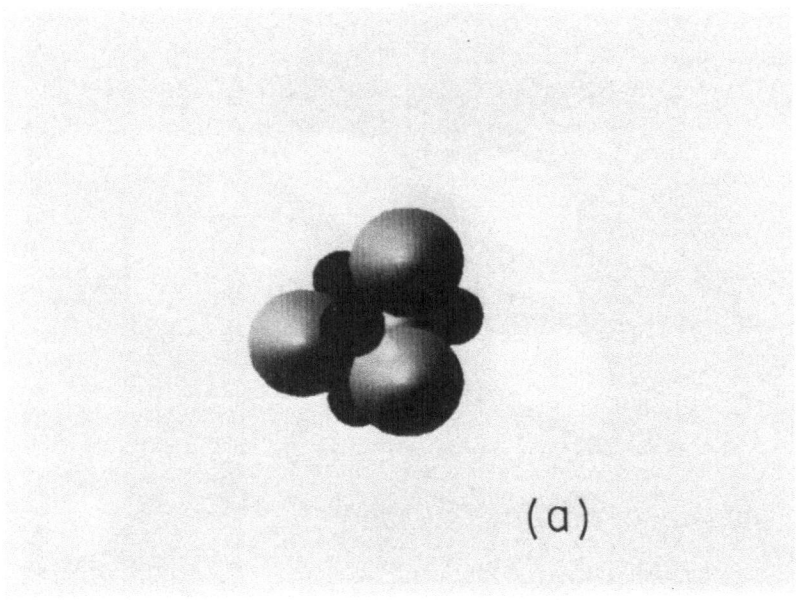

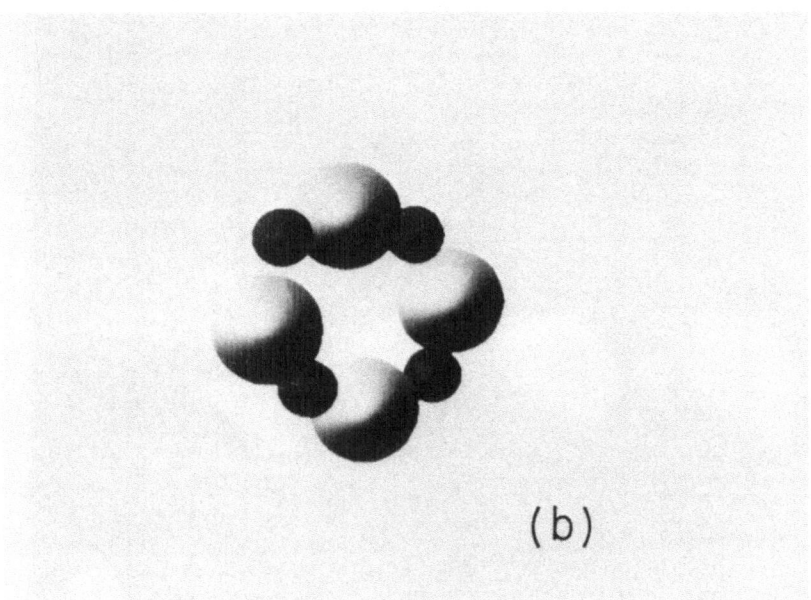

Figure 4: Configurations of Na_4Cl_4 at T<700°K (a) and at T>700°K (b).
Note the similarity between the cubic configuration in (a) and the low
temperature configuration of $Na_4Cl_4^-$ (Fig. 1(a)).

between isomers prevails. At 750°K a boat like configuration with the
electron beads in the center (Fig. 2) and a bent chain isomer are
encountered in coexistence. At T = 1000°K, an elongated chain isomer
is in coexistence with a planar ring configuration (Fig. 3).

3.1 Cluster Configurations

Electron attachment induced configurational changes in the small AHC at
relatively low temperatures. The ionic equilibrium configurations of
the electron–AHC do not necessarily correspond to the corresponding
equilibrium configurations of the neutral AHC. A molecular dynamics
(MD) study of the neutral Na_4Cl_4 cluster[4] showed that below T = 700°K
only a single configuration is stable namely a cubic one (Fig. 4). In
the temperature range 700°K < T < 875°K, a cubic and a ring isomer coexist, while at 875°K < T < 1250°K only the ring configuration is
stable. Above T = 1250°K an open chain isomer is in coexistence with
the ring and for T > 1400°K the open chain is the only stable configuration. The presence of the electron induces two types of configurational modifications in the $Na_4Cl_4^-$, different from those in the neutral
Na_4Cl_4 cluster: (i) the localized electron can assume the role
of a negative ion thus leading to a new nuclear configurations of the
ions which have no counterpart in the neutral cluster. We have found
that the new configuration which is exhibited at T ≈ 575°K has
similar counterpart exhibited in the MD study of $Na_4Cl_5^-$ over a tempera-
rature range 50°K ≤ T ≤ 1300°K (Fig. 5). (ii) partial neutralization of
the cations enables the appearance of high temperature configurations of
the parent neutral cluster at substantially lower temperatures for $Na_4Cl_4^-$.
This effect is found at all temperature domains.

3.2 Energetics

From a comparison between the energies of the ions in the neutral Na_4Cl_4
cluster and in the charged $Na_4Cl_4^-$ cluster (Table I) it is apparent that
E_{ion} for the charged cluster is substantially higher than the corresponding energy in the neutral cluster at all temperatures. This
overall increase in the ionic energies can be attributed to two
effects: (a) the inter–ionic binding interactions in $Na_4Cl_4^-$ decrease
relative to Na_4Cl_4 due to partial neutralization of the positive charges
by the excess electron. (b) The modification of the cubic structure at
low temperatures results in major modifications of the ionic potential
energy upon electron attachment. The configurational modifications
of the ionic structure in $Na_4Cl_4^-$ relative to that of the Na_4Cl_4 are
the source of the reorganization energy, E_c.
       The electron affinity, EA, of Na_4Cl_4 can be estimated from the low-
temperature (50°K) data

$$EA = QE - E_{ions} (Na_4Cl_4) \ . \tag{3.1}$$

We have found EA = (0.8 ± 0.2) eV. This value is considerably lower than
the EA of positively charged alkali–halide clusters[5].

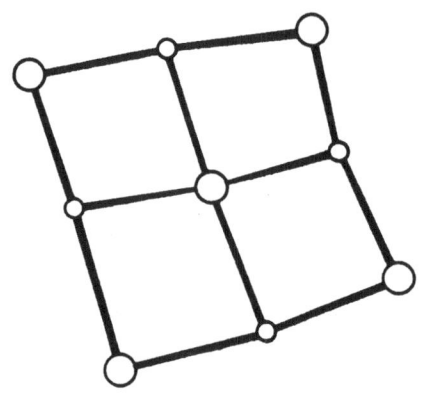

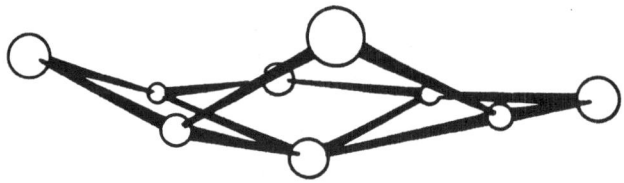

Figure 5: The stable, umbrella shaped configuration of $(Na_4Cl_4)Cl^-$, obtained via simulations in the temperature range 50° to 1000°K, shown from two different perspectives. Small and large balls, Na^+ and Cl^-.

The cluster reorganization energy is the difference between the energies of the ionic system in the neutral and the charged clusters

$$E_c = E_{ions} \ (Na_4Cl_4) - E_{ions} \ (Na_4Cl_4^-) \quad . \qquad (3.2)$$

E_c increases with increasing temperature from $E_c \sim 1.5eV$ at 50°K to $E_c \sim 3eV$ at 1000°K. The increase in the total energy QE as a function of temperature (Table I) reflects the increase of E_c.

3.3 Excess Electron Charge Distribution

The variations in the charge distribution associated with the excess electron as a function of temperature can be demonstrated by examining the following quantities (table II):
(a) The average distance between the center-of-mass (CM) of the electron necklace and the CM of the ions

TABLE I

Energies (in Hartrees) of the neutral and of the charged clusters. The energy of the ionic system, E_{ions}, corresponds to the first two terms in Eq. (2.10) for $Na_4Cl_4^-$ and the total energy of Na_4Cl_4. QE is the total energy of the cluster $Na_4Cl_4^-$. The standard deviations are shown in parentheses.

	50°K	575°K	750°K	1000°K
Na_4Cl_4				
E_{ions}	-1.02440	-1.01008	-1.00496	-0.9976
	(1×10^{-6})	(1×10^{-6})	(1×10^{-6})	(1×10^{-6})
$Na_4Cl_4^-$				
E_{ions}	-0.977	-0.93	-0.91	-0.87
	(6×10^{-3})	(1×10^{-2})	(2×10^{-2})	(2×10^{-2})
QE	-1.053	-1.00	-0.99	-0.96
	(8×10^{-3})	(2×10^{-2})	(2×10^{-2})	(3×10^{-2})

$$D_{CM} = < \left| \frac{1}{P} \sum_{i=1}^{P} \vec{r}_i - \frac{\sum_I M_I \vec{R}_I}{\sum_I M_I} \right| > \tag{3.3}$$

(b) The quantum character of the excess electron is defined by

$$QC = K_{int}/(K_{int} + K_{free}) \quad , \tag{3.4}$$

where K_{int} and K_{free} were defined in Eq. (2.11).
(c) The radius of gyration of the electron bead distribution which is given by the second moment with respect to the CM

$$R_g = [\frac{1}{2P^2} < \sum_{i,j=1}^{P} (\vec{r}_i - \vec{r}_j)^2 >]^{\frac{1}{2}} \tag{3.5}$$

TABLE II

Characterization of the excess electron in $Na_4Cl_4^-$: D_{CM} is the distance between the CM of the electron and the CM of the ions; QC is the fraction of the quantum contribution to the kinetic energy of the excess electron; R_g is the gyration radius of the electron necklace. All distances are in atomic units.

T	50°K	575°K	750°K	1000°K
D_{CM}	0.4 ± 0.2	6 ± 1	5 ± 2	4 ± 3
QC	99%	95%	93%	90%
R_g	4.2 ± 0.2	4.9 ± 0.5	4.8 ± 0.5	4.5 ± 0.6

The substantial configurational changes which occur upon a transition from the low temperature domain to the intermediate temperature domain in $Na_4Cl_4^-$, are well reflected in an abrupt change in D_{CM}. While at low temperatures, the CM of the electron necklace almost coincide with the CM of the ions, already at 575°K the corresponding two CMs are well separated in the distorted planar umbrella-like configuration. The small standard deviations in D_{CM} at 575°K suggests the presence of a single, well defined, configuration. In the high temperature domain, the high standard deviations in D_{CM} correspond to a mixture of more than one isomer, as we indeed find.

As expected, the quantum character of the excess electron is pronounced at all temperatures, slightly decreasing at higher temperatures. The gyration radius which estimates the spatial extent of the wave packet of the excess electron was almost constant within the calculation error.

The variations of the excess electron charge distribution reflect a "transition" from a relatively extended electronic state at low temperatures (T=50°K) where 75% of the electron beads reside within the cut-off radius R_c of four Na^+ cations, which are equally partially neutralized, to a localized electronic state at high temperatures where 40% of the electron beads reside within R_c of one (or two) Na^+ cation. At low temperatures the mode of electron localization is determined by the neutral cluster symmetry, while symmetry breaking effects induced by large amplitude motion and configurational changes prevail at high temperatures, inducing electron localization in the vicinity of a single Na^+ cation.

ACKNOWLEDGEMENTS

This research was supported by the U.S. DOE under Grant No. FG05-86ER45234 (to U.L.) and by Grant No. 85-00361 of the US-Israel Binational Science Foundation, Jerusalem (to J.J. and U.L.).

REFERENCES

1. G. Natanson, F. Amar and R.S. Berry, J. Chem. Phys. 78 399 (1983).
2. J. Jellinek, F. Amar and R.S. Berry, J. Chem. Phys. 84 2783 (1986).
3. R.W. Hockney and J.W. Eastwood, Computer Simulations Using Particles, (Mc.Graw-Hill, N.Y., 1981) pp 488-498.
4. J. Luo, U. Landman and J. Jortner, "Proceedings of the International Workshop on the Physics and Chemistry of Small Clusters" Eds. R. Khana and P. Jena (Plenum, N.Y., 1987).
5. U. Landman, D. Scharf and J. Jortner, Phys. Rev. Letts. 54 1860 (1985).
6. D. Scharf, U. Landman and J. Jortner, Submitted to J. Chem. Phys.
7. F.G. Fumi and M.P. Tosi, J. Phys. Chem. Solids 25 31, 45 (1964).
8. T. P. Martin, Phys. Rep. 95 167 (1983).
9. R.W. Shaw, Phys. Rev. 174, 769 (1968).
10. J.V. Abarenkov and V. Heine, Phil. Mag. 12, 529 (1965).

11. M. Parrinello and A. Rahman, J. Chem. Phys. $\underline{80}$, 860 (1984).
12. D. Scharf, J. Jortner and U. Landman, Chem. Phys. Letts. $\underline{130}$, 504 (1986).
13. R.P. Feynman and A.R. Hibbs, "Quantum Mechanics and Path Integrals" (Mc.Graw-Hill, N.Y., 1965).
14. L.S. Schulman, "Techniques and Applications of Path Integrals" (Wiley, N.Y., 1981).
15. D. Chandler and P.G. Wolynes, J. Chem. Phys. $\underline{79}$ 4078 (1981).

ELECTRON LOCALIZATION BY ATOMIC AND MOLECULAR CLUSTERS

D. Thirumalai
Institute for Physical Science and Technology
and Department of Chemistry and Biochemistry
University of Maryland, College Park, MD 20742

ABSTRACT. A simple theoretical model is used to predict the minimum number of atoms needed to localize an excess electron by heavy rare gas atoms. The minimum number of atoms, N_{min}, depends on the scattering length, the polarizability of the atom and the temperature. At T=20k, N_{min} for Xe turns out to be about 55 and for Hg it is between 5-7. The latter number is in reasonable agreement with experiment. In addition, a review of path integral Monte Carlo calculation of electron localization in water clusters is discussed. It is shown that two water molecules can localize an excess electron at T=5k in agreement with the experimental work.

INTRODUCTION

There has been considerable interest in the study of equilibrium and dynamic properties of finite sized atomic and molecular aggregrates which shall be referred to as clusters.[1] Apart from providing an obvious link between one's understanding of gas phase processes and condensed phase phenomena the behavior of clusters (containing 2-100 units) may exhibit properties not otherwise observed. Furthermore, physical clustering is viewed as an important mechanism for the occurrence of several interesting phenomena in the condensed phase. The dielectric anomaly in mercury (the so-called Marburg anomaly) at densities below the insulator-metal transition density,[2] electron localization in simple polarizable fluids[3] are just a couple of examples where clustering is thought to play a dominant role. The common aspect in the two examples cited is that a cluster of atoms is assumed to localize an electron and the behavior observed in these systems (like the dielectric constant or the mobility) is a direct consequence of this process. The purpose of this paper is to review two approaches to investigate the possibility of electron localization by atomic and molecular clusters.

Recently several theoretical studies have appeared reporting the binding of an electron by finite sized systems.[4-6] The possibility

231

J. Jortner et al. (eds.), Large Finite Systems, 231–240.
© 1987 *by D. Reidel Publishing Company.*

of observing this phenomena in molecular beam experiments has provided the major impetus for these investigations. Because of the
nonzero beam temperatures it proves to be less convenient to perform
the numerical calculations using traditional quantum mechanical
methods. Alternative methods employed in quantum statistical
mechanics have proved quite useful in probing the possibility of
electron localization in finite size systems. In this short report
we focus on two problems: (a) the formation of electron clusters by
heavy rare gas atoms using a theory due to Iakubov and Khrapak.[7] A
simple scattering mechanism combined with the notion of potential
fluctuations experienced by the electron due to the thermal motion of
the atoms allows for a simple calculation that can be used to predict
qualitatively the number of atoms needed to bind an electron. The
parameters that enter into the theory are two experimentally known
quantities, namely the scattering length, a, and the atomic polarizability α. This is presented in Section II. (b) Motivated by recent
experiments by Haberland[8] and coworkers on electron attachment to
water clusters, we performed a path integral Monte Carlo calculation
to investigate the smallest number of water molecules to localize an
electron. The calculations are based on pseudopotentials to characterize the electron water interaction. A modified central force
potential is used for the interaction betweem two water molecules.
The path integral Monte Carlo calculation and the results are presented in Section III. The paper is concluded with a discussion in
Section IV.

     Before concluding this introduction we comment on the meaning of
localization of an electron by finite sized systems. The detection
of such a species is most easily done by injecting low energy electrons into a supersonic expansion of the gas (be it rare gas or
molecular systems). The species $e^- - (A)_n$ is then detected by mass
spectrometric techniques. Thus only those species are considered
stable that live for at least the time required to reach the detector. Typically this is about a microsecond. Experiments cannot
rule out the possibility that these species can be metastable with
extremely long life times. In the calculations we make the important
assumption that this time scale is much longer than any microscopic
relaxation times (such as electronic, vibrational or rotational) so
that the solvated species can be viewed as a well "equilibrated"
thermodynamic system. This assumption[9] allows us to discuss the stability of these systems using equilibrium statistical mechanics.
Thus the theories described here cannot account for apparent
stability on a microsecond time scale due to trapping in a metastable
well or due to dynamic fluctuations.

II. LOCALIZATION OF AN ELECTRON BY RARE GAS CLUSTERS

     The well known experiments[10] on the density dependence of the
mobility of an excess in dense He gas have been explained in terms of
"bubble" formation. This is essentially due to the effective e^- –He
repulsion that dominates the scattering mechanism. On the other hand
recent path integral Monte Carlo calculations of the equilibrium

properties of an electron in Xe[11] suggest that clustering may play a
role in determining the zero field mobility at least at low densities. Thus it is of interest to determine if a free cluster of heavy
rare gas atoms can localize an electron and if so what is the minimum
size of the cluster λ_{min}, that is needed to induce binding? In the
following a simple theory based on fluctuation phenomena is presented
which predicts that under appropriate conditions cluster of noble gas
atoms Ar, Kr, and Xe as well as highly polarizable atoms like Hg can
indeed localize an electron. This theory provides only an estimate
of λ_{min}. In order to obtain accurate values for λ_{min} one has to
appeal to experiments as well as quantum statistical mechanical
calculations.

The contents of the simple theory can be outlined as follows.
Because of electron attachment there is a decrease in binding energy
given by

$$\Delta E = k + u$$

$$= \langle\psi|p^2/2m|\psi\rangle + \langle\psi|\sum_{i=1}^{N} V(r,R_i)|\psi\rangle \qquad (II-1)$$

where ψ is an "average" localized wavefunction,[12] p is the momentum
of the electron, $V(r,R_i)$ is the interaction potential between the
electron and the ith atom and R_i is the coordinate of the ith
electron. The entropy change is estimated by making the following
assumptions (a) Because the system is at a finite temperature the
positions of atoms in the cluster fluctuate. (b) For a given
realization, which specifies the positions of atoms, the eletron is
assumed to experience a specified value of u. We implicitly assume
that the time scale of translational motion of atoms is much smaller
than that of the electron, i.e., the Bern-Oppenheimer approximation
is valid. (c) The gain in potential u is a random variable which
follows from assumptions (a) and (b). It is assumed that the
distribution of u is a Gaussian, $P(u) \sim exp(-u^2/2\langle u^2\rangle)$. This is not
reasonable as the Gaussian behavior is expected to be valued only in
an infinite system where only minor deviations from the most probable
value is expected. (d) Finally, the correlations between the atoms
is neglected which is also difficult to justify given the strong
interaction between the electron and the atom. We discuss this
further below.

Using the assumptions (a)-(d) the change in free energy can be
written as[13]

$$\Delta F(u) = k + u + u^2/2\langle u^2\rangle\beta \qquad . \qquad (II-2)$$

Minimizing $\Delta F(u)$ with respect to u yields,

$$\Delta F(u) = k - \beta \langle u^2 \rangle / 2 \quad . \tag{II-3}$$

The mean square fluctuations can be expressed in terms of the effective potential (due to scattering process) and the "average" localized wavefunction as

$$\langle u^2 \rangle = \int \frac{dq}{(2\pi)^3} V_{eff}^2(q) \, \psi^2(q) \quad . \tag{II-4}$$

Because we are interested in low densities, localization is induced by large density fluctuations. Thus the scattering is essentially dominated by large electron wavelength, λ and consequently in evaluating Eq. (II-4) only $q \lesssim \lambda$ need be considered. The effective potential in this case is expressible in terms of the scattering amplitude,[14] i.e.

$$V_{eff}(q) = \frac{2\pi h^2}{m} f(q) \tag{II-5}$$

using Eq. (II-5), $\beta \Delta F$ can be approximated by

$$\beta \Delta F(\lambda) \sim 4 \left(\frac{2\pi \lambda_T}{\lambda} \right)^2 \left[1 - \frac{4Nf^2(\lambda^{-1})}{\lambda} \lambda_T^2 \right] \tag{II-6}$$

where $\lambda_T^2 = \beta h^2 / 2m$. The minimum size of the cluster which localizes the electron is found by solving $\beta \Delta F(\lambda) = 0$. One can do this by assuming that $f(\lambda^{-1})$ is given by[14]

$$f(\lambda^{-1}) = a + \alpha \lambda^{-1} / 4a_0 \tag{II-7}$$

where a is the scattering length. Setting $\Delta F(\lambda_{min})$ one obtains the following condition for the value of N_{min} when localized states dominate,

$$N_{min} \simeq 2\alpha / \lambda_T^2 |a|^3 a_0 \tag{II-8}$$

At T=20k using the parameters for Xe ($\alpha = 27.2 \, a_0^3$, $a = -7.1 \, a_0$) the size of the cluster is predicted to be about 20 Å. A cluster this size contains approximately 55 atoms. Similar calculation for Hg ($\alpha = 45 \, a_0^3$, $a = 1.72 \, a_0$) yields a value of $N_{min} \sim 5-7$.[15]
     We now present an argument that suggests that N_{min} given by Eq. (II-8) should be an upperbound to the expected (experimentally) number of atoms that will localize (exponentially) an electron. One of the most important effects that has not been considered here is

the manybody polarization effects, i.e., the electron induces a dipole moment on each atom which is proportional to α. The induced moments interact among themselves and their precise effect has to be determined self consistently. The procedure for calculating this effect has been given elsewhere[15] and adopting that argument here suggests that the effective potential experienced by the electron becomes more attractive and consequently N_{min} should become lesser when the manybody polarization effects are taken into account. This is particularly important for highly polarizable atoms like Xe and Hg.[16]

III. ELECTRON BINDING TO WATER CLUSTERS

In recent experiments Haberland et al.[8,17] have examined the stability of $e^- - (H_2O)_n$ systems as a function of n the number of water molecules comprising the cluster. The initial experiments suggested that the electron-water cluster is stable only when $n \geqslant 11$.[8] However, when the beam was seeded with Ar atoms, which effectively lowers the beam temperature it was found that two water molecules can bind an electron.[17] In this section a review of our work[5,6] which explores the thermodynamic conditions under which $e^- - (H_2O)_n$ (for n=2 and 3) is stable is presented.

The Hamiltonian for the excess electron interacting with N water molecules can be written as

$$H = \frac{p^2}{2m} + \sum_{i=1}^{N} \sum_{j=1}^{3} \frac{P_{i(j)}^2}{2M_j} + \sum_{i<j} V(R_i,R_j) + \sum U(r,R_i) \qquad (III-1)$$

$$\equiv T + V_T$$

where T is the sum of the first two (kinetic energy) terms and V_T is the sum of the last two (potential energy) terms. The symbols r is the coordinate of the electron and p is the momentum conjugate to r, $P_{i(j)}$ is the momentum of the jth species of the ith water molecule M_j is the mass of the jth species of a water molecule with $M_1 = M_O$ and $M_{2,3} = M_H$ being the oxygen and the hydrogen makes respectively, R_i denotes the collection of the coordinates of the ith water molecule, $V(R_i,R_j)$ is the interaction potential between two water molecules, and $V(r, R_i)$ is the potential energy between the excess electron and the ith water molecule. The quantum canonical partition function Q at the temperature T is given by

$$Q = Tr\left[e^{-\beta H/P} \right]^P \qquad (III-2)$$

By letting P very large and neglecting the exchange contribution due to the indistinguishability of the water molecules, Q can be written as the limit of $P \to \infty$ of the equation[6]

$$Q_p = (\frac{mp}{2\pi h^2 \beta})^{3P/2} (\frac{M_0 P}{2\pi h^2 \beta})^{3NP/2} (\frac{M_H P}{2\pi h^2 \beta})^{3NP} \times$$

$$\int \exp(-S) \prod_{i=1}^{N} \prod_{j=1}^{P} dR_i^{(j)} dr^{(j)} \qquad (III-3)$$

with

$$S = \beta (V_{e-W} + V_{W-W}) \qquad (III-4a)$$

where

$$V_{e-W} = \sum_{j=1}^{P} [\frac{mP}{2h^2\beta^2} (r^{(j)} - r^{(j+1)})^2 +$$

$$\frac{1}{P} \sum_{i=1}^{N} V(r^{(j)}, R_i^{(j)})] \qquad (III-4b)$$

and

$$V_{W-W} = \sum_{i=1}^{N} \{ \sum_{j=1}^{P} [\sum_{\lambda=1}^{3} \frac{M_\lambda P}{2h^2\beta^2} (R_{i(\lambda)}^{(j)} - R_{i(\lambda)}^{(j+1)})^2 +$$

$$\frac{1}{2P} \sum_{k \neq i}^{N} V(R_i^{(j)}, R_k^{(j)})]\} \qquad (III-4c)$$

In the discretizing path integral formulation the task of calculating the partition function of N+1 quantum particles has been reduced to calculating the partition function of $3NP + P$ classical particles.[18] The action for the isomorphic classical system is given by Eq. (III-4). This formulation is exact only in the limit of $P \to \infty$. In practice one evaluates all the observables of the equilibrium system (like distribution function, energy, etc.) for a finite but large P such that the results do not change appreciably with further increase in P. Because of the reduction to the equivalent classical system

one may use Monte Carlo techniques to evaluate the physical quantities of internate. Alternate methods have also been used.[19]
     In order to evaluate the partition function the potentials
$V(R_i,R_j)$ and $U(r,R_i)$ need to be specified. The term $V(R_i,R_j)$ is
taken to be the modified central force potential of water. Morse
potentials are used to model the internal vibrations of the water
molecule. The electron-water monomer interaction, $U(r,R_i)$, is
assumed to consist of three parts: (a) an expoential repulsive
interaction due to the closed shell water electrons. This term
accounts for the orthogonality of the wavefunction of the excess
electron to the ground electronic wavefunction of the target molecule; (b) an anisotropic electron-dipole interaction centered in the
oxygen atom and (c) an electron-induced dipole interaction due to the
polarizability of the water molecule centered on the oxygen atom.
Parts (b) and (c) contain appropriate switching functions which turns
off these interactions when the electron gets close to the respective
centers. The details of this model potential can be found
elsewhere.[5]
     The calculations are performed by equilibrating the isomorphic
classical system consisting of 3NP + P particles moving in an effective potential given by Eq. (III-4). In this work the number of
pseudoparticles, P, for the electron was taken to be 1000 and P for
both the hydrogen and oxygen atoms was set to 100. The initial configuration of the classical system is arbitrarily chosen and it has
been verified that the results are independent of the starting configuration. Standard Metropolis algorithm is used to equilibrate the
system. The results were obtained by averaging over 13000 passes
after the system equilibrated.
     The stability of the e^- – $(H_2O)_n$ system is determined by (a) the
overall external potential energy between the electron and the water
cluster (b) the behavior of the electron-oxygen radial distribution
function $g_{eo}(r)$ given by

$$g_{eo}(r) = \left\langle \frac{1}{NP} \sum_{i=1}^{N} \sum_{t=1}^{P} \delta\left(\left| r^{(t)} - R_{io}^{(t)} \right| - r \right) \right\rangle \qquad \text{(III-5)}$$

Before presenting the results a word about the $e^-(H_2O)_3$ calculation
is in order. The water trimer can exist in several configurations
which may be classified according to the number of hydrogens of the
central molecule that participate in hydrogen bonding.[20] We confine
ourselves to one electron localization by a trimer in the single
donor linear configuration. In this conformer, one of the hydrogens
of the central molecule is engaged in hydrogen bonding while the
hydrogen of the trimer molecules are free resulting in an open linear
configuration. Elsewhere,[5] it has been established that a trimer in
the cyclic conformer does not bind an electron. The binding energy
of the $e^-(H_2O)_2$ is estimated to be between 3-6 meV while that
of e^- – $(H_2O)_3$ is found to be between 6-9 meV. In Fig. (1) a plot of

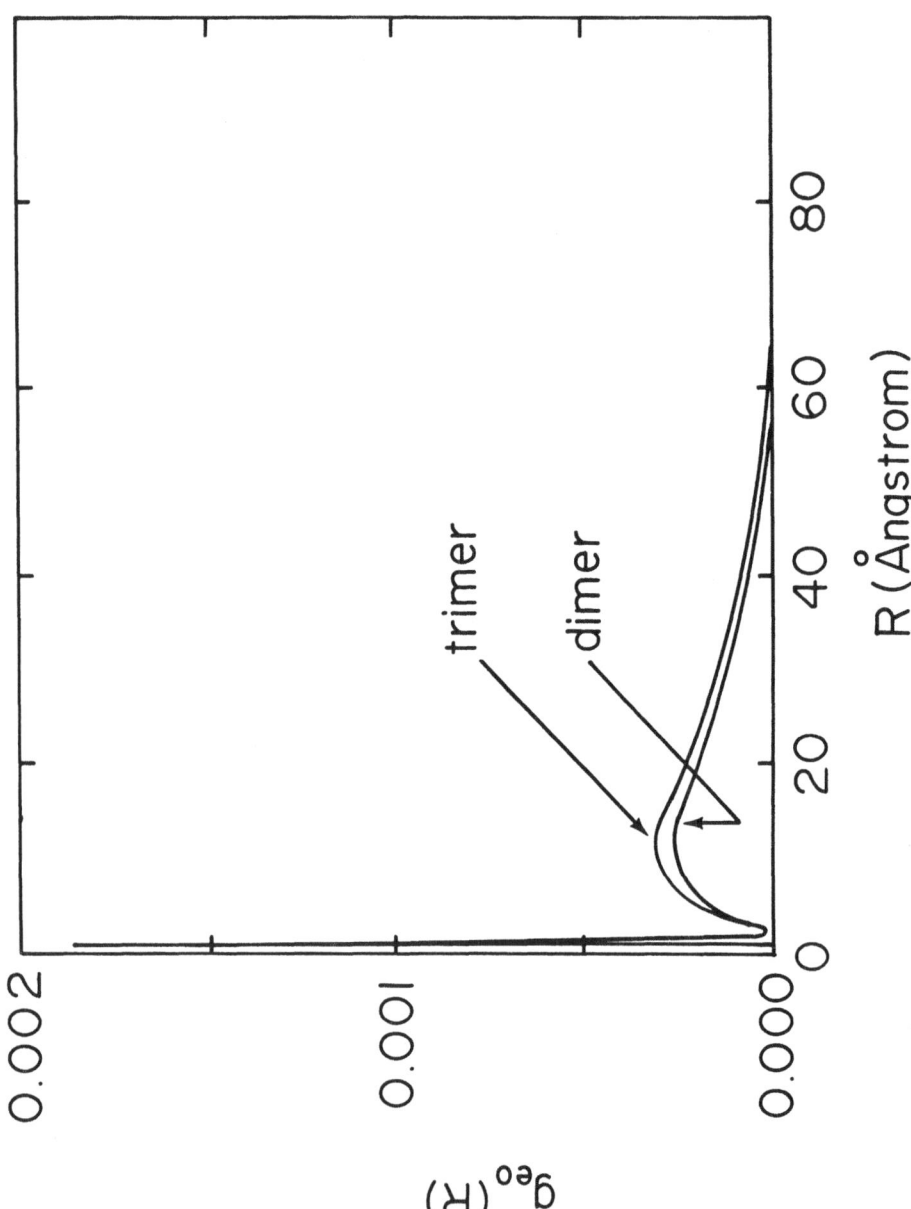

Figure 1. Electron-oxygen radial distribution functions for $e^-(H_2O)_3$ and $e^-(H_2O)_2$.

$g_{eo}(r)$ as a function of r for both the trimer and the dimer are shown. This figure shows that the peak height at r ~ 10 Å is smaller for the e^- - $(H_2O)_2$ than it is for e^- - $(H_2O)_3$. Both curves show a sharp peak at $r < 2$ Å, which is a consequence of the trapping of the electron density by the deep well present in the electron-monomer potential. In addition, there is a peak in $g_{eo}(r)$ corresponding to the electron-water trimer at r ~ 5 Å which is absent in the dimer radial distribution function. This peak corresponds to the presence of electron density near another water molecule and suggests that the electron is essentially localized by the field of two water molecules. The electron is more strongly localized by the water trimer than by the water dimer. This is consistent with the higher binding energy for electron-trimer system than for the electron-dimer system.

We now discuss the relevance of calculations to the experimental work of Haberland et al.[17] In addition to the calculations done at 5k we repeated them at 20k and the results clearly showed that the excess electron does not bind to the water dimer.[5] Thus, our calculations are in agreement with the experimental results only if the beam temperature is assumed to be low - this indeed is possible. However Haberland and coworkers estimate the binding energy for the e^- $(H_2O)_2$ to be about 17 meV which is about thrice the theoretical value obtained here. The source of discrepancy is probably due to the choice of potential energy surface. The interesting prediction of our calculation is that a trimer in the SDC configuration should bind an electron. The lack of prominence of the e^- - $(H_2O)_3$ peak in the experiment suggests that the predominant trimer conformer is the cyclic one which has been shown not to bind an electron. The seeding of the beam with Xe enhances the formation of e^- - $(H_2O)_3$ indicating that Xe may help the formation of the linear configuration. It would be interesting to confirm our prediction that the appropriate conformer of the trimer (the linear one) should localize the electron stronger than the dimer.

IV. CONCLUSION

In this report we have briefly reviewed agruments that suggest that cluster of small number of rare gas atoms can indeed localize an electron. In such a system the bulk of the electron density is expected to be confined near the location of the atoms in a cluster. The theory also suggests that small number of Hg atoms can bind an electron which is in accord with the experiments. We have also presented path integral Monte Carlo simulations which suggest that at sufficiently low temperatures two water molecules can bind an electron. The details of this work have appeared elsewhere.[5,6]

ACKNOWLEDGMENTS

The author is grateful to A. Wallqvist and B. Berne for collaborations on the work reported in Section III. I am indebted to F. Hensel and H. Haberland for useful discussions on the experimental aspects of electron localization. This work was supported in part by

National Science Foundation grant CHE-86-09722, Camille and Henry Dreyfus foundation, the Research Corporation, Alfred P. Sloan foundation, and the Presidential Young Investigators program.

REFERENCES

1. J. Jortner Ber. Bunsenges. Chem. $\underline{88}$, 188 (1984).
2. W. Hefner and F. Hensel Phys. Rev. Lett. $\underline{48}$, 1026 (1982). For a recent theoretical discussion see D. Logan and P. P. Edwards Phil Mag. $\underline{53}$, L123 (1986); R. W. Hall and P. G. Wolynes Phys. Rev. B $\underline{33}$, 7879 (1986); L. Turkevich and M. H. Cohen Ber. Bunsenges. Chem. $\underline{88}$, 292 (1984).
3. G. Ascarelli Comm. Sol. Stat. Phys. $\underline{11}$, 179 (1985).
4. U. Landman, D. Scharf and J. Jortner Phys. Rev. Lett. $\underline{54}$, 1860 (1985).
5. A. Wallqvist, D. Thirumalai and B. J. Berne J. Chem. Phys. $\underline{85}$ 1583 (1986).
6. D. Thirumalai, A. Wallqvist and B. J. Berne J. Stat. Phys. $\underline{43}$, 973 (1988).
7. I. J. Iakubov and A. G. Krapak Chem. Phys. Lett. $\underline{39}$, 160 (1976).
8. H. Haberland, H. G. Schindler and Dr. R. Worsnop Ber. Bunsenges. Chem. $\underline{88}$, 270 (1984); J. Phys. Chem. $\underline{81}$, 3472 (1984).
9. The validity of this can only be checked by doing a time dependent calculation.
10. K. Schwarz Phys. Rev. Lett. $\underline{41}$, 239 (1978); B. Springett, M. H. Cohen and J. Jortner Phys. Rev. $\underline{159}$, 183 (1968).
11. D. F. Coker, B. J. Berne and D. Thirumalai J. Chem. Phys. $\underline{86}$, 5689 (1987).
12. We assume that the wavefunction is of the form e^{-r}. If one were to perform a path integral simulation the onset of localization will be signaled by looking at conditions where appreciable electron density is in the vicinity of the cluster. Thus the theory presented here cannot account for "surface states" reported in certain clusters (see Ref. 4). This is another reason why N_{min} obtained here cannot correspond to simulation results which can identify surface states which are precursors to the truly exponentially localized states explored here.
13. L. Landau and E. M. Lifshitz, "Statistical Physics", (Pergamon Press, NY, 1977) chapter 12.
14. T. F. O'Malley Phys. Rev. $\underline{130}$, 1020 (1963).
15. Recent experiments give a value of $N_{min} = 3$ H. Haberland (Private Communication).
16. A. Wallqvist, D. Thirumalai and B. J. Berne J. Chem. Phys. $\underline{86}$, xxxx (1987).
17. H. Haberland, C. Lindewight, H. G. Schindler and D. R. Worsnop, J. Chem. Phys. $\underline{81}$, 3742 (1984).
18. D. Chandler and P. G. Wolynes J. Chem. Phys. $\underline{74}$, 4078 (1981).
19. See the review by B. J. Berne and D. Thirumalai Ann. Rev. Phys. Chem. $\underline{37}$, 401 (1986).
20. J. R. Reimers and R. O. Watts Mol. Phys. $\underline{52}$, 357 (1984).

STRUCTURAL AND DYNAMICAL PROPERTIES OF CLUSTERS ON A SUBSTRATE

Estela Blaisten-Barojas[*], I.L. Garzón[**], and M. Avalos[**]
Instituto de Física,
Universidad Nacional Autónoma de México,
[*]Ap. Postal 20–364, 01000 México D.F., México, and
[**]Ap. Postal 2681, 22800 Ensenada, Baja California, México

ABSTRACT. This talk describes an extensive computer simulation investigation of the structure, thermodynamics and phase stability of Lennard-Jones clusters adsorbed onto a surface, with special emphasis on the cluster melting regime. This investigation includes molecular dynamics simulations of the solid-like and liquid-like phases of the Lennard-Jones 13-atom cluster, of the melting and desorption process, evidence of surface wetting and mobility of the atoms within the cluster. We conclude that solid-like phase changes and wetting take place in these finite size systems and both phenomena are dependent on the ratio substrate/atom-atom interaction.

1. INTRODUCTION

Computer simulations have added a new scope to scientific research in the rapidly growing field of cluster physics and chemistry. In the last few years, numerous novel innovations in experimental cluster measurements led to the discovery of a richness of cluster geometries built up in many cases by structures with a *magic number* of atoms. The possibility to observe clusters in the liquid state has also been rised. Heterogeneous catalysis, nucleation and physisorption, charge inestabilities in electron microscopy studies are some of the phenomena that need theoretical support. Indeed, for adsorbed clusters, very little is known theoretically, although the huge number of experiments performed on supported clusters. As is common in scientific investigation, the validity of a comparison between theoretical predictions and experiments may be sometimes questioned because of the complexity of the experimental interpretation as compared to the simplicity of the theoretical model.

Furthermore, the testing of a theoretical prediction may be restricted because of limitations in the experimental state-of-the-art. In other areas of physics, *computer experiments* have alleviated these bottlenecks to the understanding of phenomena, and, hopefully, the developments in this presentation [1] will provide evidence for this point of view in the study of supported clusters.

We will consider some particular aspects of the *13-atom world* next to a surface; the phases, their transitions, and solid -state stability of a system of van der Waals atoms as determined by computer simulation experiments. Our main purpose is to give more insight on the cluster thermodynamic and structural properties as a function of the substrate-cluster interaction. This system may be considered as a prototype for rare gas clusters physisorbed on a graphite surface.

J. Jortner et al. (eds.), Large Finite Systems, 241–251.
© 1987 *by D. Reidel Publishing Company.*

Computer simulations, both molecular dynamics [2-4] and Monte Carlo [5-7] have been used to describe the thermodynamic properties of *free* Lennard-Jones clusters, with the striking observation that free clusters can exhibit phase changes analogous to the melting and freezing of bulk materials. The type of melting transition was under dispute because of apparently contradictory results giving evidence for a first-order [2-4] or second-order [5-7] transition. This contradiction has recently been conciliated [8] by recognizing that the results in the microcanonical [2-4] and canonical [5-7] ensembles need not coincide for finite systems.

The study of clusters adsorbed on surfaces has received less theoretical attention. Weissman and Cohen [9] presented a molecular dynamics study of two-dimensional(2D) clusters adsorbed on a surface. In this work, the authors showed that 2D-clusters with 7 and 19 atoms were in one stable conformation at low temperatures and changed to a set of high energy conformations when the temperature was rised. They stated that the presence of weak or strong substrates changed by little the qualitative first-order melting-like transition. Recently, we reported [10] that Lennard-Jones clusters containing 13 or less atoms could become mechanically stable 2D clusters when strong three body effects were added to the potential. Small 3D Lennard-Jones clusters do not exhibit stable 2D structures. In reference 10, we suggested that the non-pair additive interactions could be enhanced when the clusters were adsorbed on a substrate, and thus favour energetically the 2D- over 3D-clusters.

The organization of this talk is the following. First, we shall discuss details of the computational procedure. Second, we shall present the body of results divided in three parts: (i) discussion of the melting-like transition of a free cluster; (ii) results on supported clusters and description of the wetting transition; (iii) experiments on clusters in metastable states. We shall end by giving the conclusions of this work.

2. MOLECULAR DYNAMICS COMPUTER SIMULATION METHOD OF CLASSICAL STATISTICAL MECHANICS.

In this study, we have arbitrarily adopted the Lennard-Jones 12:6 potential function $v(r)$ to represent how individual atoms interact in the 13-atom cluster; it reads

$$v(r) = \epsilon\{(r_0/r)^{12} - 2(r_0/r)^6\} \tag{1}$$

where $-\epsilon$ is the minimum of the potential at the iteratomic separation $r_0 = 2^{1/6}\sigma$. This is a respectable interatomic potential for the rare gas atoms. Typical values for argon are $\epsilon/k = 120$ K and $\sigma = 3.4$ A [11], k is Boltzmann's constant.

For the interaction of each cluster atom with the solid surface we use the potential [12]

$$v_s(z) = \epsilon_s\{(2^{1/6}\sigma_s/z_s)^{12} - 2\sqrt{2}\,(2^{1/6}\sigma_s/z_s)^3\}/1.89 \tag{2}$$

where z_s is the perpendicular atom-substrate distance. This potential corresponds to a lateral averaging of the attractive van der Waals interaction similar to (1) but with different parameters ϵ_s and σ_s. The first term in both Eq. (1) and (2) represent a hard core repulsion, so the parameters σ and σ_s measure, crudely speaking, the nearest-neighbor separation between cluster atoms and the distance from the substrate to the first adsorbed *layer* in the cluster. We considered in this work $\sigma = \sigma_s$, noting that the core power-law dependence in Eq. (2) is stiffer than in Eq. (1).

The molecular dynamics simulation technique yields the motion of a small number of atoms governed by their mutual interatomic interactions, this being calculated by numerical integration of Newton's equations of motion [13]. In this conventional molecular dynamics, the total energy E for a fixed number of atoms N in a fixed volume Ω is a constant of the motion as the dynamics of the system evolves along its trajectory in phase space. The time average of any property along the trajectory is an approximate measure of the microcanonical ensemble average of that property for a thermodynamic state N, Ω, E. A time step of 0.01τ, where $\tau = (m\sigma^2/\epsilon)^{1/2}$ and $m =$ atomic mass, was used in the numerical integration of Newton's equations of motion. For argon τ is of the order of 2×10^{-12} sec. The energy and distance units adopted were ϵ for energies and r_0 for distances. Temperature is given in these units as $T^* = kT/\epsilon$.

To follow the effect of the substrate on the phase changes of the cluster we considered various strengths $\epsilon_s^* = \epsilon_s/\epsilon$ of the substrate-cluster potential as to simulate weak (0.2), intermediate (0.5 to 0.8) and strong (> 1.0) interactions. The process of heating and cooling the clusters was always carried out in presence of a frozen substrate at $T = 0$. The temperature of the system was changed by scaling the velocities of an equilibrated and previous calculation. Every heating (cooling) process was done in a steplike way [14], allowing in each step of the process for and equilibration time of 20τ and a further run of 50τ to average the calculated quantities. Clusters were heated slowly at a rate of approximately $\Delta T^*/\Delta t = 1.7 \times 10^{-4}\tau^{-1}$. In most cases the initial configuration of the cluster atoms was an icosahedron with potential energy -44.44ϵ. The substrate was kept at a fixed initial distance of $-1.8r_0$ from the closest atom in the cluster.

Several properties were monitored during the computer runs and will be discuss in the rest of the talk:

(i) Total energy per particle E as a function of the temperature T: For the free cluster the motion is referred to the cluster center of mass, so we have $T^* = 2N\langle E_{kin}^*\rangle/(3N - 3)$, where $\langle E_{kin}^*\rangle$ is the mean kinetic energy per particle. For adsorbed clusters $T^* = 2N\langle E_{kin}^*\rangle/3N$ since the center of mass motion is important in this case. In each experiment the temperature fluctuations were calulated as well,

$$\sigma_T = 2N(\langle E_{kin}^2\rangle - \langle E_{kin}\rangle^2)^{1/2}/(3N - 3) \qquad (3)$$

(ii) Mean square displacement of the atoms as a function of time:

$$\langle r^2(t)\rangle = \frac{1}{N}\sum_{i=1}^{N}\langle|\vec{r}_i(t) - \vec{r}_i(t_0)|^2\rangle \qquad (4)$$

where $\vec{r}_i(t_0)$ are the positions at an initial time during a computer run. In general we propagated the system for 100τ more from the final configurations described in (i).

(iii) Drawings of the instantaneous cluster configurations. This visual inspection was useful in discriminating between different kinds of structures.

3. THE MELTING LIKE TRANSITION OF THE FREE CLUSTER

The behavior of the equilibrium total energy as a function of temperature can be seen in Fig. 1a for the free cluster. At low temperatures the cluster is in the global minimum of the potential energy, the icosahedral structure. This region of the plot is analogous to the solid phase of bulk materials. Furthermore, the value of the slope is of the order of the classical specific heat value of a set of $3N - 6$ harmonic oscillators. In the intermediate temperature regime the cluster can be in one or two new configurations which correspond to local minima of higher potential energy, but still acquires the icosahedal configuration. When the temperature is even higher, the cluster reaches a set of high potential energy minima. During the length of our computer runs, the icosahedral structure was not observed at high temperatures. The set of high energy structures in configuration space defines the liquid-like phase of the cluster. When the liquid-like points are fitted to a straight line, it is seen that the slope is larger than in the solid-like region. This behavior is expected in bulk liquids. All together, the behavior resembles a first-order phase transition of bulk materials. The classical two-phase model can be understood as a chemical equilibrium solid $\rightleftharpoons$ liquid [8]. Then, the temperature as a function of energy in the microcanonical ensemble is

$$T_{\text{eq}}(E) = p(E)T_{\text{sol}}(E) + [1 - p(E)]T_{\text{liq}}(E) \qquad (5)$$

where $p(E)$ and $1 - p(E)$ are the fraction of time spent in the solid-like and liquid-like

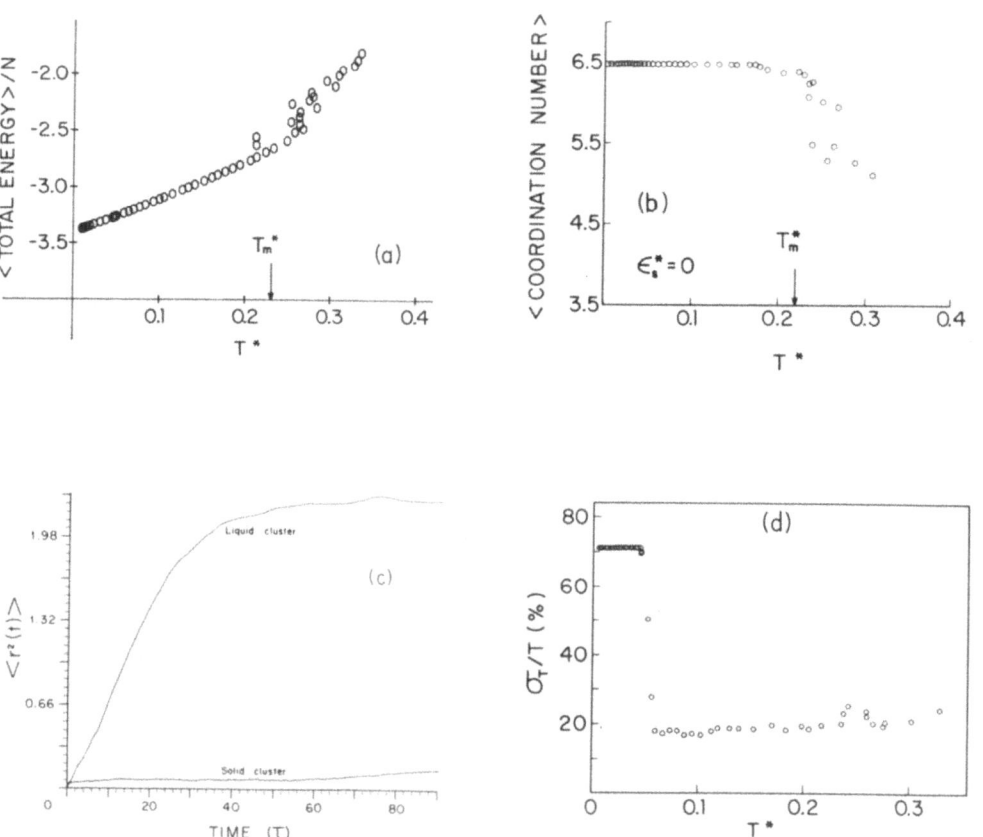

Fig. 1. Free cluster: (a) total energy per atom versus reduced temperature; (b) mean coordination number versus reduced temperature; (c) mean square displacement versus time; (d) temperature fluctuations versus reduced temperature.

phases. The coexistence of two phases is very clear when interpreted under this scope. The melting temperature T_m is defined as the mid point of the coexistence region bounded by temperatures T_1 and T_2. Our results are in agreement with previous reports, both in the solid and liquid regions [4] as in the intermediate temperature regime [8].

Further evidence of the melting like transition in the cluster is given by plots of the mean coordination number n_c as a function of temperature (Fig. 1b), and of the mean square displacement $\langle r^2(t) \rangle$ as a function of time (Fig. 1c). To obtain n_c it is necessary to calculate the pair correlation function $g(r)$ [15],

$$n_c = \rho_0 \int_0^{r_{\text{neig}}} g(r) 4\pi r^2 \, dr \tag{6}$$

where $\rho_0 = N/\Omega$ is the reference density and r_{neig} is the distance between nearest neighbor atoms in the cluster. Both, Fig. 1b and 1c exhibit a behavior very similar to what is obtained in bulk materials. The coordination number changes abruptly from a constant value in the solid phase to a lower value when the system melts. The value of the mean square displacement is very small in the solid phase and rises linearly with time in the liquid regime. After approximately 30τ, $\langle r^2(t) \rangle$ reaches a constant value consistent with the finite volume of the cluster, i.e., the deviation from the mean positions cannot be larger than a certain value if the atoms are to move in the volume occupied by the cluster.

In Figure 1d we show the change in temperature fluctuations. At extremely low temperatures, $T < 0.05$, there is a region where the system is not ergodic but harmonic, as is seen in the figure. From this threshold on, the fluctuations are of the order of $(3N-3)^{-1}$, as expected. This result gives confidence in what concerns the thermodynamic interpretation of our calculations.

4. SUPPORTED CLUSTERS AND THE WETTING TRANSITION

When the substrate is plugged into the calculations, qualitatively the total energy as a function of temperature exhibits a similar behavior as for the free cluster. This is shown in Fig. 2 and 3. The four curves in each figure correspond to heating experiments starting with the cluster in the icosahedral structure adsorbed onto the surface. All of the results show that the melting like transition is taking place while the cluster is adsorbed onto the surface. However, the melting temperature, the width of the coexisting region and the slope of the liquid line depend upon the substrate strength. Moreover, a closer inspection of the instantaneous structures indicates that a new phenomena is taking place. In Fig. 3, for strong enough substrates (b and c) the coordination number decreases in two steps. There is evidence that a set of structures with a different coordination number are reached before the cluster melts. Indeed, in the solid-liquid coexistence region two-layered structures (*rafts*) in configuration space are acquired by the system. The potential energy minima corresponding to these rafts were not present in the case of the free cluster. Several of these new structures are shown in Fig. 4 for a strong substrate.

The fact that the cluster can acquire these raft structures suggested that perhaps a surface wetting behavior could be identified in a plot of temperature versus substrate strength. In Fig. 5 we have plotted the melting temperatures as a function of substrate strength. The bars indicate the width of the solid-liquid coexistence region. Above this line the cluster is liquid; below the line the cluster is solid. The dotted curve goes through four points generated when a liquid cluster was cooled down to low temperatures. The

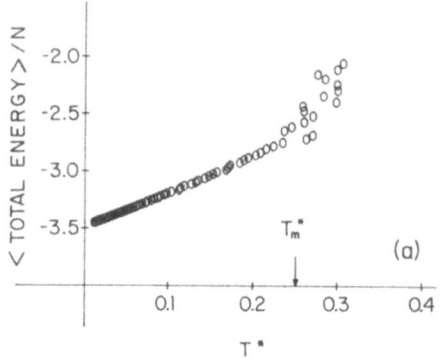
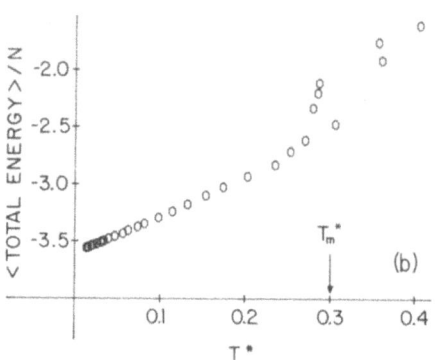

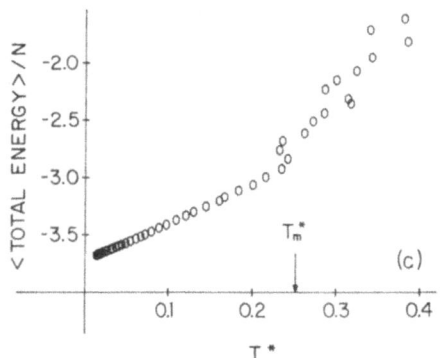
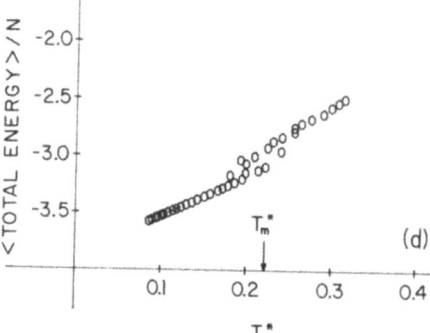

Fig. 2. Temperature dependence of the total energy for supported clusters: (a) $\epsilon_s^* = 0.2$, (b) $\epsilon_s^* = 0.5$, (c) $\epsilon_s^* = 0.8$, (d) $\epsilon_s^* = 1.1$.

point at $\epsilon_s^* = 0.2$ indicates the temperature at which the cluster ran away from the surface while freezing. For this weak substrate there is a transfer of kinetic energy from the
atoms to the center of mass, the cluster cools but abandons the surface in the icosahedral structure. The value at $\epsilon_s^* = 0.5$ corresponds to the temperature where a two-layered
structure was detected. Upon further cooling the cluster ended adsorbed onto the surface
in its icosahedral structure. The two remaining points at $\epsilon_s^* = 0.8$ and 1.1 indicate the

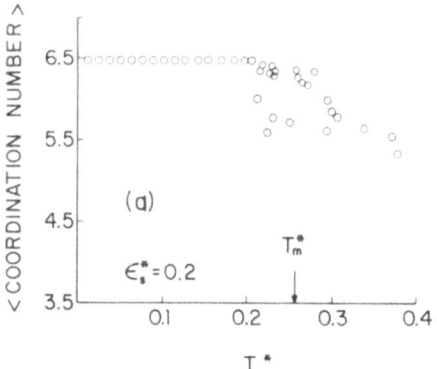

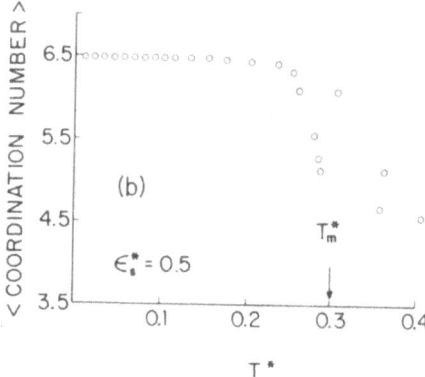

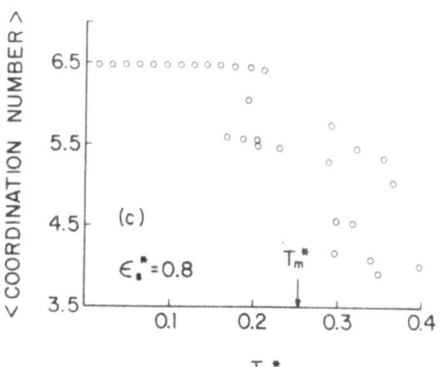

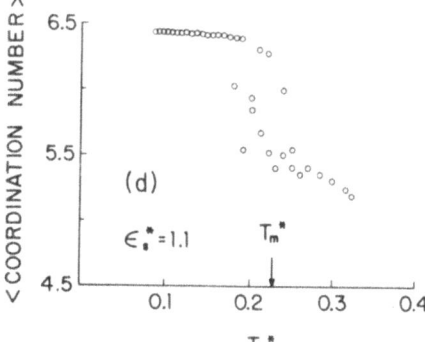

Fig. 3. Temperature dependence of the mean coordination number for supported clusters: (a) $\epsilon_s^* = 0.2$, (b) $\epsilon_s^* = 0.5$, (c) $\epsilon_s^* = 0.8$, (d) $\epsilon_s^* = 1.1$.

temperature where no more rafts were seen, and the cluster was mainly in the icosahedral structure. Of course, these temperatures are estimates, since more experiments would be necessary to determine their dispersion.

Three regions are now clearly shown in Fig. 5 by drawing the dotted line through the four points. Region I corresponds to a *dry* surface; no clusters are adsorbed. In region III the surface is *wet* and the clusters are adsorbed as rafts. In the intermediate region, $\epsilon_s^* < \epsilon_w^*$ there is *partial wetting* and the clusters are adsorbed in the icosahedral structure.

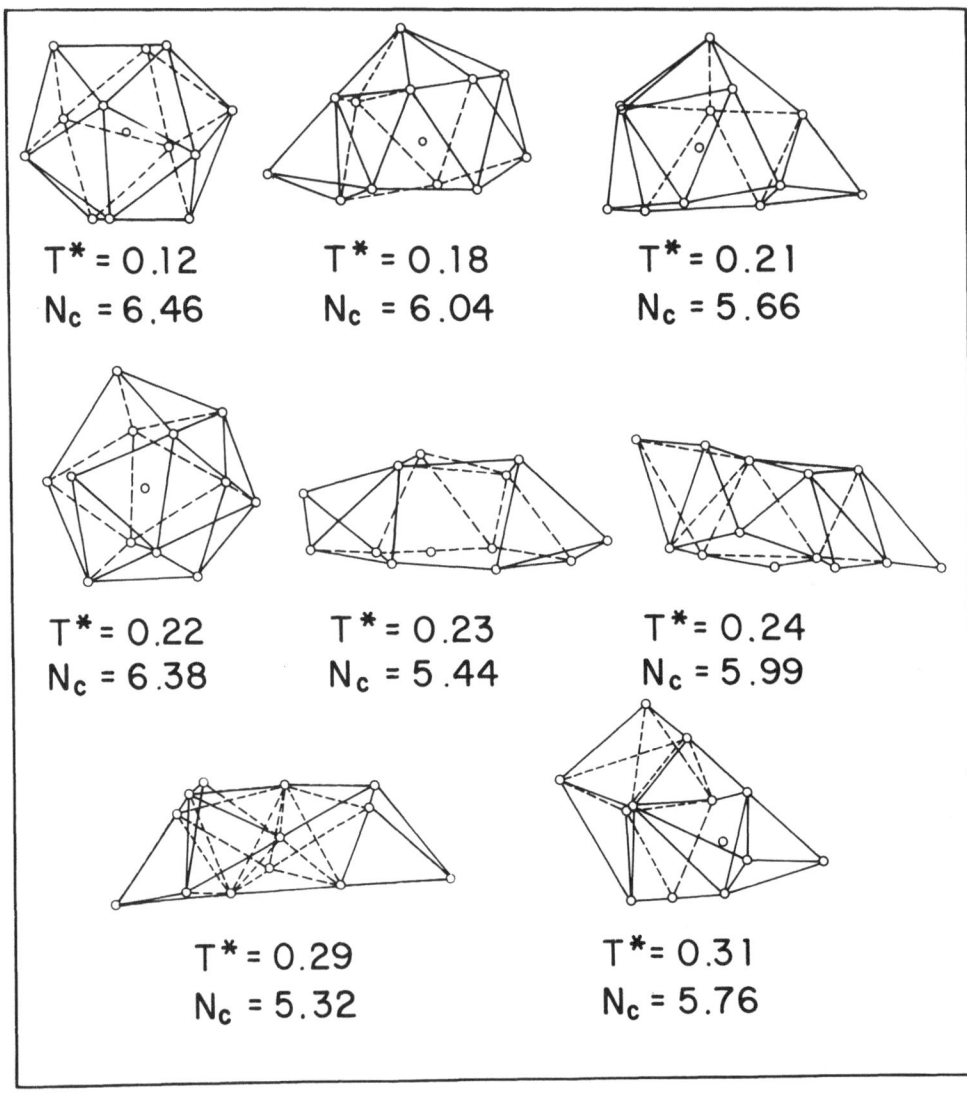

Fig. 4.   Instantaneous structures of the adsorbed cluster ($\epsilon_s^* = 1.1$) during the heating experiment. N_c is the coordination number calculated from the atomic coordinates.

In region I and region II below ϵ_i^*, the cluster desorbes as a whole. In region II above ϵ_i^*, and in region III the cluster desorbes atom by atom. When a cluster is cooled down in region III, it looses atoms towards the surface, i.e., lateral evaporation from the cluster takes place with evaporated atoms adsorbed on the surface.

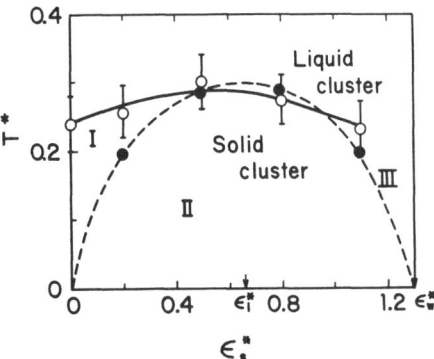

Fig. 5. The temperature versus substrate strength plane.

5. CLUSTERS IN METASTABLE STATES

Upon very fast cooling processes it was possible to trap the clusters in metastable states at low temperatures. Two situations were studied in detail. In the first, the free cluster was quenched into a structure corresponding to a local minimum of the potential energy surface. This structure corresponded to a potential energy value of -3.2ϵ, above the icosahedron value. From this structure an extremely low heating process was carried out by scaling the velocities in 3% at each step. The total energy per atom versus temperature plot is shown in Fig. 6a. The cluster started to melt, but evaporation took place and no

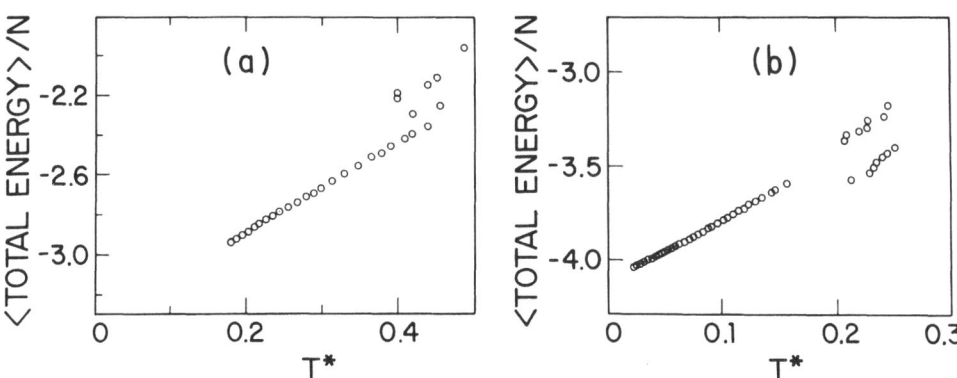

Fig. 6. Clusters in metastable states. (a) total energy versus temperature for the free cluster initiated in a state with potential energy $= -3.2\epsilon$; (b) total energy versus temperature for a raft adsorbed on a $\epsilon_s^* = 2.0$ substrate.

liquid branch was obtained. The slope of the solid-like portion of the curve is lower than in Fig. 1a. In the second situation, a raft on a strong substrate $\epsilon_s^* = 2.0$, was trapped at low temperatures. Heating from this initial configuration produced the plot of Fig. 6b. The raft underwent a discontinous transition before evaporation occurred. The high energy branch corresponds to raft structures whereas the lower branch are the usual set of disordered structures characteristic of the liquid phase.

For strong substrates, $2.0 < \epsilon_s^* < 3.5$, it was possible to begin heating experiments from the icosahedral structure. Upon heating, once more the cluster changed into raft structures before evaporating one atom on the substrate. When the substrate was even stronger, $\epsilon_s^* = 4.0$ and 5.0, the icosahedron was no longer stable and changed into a raft after 70τ run. The $\epsilon_s^* = 4.0$ substrate led to rafts with 11 atoms in the first layer. Further temperature increments produced side evaporation of one atom on the surface. The $\epsilon_s^* = 5.0$ substrate was capable to hold an adsorbed monolayer of 13 atoms, although the atoms were not all at the same distance from the surface. If this configuration was led to evolve 200τ more, one atom evaporated laterally on the surface.

It is possible that in a real experiment in the laboratory, the production mechanism of supported clusters yields clusters trapped in various metastable states. Indeed, the substrate could help to obtain a diversity of possible structures. The observation technique should then be able to detect the differences, otherwise an averaged behavior over several metastable states would mask the phenomena occurring in individual clusters in equilibrium.

6. CONCLUSIONS

We can summarize the conclusions of this talk as follows:

The 13-atom cluster trapped on the surface in the free cluster configuration of minimum energy (icosahedron) undergoes an equilibrium structural transition on the surface as the temperature is increased. It changes from the icosahedron, solid-like structure at low temperatures to a set of liquid-like structures at high temperatures. This change is analogous to the melting transition of bulk materials. The transition is smooth because the solid-like and liquid-like structures coexist over a finite range of temperatures.

For weak substrates, the solid-like branch of the total energy versus temperature plot is similar to the free cluster. The coexistence region in this plot is broad and the liquid branch is hardly present —since at high temperatures the cluster desorbs as a whole in its icosahedral structure. For stronger substrates the solid branch is again similar to the free cluster. However, in the region where solid and liquid structures coexist, typically two-layer structures (rafts) are formed before the set of liquid-like minima in configuration space could be reached by the system. Therefore, the coexistence region for these substrates is different than for the weak substrate case. Furthermore, the melting temperature and the liquid branch slope depend strongly on the substrate strength. By rising the temperature the cluster desorbs atom by atom.

This behavior indicates that a second change is taking place other than melting, a *wetting* transition. This new transition becomes apparent by analyzing the temperature versus substrate strength plane. For strong enough substrates and temperatures higher than T_m the clusters *wet* the surface before melting and are supported as rafts. For weak enough substrates the surface remains *dry* since the clusters ran away from the substrate and undergo melting away from the surface. In the intermediate substrate interaction regime both situations are present since it is at this point that the atom-atom and atom-substrate attractions become comparable and, therefore, competitive. There is *partial wetting*. The clusters undergo the melting or freezing transition while adsorbed on the surface. The final situation at low temperatures is always the cluster adsorbed in the icosahedral structure.

When the clusters are heated (or cooled) too rapidly, the cluster jumps too fast among the various structures and a simulation can end in a metastable state. Thus, free clusters can be quenched in structures different than the icosahedron by fast cooling. Heating from this initial configuration leads to qualitatively the same melting like transition, but the slope of the solid branch is definitely smaller than in the situation described initially. For supported clusters on weak substrates, adsorbance of hot clusters by fast cooling was not obtained because the drying region is approched too fast. At this point there is an internal energy transfer within the cluster and the available kinetic energy goes into the translational motion of the cluster center of mass away from the surface. For supported clusters on strong substrates it was possible to quench metastable states, such as a raft structure. Heating from this state gave a discontinuous melting transition. Finally, for supported clusters on very strong substrates, two-layered structures could not be quenched with all of its atoms because lateral evaporation of atoms on the surface took place upon fast cooling. In this case the clusters always wet the surface .

In these computer experiments we have shown the large variety of structures that a supported cluster can acquire depending upon the mechanism of adsorbtion and upon temperature. Hopefully, these parameters could be equally controlled in the laboratory during a real experiment looking for equilibrium data.

We are now extending this work for larger clusters with the hope of giving a more precise description of the wetting transition with identification of possible triple points.

ACNOWLEDGMENTS

We would like to thank Dr. Alberto Robledo for illuminating dicussions. This work was supported by Consejo Nacional de Ciencia y Tecnología, México, under grant PCCBBNA-022643.

REFERENCES

1. E. Blaisten-Barojas, I. L. Garzon, and M. Avalos, to be published.
2. C. L. Briant and J. J. Burton, *J. Chem. Phys.* **63**, 2045 (1975).
3. E. Blaisten-Barojas and H. C. Andersen, *Surf. Sci.* **156**, 548 (1985).
4. J. Jellinek, T. L. Beck, and R. S. Berry, *J. Chem. Phys.* **84**, 2783 (1986).
5. R. D. Etters and J. Kaelberg, *J. Chem. Phys.* **63**, 2045 (1975).
6. R. D. Etters and J. Kaelberg, *Phys. Rev. A* **11**, 1068 (1975).
7. J. B. Kaelberer and R. D. Etters, *J. Chem. Phys.* **66**, 3233 (1977).
8. J. D. Honeycutt and H. C. Andersen, to be published.
9. M. Weissmann and N. V. Cohan, *J. Chem. Phys.* **72**, 4562 (1980).
10. I. L. Garzon and E. Blaisten-Barojas, *Chem. Phys. Lett.* **124**, 84 (1986).
11. A. Rahman, *Phys. Rev. A* **136**, 405 (1964).
12. C. Ebner and W. f. Saam, *Phys. Rev. Lett.* **38**, 1486 (1977).
13. L. Verlet, *Phys. Rev.* **159**, 98 (1964).
14. E. Blaisten-Barojas and D. Levesque, *Phys. Rev. B* **34**, 3910 (1986).
15. E. Blaisten-Barojas, *Kinam* **6**, 71 (1984).

PHOTODISSOCIATION OF METHYL IODIDE CLUSTERS

D. J. Donaldson, S. Sapers, and V. Vaida
Department of Chemistry, University of Colorado, Boulder, CO
80309-0440, USA

R. Naaman*
Department of Isotopes Research, Weizmann Institute of
Science, Rehovot, Israel.

ABSTRACT. The dissociation of methyl iodide molecules "solvated" in
clusters has been investigated using both direct absorption spectroscopy and multiphoton ionization methods. Clusters of both neat
methyl iodide and of methyl iodide with rare gases were studied in a
molecular jet. It was found that dimerization slows the predissociation rate from the Rydberg states of CH_3I, whereas in large clusters
the direct dissociation from the valence state is slowed. We present
a model which explains the effect of CH_3I dimer formation on the predissociation dynamics. Evidence is also presented for electron de-
localization in higher clusters after excitation into the Rydberg
states.

1. INTRODUCTION

Studies of the dissociation of clusters and van der Waals molecules
have become an area of intense activity in recent years (1-5). Some
of this work has dealt with the dissociation of a strongly bound molecule "solvated" inside a cluster. This process is similar to dissociation in a solvent, where it is known that the solvent can "hold"
molecules together, even if the system has absorbed energy in excess
of that needed for its dissociation. Two types of mechanisms have
been proposed to explain this effect. In the first, the nascent fragments collide with the solvent shell, before they have completely
separated. Due to momentum transfer their direction of motion is
reversed, causing them to recombine. In the other mechanism, the
initially excited molecule collides with the solvent shell before it
can fragment, causing its excess internal energy to be transferred to
the solvent, leaving the molecule with energy below that needed for
dissociation. While the former mechanism is valid only in the case of
large clusters, where the excited molecule is surrounded by a solvent

*1986-87 JILA Visiting Fellow.

J. Jortner et al. (eds.), Large Finite Systems, 253–263.

shell, the second process has also been demonstrated when a single atom is attached to the absorbing molecule (5).

In this work we present the results of recent experiments which have used CH_3I as a probe for solvent effects. Methyl iodide was chosen for this investigation because of the extensive literature existing on the dissociation of the monomer (6-15), and the relatively good theoretical understanding of the process (16-18).

The first feature in the methyl iodide ultraviolet absorption spectrum is a broad continuum, centered at 260 nm, corresponding to a transition to the dissociative valence state (denoted the A-state). The next feature in the absorption spectrum is due to a Rydberg system, whose origin is at 201.2 nm (B-state), followed by a transition originating at ~180 nm (C-state). These two transitions arise from the excitation of a nonbonding electron on the iodine to a 6s Rydberg orbital, also centered on the iodine atom (19-22). Both these Rydberg states are predissociated via surface crossings with the A-valence state.

In the following, we describe the results of both jet-cooled absorption experiments and multiphoton ionization studies performed on the valence and Rydberg systems of CH_3I. From these relatively simple experiments, we have been able to draw conclusions regarding the details of the dissociation of methyl iodide solvated in clusters of different size.

The paper is divided into two parts. The first deals with the dissociation of methyl iodide dimers, while the second discusses the photophysics of bigger clusters. The results can be described by two types of solvent cage mechanism; one which we observe occurring in large clusters, which provides an example for the hard sphere "classical" cage effect, and a second new mechanism, which like the energy-transfer process, can occur in dimers. It arises from a solvent-induced shift in the potential surfaces which influence the photodissociation. The "cage" that slows the dissociation in this case is not formed by hard sphere type collisions with surrounding solvent molecules, but by an increase in the potential energy barrier to dissociation. Hence we refer to this model as a "quantum cage" effect.

2. EXPERIMENTAL

Two experimental techniques were applied in this investigation. The first involves direct ultraviolet absorption in a molecular jet, and the second uses multiphoton ionization, combined with time-of-flight mass spectrometry. All the experiments were carried out using the same commercial piezoelectric pulsed nozzle. In the absorption experiments a 1 mm nozzle was used, operated at 50 Hz. The focussed output of either a high-brightness D_2 lamp or a very high brightness Xe lamp intersected the molecular jet 0.5 to 1 cm from the nozzle. The transmitted light was dispersed by either a 0.6 m monochromator, in which 35 cm^{-1} resolution could be achieved, or a 1 m monochromator, which allowed a spectral resolution of 10 cm^{-1}.

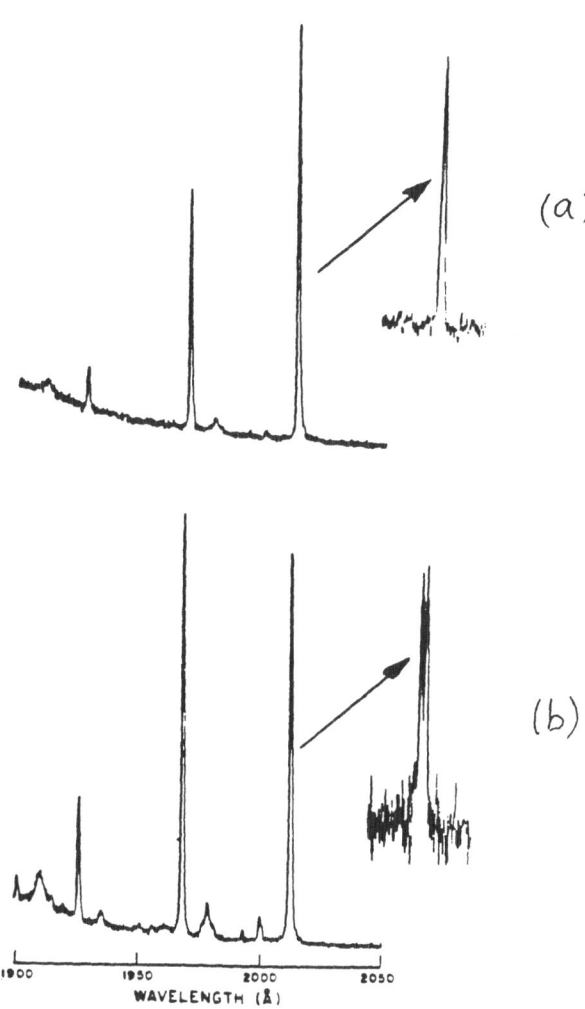

Fig. 1. Absorption spectra of CH_3I in the Rydberg B region taken at
a) 25 torr neat CH_3I behind the nozzle, and b) 80 torr of neat CH_3I
The insets show high resolution scans over the origin for each. Note
the splitting in b).

The multiphoton ionization experiment was carried out in a
differentially pumped molecular beam machine, equipped with a time of
flight (TOF) mass spectrometer. The laser intersected the molecular
beam 4.5 cm downstream from the skimmer and 5.5 cm from the 0.5 mm
diameter pulsed nozzle. An excimer pumped dye laser or excimer laser
were used, with typical 100 MW/cm^2 power density at the excitation
region.

3. RESULTS

3.1. Absorption

In the absorption studies, transition to both the valence A-state and
the two Rydberg states, B and C, were observed under expansion conditions that produced either monomers, monomers plus dimers, or large
clusters. It was found that when dimers form a substantial fraction
of the beam, the maximum of the broad peak corresponding to the X-A
transition is blue shifted by about 600 cm^{-1} with respect to that
observed for the monomer. In the X-B and X-C transitions, which show
relatively sharp features, the effect of dimer formation can easily be
quantified. In figure 1 the X-B absorption spectrum is presented for
a beam containing monomers only (a), and monomers and dimers (b). The
upper panel, (a), displays the spectrum obtained with 25 Torr of neat
three strong lines at 192.72 nm, 196.91 nm, and 201.26 nm, with weak

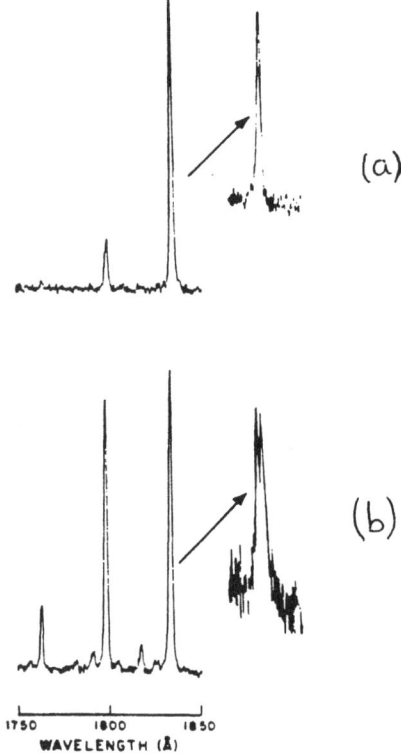

Fig. 2. Absorption spectra of CH_3I in the Rydberg C region taken at
a) 25 torr neat CH_3I behind the nozzle, and b) 80 torr of neat CH_3I.
Also presented are high resolution spectra showing a red shift of
about 10 cm^{-1} in the absorption of the dimers.

lines appearing at 191.13 nm and 197.95 nm. The strong lines were assigned to the band origin, v'=1 and v'=2 of the ν_2 umbrella mode respectively. Figure 1b illustrates the results of increasing the backing pressure to 80 Torr. The intensity ratios among the three strongest lines change dramatically. In addition, many new linesappear in the spectrum. The insets show high resolution scans over the origin region. At the higher backing pressure, we observed a splitting in the origin of ~10 cm^{-1} due to the presence of dimers and monomers in the beam. The new lines appearing in figure 1b are all assigned to CH$_3$I vibrations, except for those which lie 300 cm^{-1} to the blue of each member in the ν_2 progression. The latter have been assigned by Felps et al. (20) to transitions to the dark Rydberg state; however, the present results suggest that these lines are in- duced by the formation of methyl iodide dimers.

In figure 2 the X-C absorption spectrum is presented, taken under identical experimental conditions as those in figure 1. A high resolution expansion of the spectral origin here also reveals a splitting due to the presence of both monomers and dimers in the beam, and a red shift of about 10 cm^{-1} of the dimer lines versus the monomeric ones. Again it is evident that the higher member lines in the ν_2 umbrella mode progression are more intense in the dimer than in the monomer, and as in the case of the X-B transition, several other CH$_3$I modes appear upon dimerization.

Figure 3b shows the X-B absorption spectrum obtained when 300 torr CH$_3$I and 2.5 atm He were expanded through the nozzle. Under these conditions we expect large methyl iodide clusters to be present in the jet. Figure 3a shows the low pressure results for comparison. In the low pressure spectrum only the ν_2 progression is observed, while figure 3b displays many of the lines seen in figure 1, but contains an intense underlying continuum. The most striking feature in this spectrum is the appearance of the dips in the absorption just to the red of the band origin and the v'=1 line of ν_2. These inter- ference peaks are caused by the interaction between the sharp reso- nances and the continuum, and are observed also for methyl iodide monomers solvated in rare gas clusters. Figure 3c shows the results for 25 Torr of methyl iodide solvated in 1 atm of Ar. Under these conditions, the Rydberg spectrum looks like that of the monomer. It is interesting to note that in this latter case the absorption lines are slightly red shifted upon forming small clusters of all the rare gases with CH$_3$I, but become somewhat blue shifted in larger clusters.

3.2. Multiphoton ionization

CH$_3$I provides a convenient spectroscopic system where, by using MPI we have explored the dissociation dynamics on two very distinct electron- ic surfaces. In the ultraviolet, a one photon process will access the very broad A-state absorption occurring at 230-300 nm (λ_{max}= 260 nm). At higher energy (170-200 nm), by two photon absorption, the Rydberg system is excited. In the present study, methyl iodide clusters were excited through these two "doorway" states. Two-photon excitation

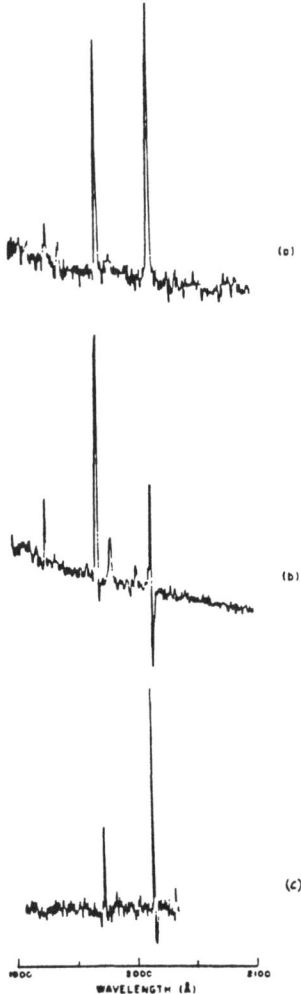

Fig. 3. The X-B absorption spectra taken at a) 35 Torr neat CH_3I
backing pressure; b) 300 torr CH_3I + 2.5 atm of He; c) 25 Torr CH_3I +
1 atm of Ar. Note the strong interference features in the origin and
ν_2 (v'=1) lines in b) and c).

into the C-state origin at 182 nm was performed with an excimer-
pumped dye laser operated at 364 nm. Single photon excitation of the
valence state was accomplished at 308 and at 248 nm using an excimer
laser. In both cases the photon density ($\approx$100 MW/cm^2) was
sufficiently large to allow subsequent photon absorption, leading to
fragmentation and ionization.

A strong ion signal was observed when methyl iodide monomers were
excited through the C-state. However, no ions were observed under
identical molecular beam conditions when the excitation was through
the valence state. Figure 4 displays TOF photofragment mass spectra
observed under different expansion conditions and excited at several

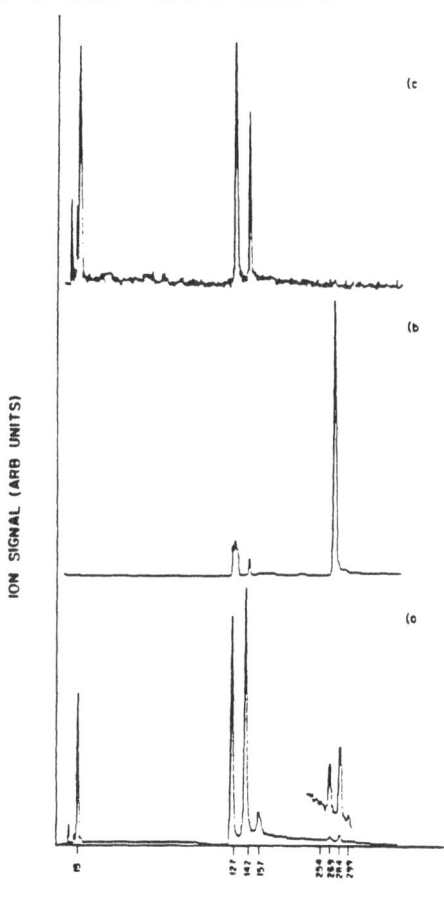

Fig. 4. Time of flight photofragmentation mass spectra of CH_3I monomers and clusters under different excitation and expansion conditions a) Excitation into the Rydberg C-state. The backing pressure is 400 torr neat methyl iodide. b) Excitation into the valence state under identical expansion conditions. c) Excitation into the valence state 400 Torr methyl iodide is expanded with 3 atm He.

wavelengths. In figure 4a we show the results obtained at 182 nm when neat CH_3I is expanded from a stagnation pressure of 400 Torr. The corresponding direct absorption experiment is shown in figure 2b. Recall that in that experiment, we observed the absorption line of the dimer, indicating a substantial fraction of dimers present in the beam. Under these conditions the photofragment mass spectrum (fig. 4a) consists of intense peaks corresponding to CH_3I^+, I^+ and CH_3^+ and somewhat weaker features corresponding to larger fragments, namely $CH_3ICH_3^+$, $CH_3I_2^+$, $(CH_3I)_2^+$ and a small signal of $(CH_3I)_2CH_3^+$. Excitation of the C-state under these conditions results in ionization of

CH_3I, $(CH_3I)_2$ and the small amount of $(CH_3I)_3$ in the beam. However, excitation of the valence state at 248 nm and 308 nm under identical expansion conditions produces mainly I_2^+ (fig. 4b) with significant smaller signals corresponding to I^+ and CH_3^+.

By adding seeding gas to change the expansion conditions, we were able to produce large methyl iodide clusters. These conditions correspond to those giving rise to the absorption spectrum obtained in figure 3b, where the growing continuum and the interference peaks indicate the formation of large CH_3I clusters. Excitation at 248 nm under these conditions generates the mass spectrum shown in figure 4c, which consists mainly of CH_3^+, I^+, and CH_3I^+. A similar mass spectrum was observed when the excitation was through the Rydberg state.

4. DISCUSSION

The discussion is divided into two parts. The first deals with the processes in dimers, while in the second, larger clusters are discussed.

4.1. Dimers

In the absorption spectra to the two Rydberg states (X-B and X-C transitions), the same basic phenomenon is observed: we find only the ν_2 (CH_3 "umbrella") mode in the monomer spectra. As the jet conditions are changed to optimize $(CH_3I)_2$ formation, more vibrations are evident, and the relative intensities in the ν_2 progression vary. At the same time as these new lines are observed, the high resolution experiment shows splittings in the origin band. The red component of the split origin is assigned to the origin of the dimer-induced spectrum. The appearance of this splitting is thus an unambiguous indication of the presence of dimers in the beam. All of the CH_3I lines we observe upon dimer formation have been reported previously and assigned to monomer absorption (20). The present results, however, show clearly that these lines appear only when CH_3I is bound in a dimer.

All the absorption data can be rationalized by assuming that dimerization causes a change in the predissociative curve crossing between the valence and the Rydberg states. As is evident from the blue shift in the valence state absorption, the A-state is less stabilized in the dimer than is the ground electronic state. This is probably due to the reduction in the dipole moment in the electronically excited molecule because of the transfer of an electron, located on the iodine, to an antibonding orbital centered on the C-I bond. Since the Rydberg states are stabilized as much as is the ground state (as evidenced by the minor shift in their absorptions), the result of dimer formation on these states is a net reduction in their energies versus that of the valence state. This implies a shift to higher energy of the crossing between these potentials. As a result, the "new" vibronic states which appear upon dimer formation are caused to dissociate more slowly in the dimer than in the monomer. This, in

turn, causes their line widths to be narrower; thus they can be observed above the background.

In the case of the ν_2 mode, the argument given above immediately
explains the change in the relative peak intensities upon dimer formation. Direct excitation of ν_2 (v'=2) provides the vibrationally
adiabatic gateway to the most probable CH_3 product vibrational state.
Hence it dissociates rapidly and as a result becomes broader and
appears in the monomer spectrum as a weak line above the continuum.
Excitation of ν_2 (v'=1) gives rise to less efficient coupling to the
dissociative state. Consequently its line width becomes narrower and
its peak intensity is thus greater than that of ν_2 (v'=2). The
changes in this mode are thus also the direct result of the higher
barrier for predissociation in the dimer.

While the MPI data indicates that excitation of the dimer through
the valence state results in formation of I_2^+ (fig. 4b), thermodynamic
considerations show that the channels leading to $CH_3ICH_3^+$ and $CH_3I_2^+$
are lower in energy by 0.25 - 0.55 eV. These lower energy channels
govern the dissociation of the dimer when excited into the higher
energy Rydberg C-state (fig. 4a). The predominance of I_2^+ in figure
4b implies that dynamics play an important role in the fragmentation
process. The first step in the process following the absorption of a
photon at 248 nm by the dimer is most likely dissociation to give
$CH_3I \cdot I$. We estimate, based on references 7 and 8, that at this
excitation energy the CH_3 and I fragments are formed with 2.0 and 0.25
eV of kinetic energy, respectively. The absorption of a second 248 nm
photon dissociates the remaining CH_3I to form I_2 which is subsequently
ionized and detected. This process provides one of the few examples
of reactions occurring by dissociation within the cluster.

4.2. Large clusters

As the backing pressure of CH_3I is increased, both the MPI and
the absorption results indicate the presence of high clusters of CH_3I
in the jet. In the absorption spectrum, a strong continuum grows in
beneath the Rydberg B and C states. At the same time, the origin and
the ν_2(v'=1) lines show red-shifted dips. Such interference effects
in absorption spectra have been documented and explained in two
limits: atomic spectra and large molecules (23-25). The effect results from the interaction of a continuum with a sharp resonance in
which both carry similar oscillator strength.

Under these same conditions, the MPI experiment gives rise to
CH_3I^+ as the sole signal carrier, independent of the specific doorway
state excited. The independence of the results on the initial excitation may be explained by a caging effect such as those described in
solution. Large clusters can form a cage about the chromophore, which
prevents fast dissociation and allows its subsequent ionization to
CH_3I^+, independent of doorway state. However, the appearance of the
monomer ion as the sole signal carrier seems counterintuitive, as one
would expect to see a distribution of fragment ions from the larger
methyl iodide clusters.

One intriguing possibility is that in the large clusters an exciton type state is formed. If this is the case, then both the absorption spectra and the MPI data may be explained. The formation of such an exciton state will definitely give rise to a continuum as is observed in the absorption. This idea is somewhat supported by the absorption spectrum obtained when CH_3I monomer is solvated in large rare gas clusters. Even though the sharp Rydberg structure observed in such clusters corresponds to that of the monomer, an intense continuum and interference dips are also apparent, as is seen in figure 3c. These effects grow more important with increasing atomic number of the rare gas. These results suggest that the continuum is associated with the cluster itself, rather than with some greatly-perturbed state of the methyl iodide.

If this model is true, then the MPI results may also be explained. If one or two UV photons suffice to excite the exciton state, subsequent photon absorption may only serve to dissociate the cluster. If a single CH_3I chromophore provides the electron which is promoted to the excitonic state (as implied by the rare gas cluster results), it may remain as a "hole", and the cluster may fragment into CH_3I^+ + (cluster)$^-$. This would then explain why no larger fragments are seen in the MPI experiment.

5. CONCLUSIONS

In this work two mechanisms which slow down dissociation in a cluster have been demonstrated. The first can be observed in clusters as small as dimers. In this case the change in the absorption spectra of the Rydberg B and C states of CH_3I upon dimer formation reflects the change in the dissociation dynamics. In the monomer those vibronic states which dissociate fast appear either as weak peaks or can not be observed at all in the spectrum, due to their very large linewidth. Upon dimerization, the crossing of the B and C states by the repulsive valence potential is shifted to higher energies. This in turn causes the rate of predissociation out of low lying-vibrations in the Rydberg system to decrease, and therefore more lines appear in the spectrum.

In studying the MPI of methyl iodide dimers, we find that the lifetime of the "doorway" state has a pronounced effect on the dissociation mechanism. In intermediate sized clusters, we observe fragmentation which is indicative of a classical cage effect. In very large clusters, on the other hand, we postulate the excitation of an exciton-type state, followed by its dissociation.

ACKNOWLEDGEMENTS

Support from the National Science Foundation is gratefully acknowledged. R. N. acknowledges support from the Fund for Basic Research administrated by the Israel Academy of Sciences, and the Minerva Foundation Munich, Germany.

REFERENCES

1. J. B. Valentini and J. B. Cross, J. Chem. Phys. **77** 572 (1982).
2. F. G. Amar and B. J. Berne, J. Phys. Chem. **88** 6720 (1984).
3. M. L. Alexander, M. A. Johnson and W. C. Lineberger, J. Chem. Phys. **82** 5288 (1985).
4. a) F. G. Celii and K. C. Janda, Chem. Rev. **86** 507 (1986).
   b) K. C. Janda, Adv. Chem. Phys. **60** 201 (1985).
5. a) D. H. Levy, Adv. Chem. Phys. **47** 323 (1981).
   b) D. H. Levy, Annu. Rev. Phys. Chem. **31** 197 (1980).
6. G. N. A. van Veen, T. Baller, A. Z. De Vries and N. J. A. van Veen Chem. Phys. **87** 405 (1984).
7. M. D. Barry and P. A. Gorrey, Mol. Phys. **52** 461 (1984).
8. S. J. Riley and K. R. Wilson, Discuss. Faraday Soc. **53** 132 (1972).
9. S. L. Baughcum and S. R. Leone, J. Chem. Phys. **72** 6531 (1980).
10. R. K. Sparks, K. Shebatake, L. R. Carlson and Y. T. Lee, J. Chem. Phys. **75** 3838 (1981).
11. H. W. Hermann and S. R. Leone, J. Chem. Phys. **76** 4759 (1982).
12. M. O. Hale, G. E. Galica, S. G. Glogover and J. L. Kinsey, J. Phys. Chem. **90**, 4997 (1986)
13. M. Dzvonik, S. Yang, and R. Bersohn, J. Chem. Phys. **61** 4408 (1974).
14. J. L. Knee, L. R. Khundkar and A. H. Zewail, J. Chem. Phys. **83** 1996 (1985).
15. Y. Jiang, M. R. Giorgi-Arnazzi and R. B. Bernstein, Chem. Phys. **106** 171 (1986).
16. M. Shapiro, J. Phys. Chem. **90** 3644 (1986) and references cited therein.
17. a) R. L. Sunberg, D. Imre, M. D. Hale, J. L. Kinsey and R. D. Coalson, J. Phys. Chem. **90** 5001 (1986).
    b) S. Y. Lee and E. J. Heller, J. Chem. Phys. **76** 3035 (1982).
18. S. K. Gray and M. S. Child, Mol. Phys. **51** 189 (1984).
19. G. Herzberg and G. Scheibe, Trans. Faraday Soc. **25** 716 (1929); Z. Physik. Chem. **B7** 390 (1930).
20. a) W. S. Felps, P. Hochmann, P. Brint and S. P. McGlynn, J. Mol. Spectrosc. **59** 355 (1976).
    b) J. D. Scott, W. S. Felps, G. L. Findley and S. P. McGlynn, J. Chem. Phys. **68** 4678 (1978).
21. W. C. Price, J. Chem. Phys. **4** 539 (1936).
22. a) R. S. Mulliken and E. Teller, Phys. Rev. **61** 283 (1942).
    b) R. S. Mulliken, Phys. Rev. **61** 277 (1952).
23. U. Fano, Phys. Rev. **124** 1866 (1961).
24. J. Jortner and G. C.Morris, J. Chem. Phys. **51** 3689 (1969).
25. R. Scheps, D. Florida and S. A. Rice, J. Chem. Phys. **56** 295 (1972).

THE YIELD OF EXO-ELECTRONS FROM CLUSTER BOMBARDMENT OF METALLIC SURFACES

Uzi Even, Pieter J. de Lange, Peter J. Renkema and Jan Kommandeur
Laboratory for Physical Chemistry
Nijenborgh 16
9747 AG Groningen
The Netherlands

Abstract

Electron emission is observed when a neutral supersonic cluster beam collides with a metallic surface. Using a very simple theoretical model, based on thermoionic emission, we compare calculated electron yields with measured yields for several cluster distributions. This is done for two different collision models. Surprisingly, we have to use a very low value of the heat conductivity to fit the experimental result of reaching surface temperatures of about 5000 K.

1. Introduction

The study of the properties of molecular clusters is exceedingly interesting, because they form aggregates which can be thought of as intermediate between the solid, the liquid and the gaseous phase. After discovering the convenient way of preparing clusters of almost all compounds in a supersonic expansion, many experiments have been performed with them (1). Recently, electron emission was observed when a supersonic cluster beam collides with a metallic surface (2,3). We have called these electrons exo-electrons. The kinetic energy of clusters which are seeded into helium is so high that, when part of this energy is transferred to a metallic surface, an area becomes locally so highly excited that electron emission may result. The number of exo-electrons emitted depends strongly on the cluster size distribution and increases very fast when the distribution shifts to larger clusters. No electron emission at all was observed when argon was used as carrier gas. Although the cluster distribution shifts to larger clusters using argon, because of the more effective cooling of the larger carrier gas atoms, the velocity of the beam decreases so much that the kinetic energy of the clusters becomes too low to induce electron emission.

By measuring the kinetic energy distribution of the exo-electrons it is possible to estimate the surface temperature of the metal. This was done for the bombardment of various metals (copper, aluminium,

J. Jortner et al. (eds.), Large Finite Systems, 265–275.

nickel) with CCl_4 clusters seeded in helium (2). A local surface temperature of about 5000 K was derived from the experiments, essentially independent of the metal used as a target.

In this paper we try to find a relation between the number of exo-electrons which are ejected and the cluster distribution. In the experimental section we describe the time of flight mass spectrometer with which we were able to measure cluster distributions. In the theoretical part we try to calculate the number of exo-electrons we expect from the collision of one cluster of size m with the metal, assuming the electron ejection can be described as thermoionic emission. Using this result and a measured cluster distribution we can calculate the total number of exo-electrons we expect for this distribution. Comparing this theoretical number with the experimentally measured number we can test our theory.

2.Experimental

A scheme of the experimental set-up is given in fig. 1. It has also been described elsewhere (3). It basically consists of a cluster formation part and a time of flight mass spectrometer. The cluster formation takes place by expanding triethylamine (TEA) seeded into helium carrier gas (2 bar) through a conical quartz nozzle into a vacuum chamber. The conical design increases cluster formation considerably by increasing the beam pressure through its focusing effect (4,5). Before reaching the T.O.F. mass spectrometer the beam passes 1 cm downstream from the nozzle through a conical skimmer to the second chamber and 8 cm further downstream through a similar skimmer to the third chamber. The three chambers are differentially pumped, the pressures under beam-on conditions were respectively 10^{-3}, $2*10^{-5}$ and 10^{-6} Torr. Between the nozzle and the first skimmer there is a possibility to chop the beam. This is performed by a rotating aluminium wheel perforated with 16 holes. The wheel is driven by a 6V Graupner Marbuchi RS-380 S toy electromotor, its opening and closing time is measured by a LED coupled to a photodiode. In practice we worked with an opening time of about 150 µs. The ionization part of the T.O.F. mass spectrometer consists of a 5 bar argon discharge lamp running at a frequency of about 5000 Hz. The lamp is separated from the vacuum by a CaF_2 window (hv<8.5eV). The light is focused by a CaF_2 lens between the first two of three parallel acceleration plates with central holes. After passing the third plate the ions enter a field free drift tube and after 25 cm free flight enter the detection part consecutively, delayed according to their mass. In the detection part positive ions are accelerated to and stopped by a cylindrically shaped aluminium cup at -15 kV. Secondary electrons formed in these collisions are accelerated out of the cup through a hole in the side by the high negative voltage. They impinge on a scintillating material covered with a thin layer of aluminium which is at ground potential. Photons which arise through the scintillation are detected by a photomultiplier. Neutral clusters also arrive in the cup and exo-electrons obtained by the collision are detected as well. This signal was seperated from the signal of the ions by chopping the beam,

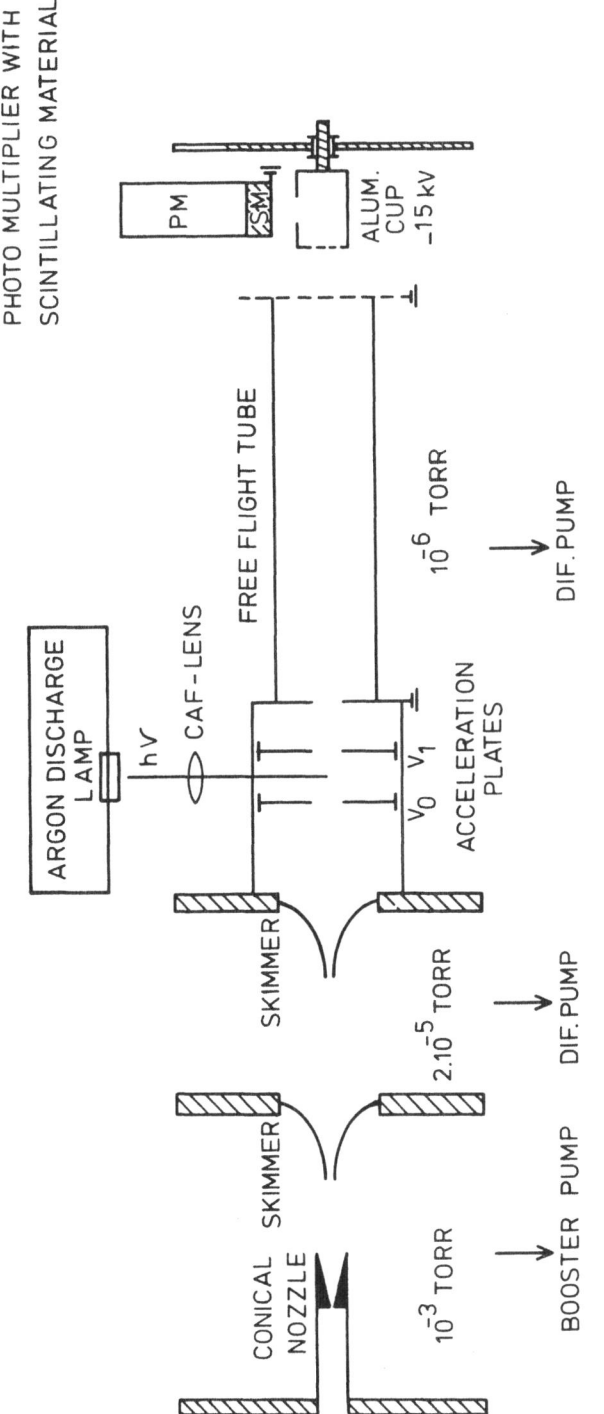

Fig. 1 : Experimental set-up, consisting of a cluster source and a time-of-flight mass spectrometer

through which the ionized and neutral clusters arrive seperated in time at the aluminium cup. Moreover, chopping the beam gives an easy way of measuring the beam velocity.

3. Theoretical

Assuming the collisions of clusters with a metallic surface to be independent of eachother, the total number of exo-electrons (I_{ex}^{tot}) we expect for a given distribution is

$$I_{ex}^{tot} = \int_0^\infty N(m) \, I_{ex}(m) \, dm \tag{1}$$

where $N(m)$ is the number of clusters of size m and $I_{ex}(m)$ the number of exo's we get when a cluster of size m collides with the surface. The assumption of independent collisions seems to be correct because of the fast cooling of a hot spot on a metallic surface and the relatively low collision frequency. Eq. (1) can be rewritten to

$$I_{ex}^{tot} = \int_0^\infty N(m)*m*\frac{I_{ex}(m)}{m} \, dm \tag{2}$$

This is done because we do not measure $N(m)$ in our experiment, but $N(m)*m$, which is caused by the ionization-efficiency of the clusters, which increases linearly with m. For weakly bound clusters (van der Waals bound) we expect an ionization-efficiency which is proportional to the cluster size m. Therefore, what we want to know is $I_{ex}(m)/m$, the number of exo-electrons per monomer as a function of the cluster size. We will try to calculate this quantity for two extreme cases.

I) The cluster spreads totally over the metal and heats the whole surface eventually covered. In this case (we will refer to it as case I), the heated surface area is proportional to the clustersize m and the heat per unit surface q is independent of m.
II) The cluster spreads, but the heat (kinetic energy) is transferred only to the projection of the cluster on the surface. Now (we will refer to it as case II) the heated surface area is proportional to $m^{2/3}$ and q is proportional to $m^{1/3}$.
For both cases it is possible to calculate the temperature of the metal as a function of time, radial distance, depth and cluster size using a numerical heat conduction program. From the surface temperature we calculate, with the expression for thermoionic emission the number of electrons ejected:

$$I_{ex}(m) = C \int \int T_S^2(m) \exp \left(\frac{-W}{kT_S(m)} \right) dAdt \tag{3}$$

in which C is a constant depending on the metal and the detection efficiency, T_S the surface temperature and W the work-function of the metal. The integral has to be carried out over the surface A and the time t.

4. Results and discussion

a) Experimental results:

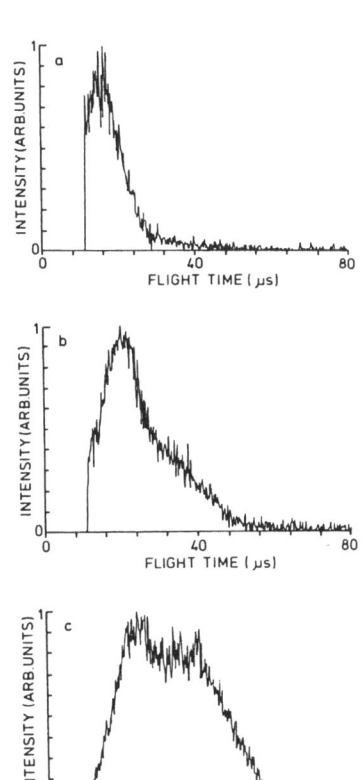

Fig. 2:a) TOF spectrum of triethylamine seeded into helium (2 bar) on a 80 µs timescale (detection up to 60,000 a.m.u.) Relative measure of the number of exo-electrons experimentally detected with this distribution: 1

b) Same as a) but with higher ratio of T.E.A. vapour to heliumgas. Relative measure of the number of exo-electrons: 23

c) Same as a) but with highest ratio of T.E.A. vapour to heliumgas. Relative measure of the number of exo-electrons: 190

In figs. 2 a,b,c we show three time of flight spectra of triethylamine (T.E.A.) seeded into helium carrier gas on a timescale of 80 µs. With the acceleration voltage used (2560 V) this means detection of clusters up to a mass of 60,000 a.m.u. The exact ratios of T.E.A. vapour to heliumgas are not known, but ratio (a) < ratio (b) < ratio (c). The expansion pressure was in all cases 2 bar. The resolution on this timescale is not high enough to resolve different mass peaks, but for the present purpose there is no need for this. For accuracy, masses below 500 a.m.u. (including the monomer and fragments) were filtered out by a pulse generator working as an AND-gate.

The "double hump" we measured for broad cluster distributions (see fig. 2c for instance) is a bit of a problem. We found it not only for T.E.A. but also for benzene, acetone and other organic compounds. Because the first maximum is at about half of the mass of the second, double ionizations would be a possibility. But reducing the light intensity by defocussing the lamp didn't change the relative intensities of the two maxima significantly. Moreover, given the low ionization efficiency, the chance of a double ionization is so low that we practically can exclude this ground for the "double hump". If it is real and also present in the neutral cluster distribution, it would be very interesting. Many groups have reported "magic numbers" (6,7) in cluster distributions, but most of the time these concern much smaller clusters. Because of the present uncertainty about its real existence or as an experimental artifact, we don't speculate further about it.

In the captions of figs 2a,b,c we give a relative measure of the number of exo-electrons detected experimentally. From the numbers it is clear that we measure a higher number per molecule when the distribution shifts to larger clusters.

In fig. 3 we give an example of a measurement of the velocity of the neutral cluster beam. Using the signal of the exo-electrons, we measured a flight time of 400 μs for 50 cm, which means a velocity of about 1250 m/s. With the simple equation for a normal supersonic nozzle:

$$v = \frac{\gamma}{\gamma-1} \left(\frac{2kT_0}{m}\right)^{1/2} \quad (8)$$

we calculate 1700 m/s, a satifactory agreement, since the latter number is not corrected for the presence of TEA.

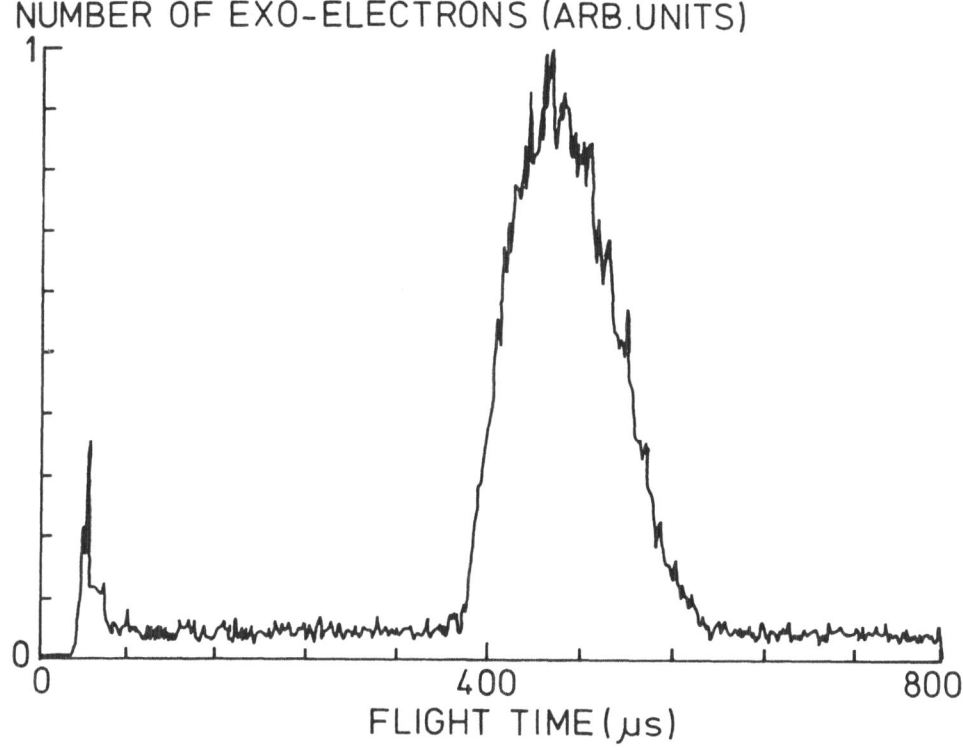

NUMBER OF EXO-ELECTRONS (ARB.UNITS)

FLIGHT TIME (μs)

<u>Fig. 3</u>: Velocity measurement of the neutral clusterbeam. A Time to
        Pulse Height Converter is started by the signal of the chopper
        and stopped by the signal of the exo-electrons.

b) Theoretical results.

      In fig. 4a we give the calculated number of exo-electrons per
monomer for case I, in fig. 4b we do the same for case II. For the
collision time we used the interaction distance between cluster and
surface divided by the velocity of the cluster, about 10^{-13} sec. The
heat capacity was approximated by the electronic heat capacity $C_v(T) =
\pi^2(kT/2E_F)k$ (where k is Boltzmann's constant, and E_F the Fermi
energy), because the electronic gas is excited by the collision and
the heat flows away so fast that there is no or hardly any time to
heat up the atomic motion. Not much is known about the heat
conductivity of metals at very high temperatures (T ≈ 5000 K). Because
of this we used such a value that we could fit the experimental result
of reaching a surface temperature of 5000 K. When we compare the value
thus derived with the value of the electronic heat conductivity for
aluminium at room temperature we see that it is a factor of 100 lower.
Apparently, the conductivity decreases enormously for such high
temperatures, otherwise thermoionic electron emission could not result
from the collision.

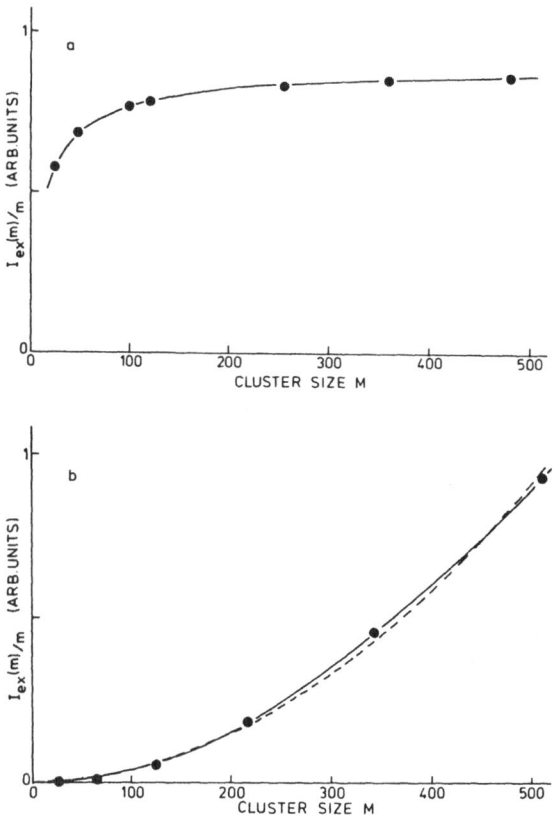

Fig. 4: a) Calculated number of exo-electrons per monomer as a
function of cluster size for heating the whole surface
eventually covered (case I)
b) Same as a) but for heating only the projection of the
cluster on the surface (case II). Dotted line is a quadratic
best fit.

The curve for case I can be understood very easily. If the
cluster size grows the relative radial heat loss decreases and because
the entrant heat per surface unit is constant, the number of exo-
electrons per monomer goes to a constant value. In case II the entrant
heat per surface unit increases with cluster size, because of this
higher temperatures are reached and the yield per monomer increases
with cluster size. In a wide range the curve can be fitted with a
quadratic function (dotted line).

c) Combined Results

In fig. 5 we have plotted the measured number of exo-electrons
against the calculated number (using eq. 2) for several cluster
distributions. Because we only know relative numbers, we ideally
expect a linear relation. For case I (fig. 5a) we approximated the

curve of fig. 4a with $I_{ex}(m)/m$ = constant. This approximation doesn't disturb the results too much because the relative weight of the small clusters is low in the total exo-integral. For case II (fig. 5b) we approximated the curve of fig. 4b with
$I_{ex}(m)/m = C*m^2$.
         Comparing the results for both cases it is clear that we find the best linear relation in the second case. Therefore, we favour the model in which the colliding cluster heats only its projection on the surface.

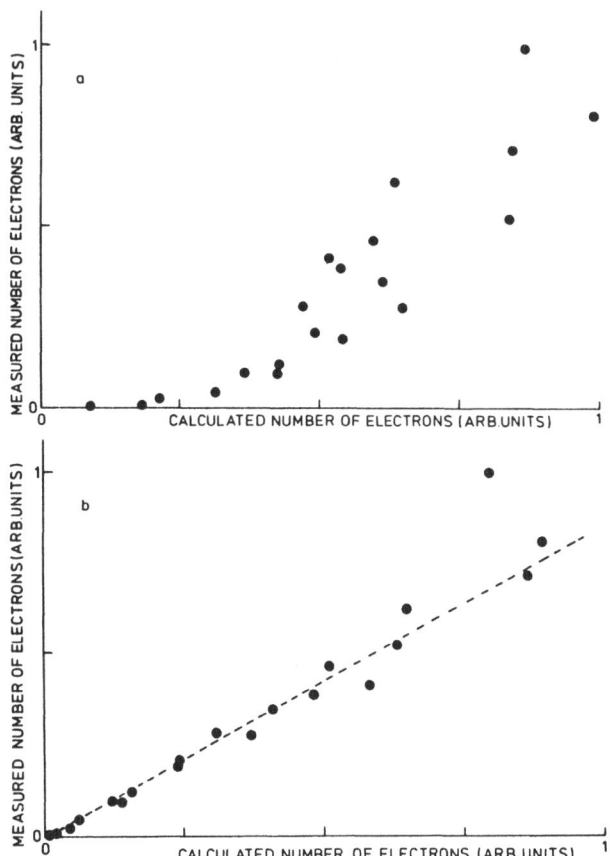

Fig. 5: a) Plot of the measured number of exo-electrons against the
        calculated number for case I
     b) Same as a), for case II

5 Conclusions

        Using measured cluster distributions together with their relative exo-yield and a very simple theoretical model we have tried to establish how the kinetic energy of a cluster is transferred to a metallic surface. Given the simplicity of the theory the results are not disappointing and we certainly can give a jugdgement about the two different models we have used. If a cluster would excite the total surface eventually covered the yield per monomer would go to a constant value for large clusters. This must hold more generally and not only for our theoretical model. But it is in contradiction with our measurements. To fit the measurements, the energy input per surface unit has to rise with increasing cluster size, then the number of exo-electrons per monomer increases with cluster size. The relation calculated for heating only the projection of the cluster gives satisfactory agreement, but of course other theories in which the energy input per unit surface increases with cluster size would also work.

We have to use a surprisingly low value of the heat conductivity to fit the experimental result of reaching a surface temperature of 5000 K. This is an indication that the heat conductivity decreases enourmously at such high temperatures. Otherwise it would be impossible to observe thermionic electron emission when a metal is bombarded with neutral clusters.

Acknowledgement

We express our thanks to Dr. Jonkman for the use of his computer program.

This work was supported by the Netherlands Foundation for Chemical Research with financial aid from the Dutch Organization for the Advancement of Pure Research (Z.W.O.)

References

1) J.C. Philips, Chem. Rev. 86, 619 (1986)

2) U. Even, P.J. de Lange, H.T. Jonkman and J. Kommandeur, Phys. Rev. Lett. 56, 965 (1986)

3) P.J. de Lange, P.J. Renkema and J. Kommandeur, (to be published)

4) O.F. Hagena and W. Obert, J. Chem. Phys. 56, 1793 (1972)

5) H.T. Jonkman, U. Even and J. Kommandeur, J. Phys. Chem. 89, 4240 (1985)

6) a) H.W. Kroto, J.R. Heath, S.C. O'Brien, R.F. Curl and R.E. Smalley, Nature 318, 162 (1985)
   b) M.Y. Hahn, E.C. Honea, A.J. Paguia, K.E. Schriver, A.M. Camarena and R.L. Whetten, Chem. Phys. Lett. 130, 12 (1986)

7) O. Echt, A. Reyes Flotte, M. Knapp, K. Sattler and E. Recknagel, Ber. Bunsenges. Phys. Chem. 86, 860 (1982)

8) G. Ter Horst, thesis, University of Groningen (1982)

THE ROLE OF LARGE OLIGOMERS IN DESIGNING MATERIALS OF INTEREST FOR THEIR ELECTRICAL AND OPTOELECTRONIC PROPERTIES.

J.M.André
Facultés Universitaires ND de la Paix
Laboratoire de Chimie Théorique Appliquée
61, rue de Bruxelles
5000 Namur
Belgium

ABSTRACT. In this paper, we develop the links between clusters (or aggregates) and infinite systems by examples chosen in the field of quantum chemistry of polymers. On a methodological basis, we stress that electron properties in band terms are equivalent to molecular results in terms of molecular orbitals for energies, topologies and electron densities.The role of exchange, correlation and localization is discussed and illustrated by examples on linear chains of hydrogen atoms and polyenes. As a practical application, attention is paid to the question of electronic response to external electric fields as exemplified by calculations on chains of hydrogen atoms. The pattern of saturation for polarizability properties of long conjugated chains is discussed in the case of polyenes.

1. INTRODUCTION

The present book is concerned with large but finite systems. Those systems form, in organic chemistry, a somewhat new class having characteristics intermediate between those of molecules and polymers. It is tempting to establish a correlation between the strict division into molecules and polymers and that which has historically existed for a long time between *molecular* quantum chemistry and *solid state* physics. Until recent days, it has been sometimes assumed that there was a continuous link between the properties of *real* molecules of increasing but finite sizes and those of the *ideal* infinite solid. However, if it is true that some properties extrapolate to those of the infinite system (independently of its dimensionality), it is also true that new concepts do appear.

In this paper, we would like to contribute by some aspects we have met in applying the methods of molecular quantum chemistry and solid state physics to the study of the electronic structure of polymers, a field now commonly referred to as *polymer quantum chemistry*. Throughout this paper, we restrict ourselves to linear chains. The paper is organized as follows: after this short introduction, we discuss in part 2, how basic concepts link standard molecular orbital theory and standard band theory, emphasizing their interrelation but also domains where they deviate from each other when we are concerned with large finite oligomers. In part 3, we illustrate those concepts by applications in the molecular design of materials of interest for their electrical or optoelectronic properties. Finally, part 4 displays our conclusions.

J. Jortner et al. (eds.), Large Finite Systems, 277–288.

2. MOLECULAR QUANTUM CHEMISTRY versus SOLID STATE PHYSICS

2.1. Molecular Orbitals and Bands in Polymeric Molecules

In molecules and polymers, the standard theory is based on the Hartree-Fock theory which is an independent model where a single electron moves in the field of the nuclei and in the mean coulombic and exchange field of all the electrons. A set of molecular orbitals (MO's) is obtained to describe the occupied and unoccupied one-electron wave functions. In molecular quantum chemistry, it is usual to represent the molecular orbitals by single levels which can be at most doubly occupied by a pair of electrons of opposite spins.

If he is interested in evaluating the electronic properties of a polymer or of a large oligomer, the molecular quantum chemist will probably start by an extrapolation study like that summarized in Figure 1. That figure is an adaptation to the case of polyacetylenes of a classical figure explaining the properties of planar graphite given in the pioneering book by the Pullmans on "quantum biochemistry"(1) and has been frequently used since for the polyethylene (methane, ethane, propane,...) series (2,3) and even previously for the polyene-polyacetylene series (ethylene, butadiene, hexatriene,...) (4-6) in the framework of the Hückel method. In its leftmost part, Figure 1 plots the MO levels as a function of the number of carbon atoms in the oligomeric chain. It can be noticed that as the number of carbon atoms increases the distance between the energy levels diminishes in order to form, in the infinite limit, occupied energy regions (the internal and valence bands), unoccupied energy regions (the conduction bands) and forbidden energy regions (the forbidden bands or energy gaps).

On the other hand, the solid state physicist makes full use of the possible translation symmetry of the lattice and uses the language of Brillouin zones introduced in the classical work by Bloch in 1928. In this theory, the so-called Bloch functions (molecular orbitals for an infinite 1D chain) are eigenfunctions of a translation operator and as such express in terms of a wave number, k, defined in the reciprocal space. For systems having one-dimensional periodicity (and an electronic density periodic between sites separated by a, the length of the direct cell) , the Bloch- or polymeric orbitals are functions of that wave number k:

$$\phi(k,\mathbf{r}+ja) = \exp(ikja)\ \phi(k,\mathbf{r})$$

Thus, the associated "orbital" energies are also functions of k. The representation of those dispersion curves is called an energy band. Standard theorems of solid state physics demonstrate that those energy bands are periodic in the reciprocal space. Thus, their search must be performed only on a length equivalent to a single unit cell of the reciprocal lattice. Further theorems state that the energy bands are symmetric with respect to k=0. The search is simplified if we use a symmetrized part of the reciprocal space, the so called first Brillouin zone ranging from $-\pi/a$ to $+\pi/a$ which must be only explored from k=0 to k=$+\pi/a$ (half the first Brillouin zone). In the selected example of the center part of Figure 1 (polyacetylene), the translation unit cell contains a -CH=CH- unit and thus 2 π-electrons. The first Brillouin zone contains the corresponding 2 valence bands. The rightmost representation is the easiest one for experimental comparisons; it is the density-of-states (DOS) which plots the number of available energy levels as a function of

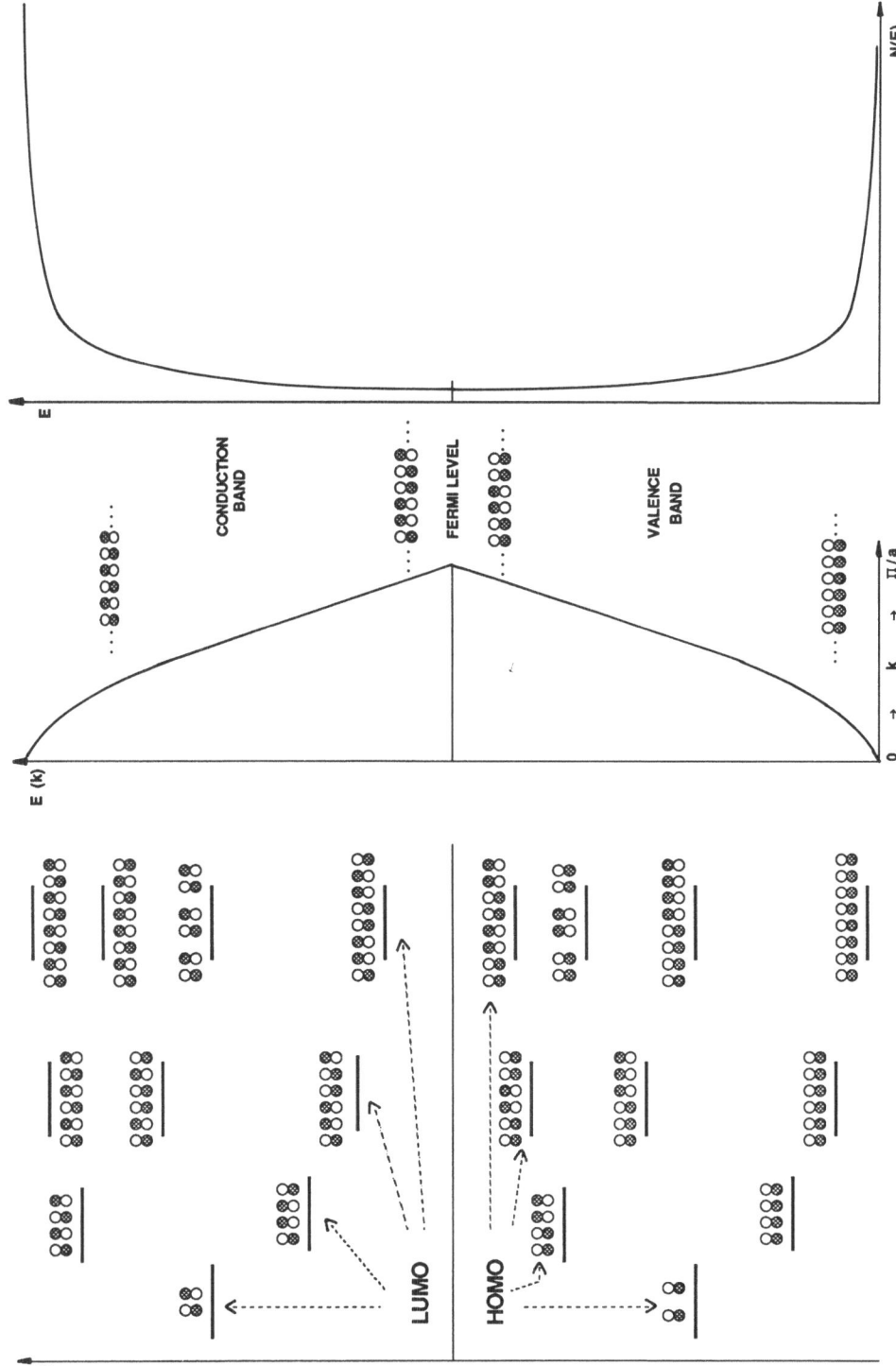

Figure 1: Hückel π energy levels and molecular orbitals of ethylene, butadiene, hexatriene and octatetraene (left),

Hückel π band structure (center) and Hückel π density of states (right) of polyacetylene

energy for infinite systems. It is usually normalized to the number of electrons per unit cell.

If, at first sight, the three representations look quite different, in practice, they are equivalent. This point is more clearly seen when schematizing the forms of the MO's in a LCAO expansion in terms of atomic orbitals (AO). The number of MO's obtained by the procedure is equal to the number of AO's included in the expansion. Due to first principles, the number of nodes increases with orbital energy. The bonding and antibonding π orbitals of ethylene are easily recognized at the first point as well as the four orbitals of butadiene popularized by the work of Woodward and Hoffman (7). The situation is similar in band theory. For high symmetry points of the first Brillouin zone the polymeric orbitals are real and can be plotted easily. The lowest orbital has no node (symmetric combination of AO's, $\Sigma\exp(ikja)=\Sigma 1$ if k=0) while the highest has the maximum number of nodes (most antisymmetric combination, $\Sigma\exp(ikja)=\Sigma(-1)^j$ if k=π/a), intermediate situations occur at the Fermi level where, in the case of a regular chain of polyacetylene, near degeneracies appear in the structure of the Highest Occupied Molecular Orbital (HOMO) and the Lowest Unoccupied Molecular Orbital (LUMO). For other points of the first Brillouin zone, the polymeric orbitals have also imaginary components but still obey the same principles. Qualitative molecular orbital techniques are now commonly used to construct π-band structures of conjugated polymers (8,9).

Figure 1 appears to be a crossing link between molecular quantum chemistry and solid state physics. The interpolation properties between the finite oligomers and the infinite polymer have also been emphasized by compact formulas giving the orbital energies (or possibly their structure) as a function of the number of carbon in the chains. Examples of such formulas are found in many textbooks (10,11,12). A rather complete analysis in the framework of Hückel methodology is available by the general technique of finite differences (13). For regular chains of N carbon atoms, the orbital energies are given in terms of the Hückel parameters α and β by:

$$\varepsilon_j = \alpha + 2\beta\cos(j\pi/N+1) \qquad \text{for linear chains}$$

$$\varepsilon_j = \alpha + 2\beta\cos(2j\pi/N) \qquad \text{for cyclic chains}$$

They correspond to an evolution of the energy gap like:

$$\Delta E = -4\beta\sin(\pi/2(N+1)) \qquad \text{for linear chains}$$

$$\Delta E = -4\beta\sin(\pi/N) \qquad \text{for cyclic chains}$$

For alternant chains of carbon atoms, the mathematical expressions are more involved and solutions are, as far as we know, only available for cyclic chains of N atoms and linear chains containing an odd number of N-1 atoms. They read, in both cases, as

$$\varepsilon_j^{(1)}, \varepsilon_j^{(2)} = \alpha \pm \sqrt{\beta_1^2 + \beta_2^2 + 2\beta_1\beta_2\cos(4j\pi/N)}$$

For the linear case, there also exists one non-bonding orbital at the energy α. It is important to note (that point will be relevant to the discussion developed in the next section) that in the case of an alternant systems, ΔE does not tend to zero as $N \to \infty$ but instead to the finite value:

$$\lim_{N \to \infty} \Delta E(N) = 2 \ (\beta_1 - \beta_2)$$

Other N-dependence are cited in the literature. For example, the dependence of the physical properties of a homologous series of conjugated compound with the conjugated length n has been shown (14) to follow a *universal* equation

$$P = A + B \ (1/2)^{2/n}$$

where A and B are two parameters independent of n.

The present discussion would raise the impression that discussing the electronic properties of finite oligomers or infinite polymers in terms of molecular orbitals or electron bands does not introduce new concepts, or, in other words, the infinite polymers can be treated as large molecules. However, we will illustrate in the next sections that this is not always the case and that the electron properties, in general, do not extrapolate as ideally as could be thought from the preceding formulas.

2.2. Electronic Localization and Hartree-Fock deficiencies

A nice example of the lack of extrapolation of data between the finite range and the infinite one is found in the UV transitions of polyene molecules. Historically, this point has been related to chain geometry by the pioneering work of Kuhn (15). Kuhn has shown that in polymethine dyes $=N\text{-}(CH=CH)_n\text{-}C=N^+\text{-}$, a simple free electron model correctly reproduces the first UV transition with a metallic extrapolation for the infinite system. In the polyene series, $CH_2=CH\text{-}(CH=CH)_n\text{-}CH=CH_2$, he had to perturb the constant potential by a sinusoïdal potential in order to cover the experimental trends. The role of the sinusoïdal potential is to take into account the structural bond alternation between bond lengths of single and double bond character. When applied to the infinite system, in such a perturbed free electron model or Hückel-type theory, a non-zero energy gap is obtained (about 1.90 eV in Kuhn's calculation). From that work, it became clear that the metallic character of the infinite chain would strongly depend upon geometrical parameters, an alternating chain being semiconductive, a regular one being possibly metallic.

A renewed interest for that question appeared in the late 1970's, when conductive films of doped polyacetylenes were synthesized by Shirikawa (16) and Heeger & McDiarmid (17). As pointed out above, a first order explanation should imply geometrical rearrangements. We made a model study (18) of a doped chain of hydrogen atoms (similar to polyacetylene by having a single electron per site and exhibiting bond alternation when undoped). That calculation showed that the electron transfer has an important effect on the electron density and considerably perturbs the one dimensional bond network. In particular, the tendency of bond alternation to decay upon doping was proven but, even if that was a support to the structural transition hypothesis, the band gap was not completely removed in the Restricted Hartree-Fock (RHF) scheme. The same results were obtained on more realistic polyacetylenic models (6,19). In all calculations, the main effect of doping was to pull up the top of the occupied bands and push down the bottom of the empty ones in order to close the forbidden region. However, ideal metallic situations are not observed, no degeneracy of the HOMO and LUMO states is obtained. That alteration of the chain geometry upon metallic conditions implies a strong electron-lattice coupling and, consequently, tends to localize the optical excitations.

The metallic behavior of doped conjugated systems has been rationalized in terms of the polaron-bipolaron model (20) but here we will pay more attention to the question of the localization of excitations in relation with correlation effects. Early calculations (21,22) of the $^1B_{1u}$ transition by semi-empirical Pariser-Parr-Pople-like (PPP) calculations are represented in Figure 2. The figure plots the theoretical first UV transition with respect to the number (N) of carbon atoms in the polyene chain. Experimental points are represented by the dots. It is surprising to see that the agreement is perfect until N≈26 and that the theoretical curve has a spurious minimum not observed in standard experiments. This is an artefact of the methodology and an interesting example of the non-correspondence between finite and infinite results.

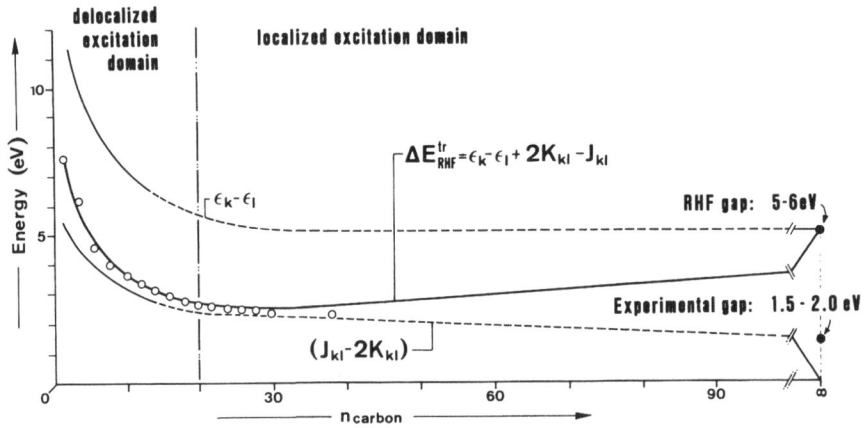

Figure 2: First singlet transition energies ($^1\Delta E$), LUMO -HOMO energy difference (ϵ_k-ϵ_l), Coulomb and exchange interactions (J_{kl} - $2K_{kl}$) as a function of the number of carbon atoms in polyenic chain. Energies in eV.

In the RHF scheme, first singlet and triplet transition energies are expressed by the LUMO (ϵ_k)-HOMO (ϵ_l) energy difference and by Coulomb (J_{kl}) and exchange (K_{kl}) modifications between the two one-electron states:

$$^1\Delta E = \epsilon_k - \epsilon_l + 2K_{kl} - J_{kl}$$

$$^3\Delta E = \epsilon_k - \epsilon_l - J_{kl}$$

In those previous expressions, many energy terms can be experimentally evaluated by usual approximations of quantum chemistry. In particular, ε_k is the electron affinity of the molecule, ε_l is the Koopmans' ionization potential and K_{kl} is half the singlet-triplet energy difference. The calculations show a perfect agreement with the available experimental data. Keeping in mind molecular arguments, the integrals J_{kl} and K_{kl} tend to zero for infinite chains and fully delocalized excitations take place so that the values of the first singlet transition energy, the first triplet transition energy and the energy gap become identical.The theoretical values start to increase to get an overestimated value of the energy gap of about 5 eV. That typical deficiency of the Hartree-Fock model is explained by the fact that in the model the electrons are spread out over the whole molecule and thus the (J-2K) exchange and Coulomb terms quickly vanish. The origin of the theoretical minimum is thus due to a poor balance between the HOMO-LUMO energy difference and the $(2K_{kl} - J_{kl})$ interaction terms.

In practice, one does not observe fully delocalized excitations but instead a localization in the middle of the molecule. That localization is proved by making PPP Configuration Interaction calculations on the first excited $^1B_{1u}$ state (23). An important flux of electronic charges from the outer part of the molecule towards the middle of the molecule is observed in the excited state. Quantitatively, for a chain of 40 atoms, the charge gain is 0.36 e$^-$ on the inner 16 atoms. That localization of the excitation in the middle of the molecule can be viewed as exciton or correlated electron-hole pair formation. This optical excitation is sometimes referred to as a self-trapped singlet exciton (24). The accompanying localization of the excited charge distribution results in a non-negligible (J-2K) term and a significant decrease of ΔE. The detailed calculations correlates nicely with well-known facts on localization effects; the RHF gap are smaller for regular chains than for alternant chains, the extent of localized excitations is in the range 10-15 double bonds.

The argumentation can be developed further. If as just demonstrated there exists a relation between correlation and energy gaps and if, as popularized by the $\rho^{1/3}$ exchange + correlation potentials of solid state physics, there exists some links between correlation and exchange interactions, there must exist some relations between the importance of exchange effects and the size of energy gaps.

That question has been investigated independently in our analysis of the properties of an hydrogen chain (25) and more formally by Monkhorst & Kertesz (26). In our study of hydrogen chains, we have modified the size of the energy gap by acting on the geometry between a symmetry fixed regular chain (metallic situation) and a strongly bond alternant form (insulating form) and calculated the range of the exchange potential by measuring the ratio of the exchange terms in a cell distant by five units and in the first unit cell. The calculations have shown that, for metallic-like situations, when the gap tends to zero, the exchange potential has a long-range behaviour [$E_{exch}(j=5)/E_{exch}(j=1)= 0.1$ $10^{-3}/0.3\ 10^{-1}$] and the correlation corrections should play an important role as indicated by the existence of Hartree-Fock instabilities. For insulating systems, on the other side, we observe a significant gap, the exchange interactions are rather localized [$E_{exch}(j=5)/E_{exch}(j=1)= 0.3 10^{-10}/0.5\ 10^{0}$] and the Hartree-Fock description is rather good since we have no indication of Hartree-Fock instabilities.This clearly demonstrates that the short-range nature of exchange is a function of the size of the energy gaps.

That work has been nicely extended by Calais & Delhalle (27,28). They have investigated the exchange lattice sums in Hartree-Fock theory and have shown that the absolute decay of the LCAO density matrices with respect to distance (as measured by its cell index j) is of exponential character for nonmetallic situations while they fall like $|j|^{-1}$ in metallic situations. Also, they proved that a metallic situation is unlikely to exist in a

RHF description except if symmetry constrained. Any active infinitesimal perturbation should remove this *artificial* metallic state. In connection with the previous points concerning exchange, they have also demonstrated in a direct space approach, the generality of the vanishing of RHF density of one-electron states at the Fermi level for systems having one-dimensional periodicity. This *pathological* behaviour results from the combination of the slowly decaying elements of the LCAO density matrices and the long-range nature of the Coulombic interaction.

This point is related to the definition of the softness (reciprocal of the hardness) introduced by Parr as the density of states at the Fermi level (29).

Monkhorst (30) was the first to provide a general proof of that vanishing of the RHF density of states at the Fermi level but in a reciprocal space approach.The direct space approach used by Calais & Delhalle is intuitively superior to analyze the origins of similar pathologies in long but finite chains (oligomers). This Hartree-Fock *pathology* raises the questions as to the basis of methods (SAMO, LCLO, VEH,..) simulating ab initio results by constructing effective matrix elements from standard (i.e. including the full exchange) ab initio treatments of limited extension. Such pathological behaviours should also occur in approximation logically descending from full ab initio framework (CNDO and the like) when the lattice sums are carried out to their bitter limits.

3. ROLE OF STUDIES OF LARGE OLIGOMERS IN NON-LINEAR OPTICS.

The organic solid state has recently gained much interest in the fields of highly conducting polymers and non-linear optics. In the latter, their interest over inorganics is the occurrance of much greater effects due to higher optical damage thresholds, purely electronic effects inducing quasi-instantaneous response and ultra fast signal processing. Their excellent mechanical and molding properties added to the virtually unlimited potentialities of organic synthesis have suscitated many studies and the development of a new physics. They raise interesting questions that will be exemplified by a model study of the linear electric polarizability of a linear chain of hydrogen atoms in the presence of an external field and generalized to polyacetylene chains.

In the study of the perturbation due to the switching of an external electric field, it is anticipated that the polarizability, normalized to the monomeric units, tends to reach an asymptotic limit which should grow when the systems exhibit increased geometrical regularity (metallic situation). For complex systems, this limit will soon be out of reach from studies on chains of increasing length. Thus, it would be very useful to be able to estimate this limit from calculations on infinite chains. One could consider as being rather trivial replacing field-dependent MO's by field-dependent Bloch polymeric orbitals and assuming the usual periodicity properties. However, two types of questions are raised.

On the one hand, as shown by Churchill and Holmstrom (31,32), serious difficulties arise in imposing realistic boundary conditions to solve the one-electron eigenvalue equations; under the boundary conditions commonly used in treating the zero-field case (e.g., Born- Karman boundary conditions), this equation either leads to physically inconsistent results or, still worse, has no solution at all. This strange behaviour is a consequence of the pathological nature of the perturbing term, $e\mathbf{F}.\mathbf{r}$, due to the external electric field $\mathbf{F}$ which becomes undetermined in the limits as $\mathbf{F}{\rightarrow}0$ and $\mathbf{r}{\rightarrow}\infty$.

On the other hand, the periodic character of the perturbation is not guaranteed under the non-periodic linear external perturbation which would rule out the use of field-perturbed Bloch orbitals. That point has been investigated by finite field calculations over a chain of 24 hydrogen atoms. The results are summarized in figure 3 which presents the net charge in the unperturbed and perturbed systems. It is seen that the response appears at first sight to be periodic (at least in the middle of the molecule) even if the resulting potential deviates from the ideal e**F**.**r** behaviour. An ab initio study of the asymptotic limit of the infinite one-dimensional periodic chain is being developed in our laboratory by Barbier (33) using the SOS (Summation-Over-States) perturbative methodology of Genkin & Mednis (34).

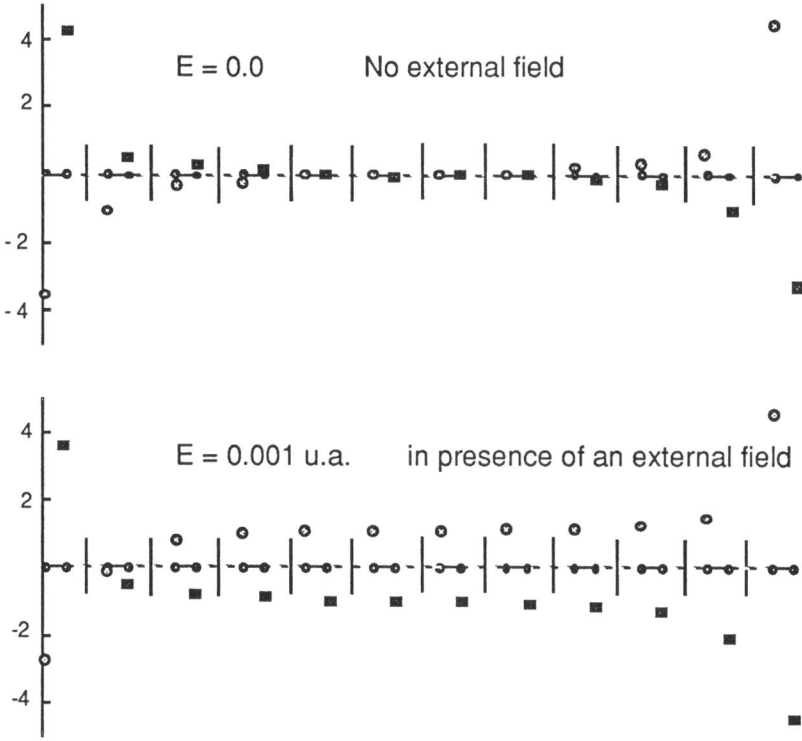

Figure 3: Net charges (x 10^{-2}) of a linear chain of 24 hydrogen atoms in the absence (top) and in the presence (bottom) of an external electrical field. 4-31G calculations.

The size-dependency of the electronic polarizabilities can be illustrated in the polyene series. For regular polyene chains, both free-electron and Hückel theories predict a behaviour of the longitudinal polarizability proportional to L^3 (where L is the total chain length). Accordingly, the longitudinal polarizability per unit cell grows as the second power of the chain length, L^2, and diverges in the limit of an infinite chain. However, both free-electron and simple Hückel models do not take into account Coulombic interactions explicitly. An investigation of the effect of size and bond alternation on oligomers of various types has been explored to get an indication on the saturation pattern of (hyper)polarizabilities by a more elaborate methodology which includes electron-electron interactions in an average way and allows for the electron charge distribution to relax self-consistently as the size of the system increases.

The Finite-Field (FF) STO-3G results (35,36,37) are schematized in Figure 4 for several bond alternation degrees and sizes of the main chain. The investigation of the dependence of the longitudinal (FF) polarizability on bond- and chain-length modifications shows an increase of the polarizability by unit cell with respect to size and a decrease with respect to larger bond alternation. Similar trends are observed in the SOS calculation of the longitudinal cubic hyperpolarizability.

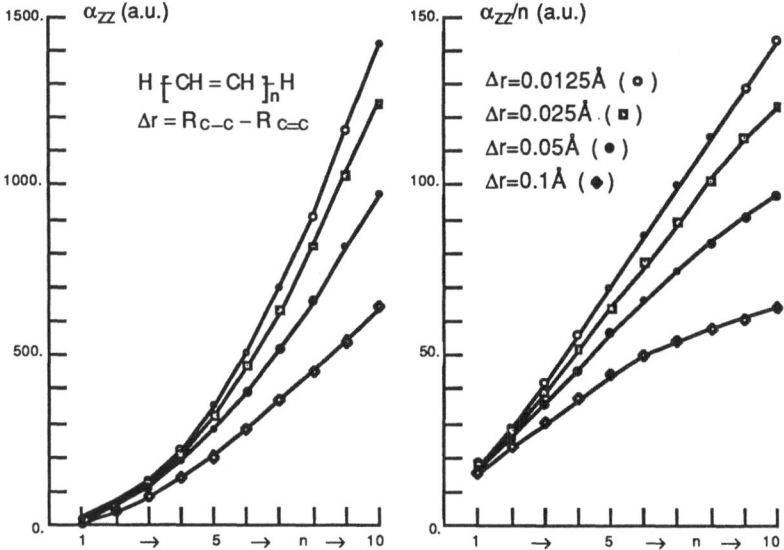

Figure 4: Longitudinal polarizabilities α_{zz}(total [left] and per CH=CH unit [right]) of polyenes as a function of the number (n) of CH=CH units

A detailed analysis of the results shows that structural alterations, enforced chemically or mechanically, and ranging from 0.05 to 0.12 Å are predicted to yield important changes in α_{zz} and suggests the possibility of enhancing the value of not only the polarizability α but also the hyperpolarizabilities β and γ by controlled tuning of the geometry. It also raises some warnings in the use of bond additivity schemes when applied to (hyper)polarizabilities considering their high sensitivity upon relatively small bond length differences.

4. CONCLUSIONS

This paper has illustrated the links existing between the one-electron theory of infinite systems (band theory) and limited molecules (molecular orbital theory). It has been shown that some difficulties arise in the extrapolation properties from limited molecules to infinite systems. When discussing the properties, a strict attention has to be paid on localization effects, on some of the Hartree-Fock pathological behaviours and thus on correlation and its accompanying size-consistency effects.

ACKNOWLEDGEMENTS

This paper is a contribution presented at the twentieth Jerusalem Symposium in Quantum Chemistry and Biochemistry on large finite systems organized under the auspices of the Israël Academy of Sciences and Humanities and of the Hebrew University of Jerusalem. The author is very grateful to Profs.B.PULLMAN and J.JORTNER for allowing him to attend that stimulating experience. He acknowledges the support of the Institute of Biologie Physico-Chimique (Fondation Edmond De Rotschild, Paris) for making his stay possible. He is also indebted to Prof.Joseph DELHALLE and Drs. Joseph FRIPIAT, Jean-Luc BREDAS and Christian BARBIER for frequent and helpful discussions on the topics developed in this paper. All calculations reported here have been made on the Namur-Scientific Computing Facility (Namur-SCF), a result of a cooperation between the Belgian National Fund for Scientific Research (FNRS), IBM-Belgium and the Facultés Universitaires Notre-Dame de la Paix (FUNDP).

REFERENCES

[1]   B.Pullman, A.Pullman, **Quantum Biochemistry**, Interscience Publishers (1963), see pp.304 & 305
[2]   J.M.André, **Etude Théorique de la Structure de Bandes des Systèmes Périodiques**, Ph.D.Thesis, University of Louvain (UCL), Belgium, (1968) see p.131 et sq.
[3]   J.M.André, in **Electronic Structure of Polymers and Molecular Crystals**, J.M.André, J.Ladik (Eds.), 1, Plenum Press (1975)
[4]   J.M.André, J.Delhalle, in **Quantum Theory of Polymers**, J.M.André, J.Delhalle, J.Ladik (Eds.), 1, D.Reidel Publishing Company (1978)
[5]   J.M.André, Advan.Quantum Chem. **12**, 65 (1980)
[6]   J.M.André, in **Current Aspects of Quantum Chemistry**, R.Carbo (Ed.), 273, Elsevier (1982)
[7]   R.Hoffmann, R.B.Woodward, J.Amer.Chem.Soc. **87**, 2046 (1965)
[8]   J.P.Lowe, S.A.Kafafi, J.P.LaFemina, J.Phys.Chem., **90**,6602 (1986)
[9]   K.Tanaka, S.Yamashita, H.Yamabe, T.Yamabe, Synth.Metals, **17**, 143 (1987)
[10] W.Kauzmann, **Quantum Chemistry**, Academic Press (1957) see p.682
[11] F.L.Pilar, **Elementary Quantum Chemistry**, McGraw-Hill (1968) see p.593 et sq.
[12] Th.A.Albright, J.K.Burdett, M-H.Whangbo, **Orbital Interactions in Chemistry**, J.Wiley & Sons (1986) see p.229 et sq.
[13] T.K.Rebane, in **Methods of Quantum Chemistry**, M.G.Veselov (Ed.), 147, Academic Press (1965)
[14] Y.Cao, D.Guo, M.Pang, R.Qian, Synth.Metals, **18**,189 (1987)

[15] H.Kuhn, J.Chem.Phys., **17**, 1198 (1949)
[16] T.Ito, H.Shirakawa, S.Ikeda, J.Polymer Sci.Polymer Ed., **12**, 11 (1974)
[17] C.K.Chiang, C.R.Fincher, Y.W.Park, A.J.Heeger, H.Shirakawa, E.J.Louis, S.C.Gau, A.G.MacDiarmid, Phys.Rev.Letters, **39**,1089 (1977)
[18] J.Delhalle, J.L.Brédas, J.M.André, Chem.Phys.Letters, **77**, 93 (1981)
[19] J.L.Brédas, B.Thémans, J.M.André, R.R.Chance, D.S.Boudreaux, R.Silbey, Journal de Physique, **44**, C3-273 (1983)
[20] J.L.Brédas, B.Thémans, J.G.Fripiat, J.M.André, R.R.Chance, Phys.Rev.,**B29**, 6761 (1984)
[21] J.M.André, G.Leroy, Theoret.Chim.Acta, **9**, 123 (1967)
[22] J.M.André, G.Leroy, Ann.Soc.Scient.Brux., **84**, 133 (1970)
[23] J.L.Brédas, B.Thémans, J.M.André, L.Piela, Bull.Soc.Chim.Belges, **92**, 1 (1983)
[24] E.J.Mele, G.W.Hayden, Synth.Metals, **17**, 107 (1987)
[25] L.Piela, J.M.André, J.G.Fripiat, J.Delhalle, Chem.Phys.Letters, **77**, 143 (1981)
[26] H.J.Monkhorst, M.Kertesz, Phys.Rev., **B24**, 3015 (1981)
[27] J.L.Calais, J.Delhalle, in **Understanding Molecular Properties**, 511, J.Avery et al. (Eds.), D.Reideil Publishing Company (1987)
[28] J.Delhalle, J.L.Calais, J.Chem.Phys., **85**, 5286 (1986)
[29] W.Yang, R.G.Parr, Proc.Natl.Acad.Sci.USA, **82**, 6723 (1985)
[30] H.J.Monkhorst, Phys.Rev., **B20**, 1504 (1979)
[31] J.N.Churchill, F.E. Holmstrom, Amer.J.Phys., **50**, 848 (1982)
[32] J.N.Churchill, F.E. Holmstrom, Physica B.,**123**, 1 (1983)
[33] Ch.Barbier, to be published
[34] V.M.Genkin, P.M.Mednis, Sov.Phys.-JETP **27**, 609 (1968)
[35] V.P.Bodart, J.Delhalle, J.M.André, Springer Series in Solid State Sciences, **63**, 191 (1985)
[36] J.M.André, C.Barbier, V.P.Bodart, J.Delhalle, in **Nonlinear Optical Properties of Organis Molecules and Crystals**, D.S.Chemla & J.Zyss (Eds), Vol.2, 137, Academic Press (1987)
[37] J.M.André, J.O.Morley, J.Zyss, **Volume in tribute of Prof.R.Daudel**, in press (Reidel Publishing Company)

ISOLATED, BARE METAL CLUSTERS: ABUNDANCES AND IONIZATION

Ernst Schumacher and Manfred Kappes
Institute for Inorganic, Analytic, and Physical Chemistry
University of Berne
CH-3000 BERNE 9
Switzerland

ABSTRACT: Metal cluster ensembles serve as ingredients for modelling disordered (glassy, liquid) metals. They are also the first manifestations of bonding interactions of metal atoms. Some of their properties make them new chemical entities which merit research in their own right. Clusters can now be made and studied for all metals as size distributions. There is no agreement yet, what constitutes a reproducibly "correct" size-abundance distribution, nor do we know how to port results from one cluster source to the next. All inferences drawn from dominant species in a mass spectrum must, therefore, be scrutinized for a number of influences before one can draw conclusions about intrinsic properties. This is exemplified with abundances and phototoinization potentials of alkali cluster beams. The scope and experimental applicability of global and detailed theoretical descriptions is discussed and problems to be solved are enumerated.

1. INTRODUCTION

Atomic clusters from metallic elements have been objects of a fast growing research activity since about 20 years [1]. During this period we have also witnessed the partial collapse of an extremely fertile field of solid state science: Physical properties of liquid and amorphous metals and semiconductors can only tenuously be described within the frame of band theory and k-space [2]. The boundary conditions imposed by an ideal periodic lattice with translational symmetry have to be replaced by those of a no longer negligible surface and proper structure of the finite entities making up those systems. The isolated metal clusters seem to be the first order objects to model e.g. glassy, liquid, and dense vapour metals. Their coupling into ensembles allows to explain many puzzling experimental data of non crystalline and quasicrystalline condensed systems from metallic elements.
Today we have the tools to make isolated clusters from every metal ranging in size from the dimer to many dozens of atoms. Most cluster sources use condensation of a high temperature vapour offering little

J. Jortner et al. (eds.), Large Finite Systems, 289–301.
© 1987 by D. Reidel Publishing Company.

control over the ensuing size-abundance distribution. The latter is to
a certain extent source specific and reflects the history of the formation processes. The 2. section addresses this complex and by no means
solved problem.

In order to gain size specific information the clusters are usually
ionized and the ions separately measured in a mass spectrometer. If
combined with a proper method of ionization - photoionization or electron attachment - ionization thresholds and relative cross sections can
be obtained. This serves as a first insight into the electronic properties. The 3. section shows the evolution of the IP from the atom to the
bulk as well as the problems encountered with the interpretations of
the finer details of this behaviour. New, higher precision measurements
are used as a basis.

A brief summary of other experimental information is relevant to our
discussion (for gas-phase isolated clusters only):

Optical spectroscopy: Very little or no information beyond M_3, where
Na_3 and Cu_3 are nice examples [3,4];

Structure: Nothing $> M_3$;

Magnetic moments: known for a few, rather small, clusters of Na,K,Al,
and Fe [5,6];

Electrical polarizabilities: for $Na_x \leq 40$ [7];

Chemical reactivity: Stoichiometric reactions of alkali and small
transition metal clusters with oxygen, sulfur, halogens, and
hydrocarbons [8,9]; catalytic(?) reactions of iron, cobalt, nickel and
platinum metal clusters [10,11,12].

The field is still at an early stage and it will be immediately
evident, why experimental progress is difficult. The interaction with
theory is, therefore, of paramount importance.

2. CLUSTER SOURCES: SIZE-ABUNDANCE DISTRIBUTIONS

Not many years ago we were happy if we could make clusters at all. This
is still true for enthusiastic newcomers to the field. As soon as a
mass spectrum shows some polymer peaks they rush to print and draw all
sorts of fancy conclusions from what they believe the data mean. A mass
spectrum *per se* contains very little immediately interpretable information. This is especially true for a system of homonuclear polymers,
where a given peak may be a genuine product of the cluster growth
process or else contain fragments from any number of larger parent
clusters either formed by collisions, unimolecular decay of hot
species, or during ionization. There is no way to distinguish the many
origins without a lot of work and clever experimentation. If one is
only interested in measuring certain intrinsic properties of clusters,
it is often not necessary to know where they come from. Any source will
do. Since the Konstanz group introduced the term "magic number" into
the field and many other groups have seen more or less reproducible
culminations or breaks in the mass spectra of metal clusters it has
become a habit to ponder about their meaning. It is customary to
identify abundant cluster sizes with local stability optima. This may
often be true but the proof is exceedingly difficult and has not been

given for a single case we are aware of.
What would this proof consist of ? We have to look at the products of
the most common cluster sources to indentify the problem. Without
going into experimental details (which have been addressed elsewhere
[13]) four main types of sources for metal clusters have to be briefly
characterized:
1) cylindrical nozzle (super)sonic beam with or without seed gas;
2) conical nozzle (super)sonic beam;
3) laser pulse evaporation with long, fast flow exit tube and
supersonic expansion;
4) gas aggregation source.
All these types exist in as many varieties as there are experimenters
using them. The published details are never sufficient to compare
operational parameters. What is the preferred cluster source which we
all could adopt and hopefully arrive at the same results with ? We try
to extract the idiosyncrasies of each experimental design using the
formation of alkali clusters as test system.

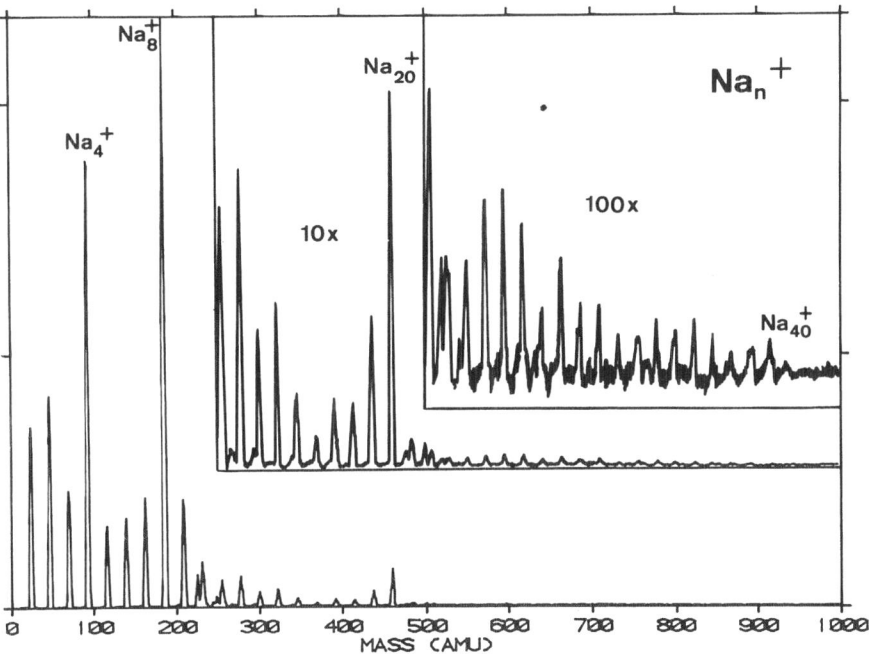

Fig 1. Representative photoionization mass spectrum obtained for a neat
sodium vapour expansion using a 30° conical nozzle with a 0.6 mm
diameter throat. Oven temperature was 800°C, corresponding to a stagnation pressure of 350 torr. Cluster ions Na_x^+, (x<86), have been observed
with this nozzle configuration at slightly higher temperatures.

Neat metal vapour expansion in source 1) results in monotonously
decreasing abundances with size, leading to < M_{20}. With an inert seed
gas very large clusters can be formed and, what was surprising when it
was observed for the first time in 1982, a multimodal distribution is
generated with local optima at 7, 19, 38, atomic clusters [14]. With
the "same" type of source but a three times higher seed gas ratio,
W.D.Knight et al. obtained in 1984 the by now famous series of abundance maxima at 8, 20, 40, and 58 [15]. Applying the standard formula
for adiabatic expansion, they concluded that their beam was colder than
ours which explained the difference. Experiments with drastically
superheated nozzles partially produced beams where 19 grew from 20 and
7 from 8. Since our nozzles were far from the temperatures of those
simulations we came recently to the inverse conclusion: Our beam was
colder than Knight's and very probably contained frozen clusters
whereas his had liquid droplets. The argument is based on several years
of experience with the use of a source 2) with a conical nozzle whose
throat and angle can be systematically varied. As Hagena has shown long
ago [16] with noble gases this nozzle has a welcome tendency to generate large clusters. It creates a regime between adiabatic and isothermal expansion and extends the "viscous zone" downstream of the orifice
thus drastically increasing the number of collisions before rarifi-
cation sets in. These beams are generally warmer than those from
sources 1). But as shown in Fig.1 the cluster ensemble switches now
over to the numbers Knight's group has found. Note also that this is a
neat expansion without the cooling benefit of a seed gas. We have been
able to observe the same preferred cluster sizes with Li, Na, K, and
mixed clusters of Li-Na, Na-K [17]. A nice observation, documented in
fig.2, has proven the conjecture of the warm clusters: During the
formation of mixed Li-Na clusters of e.g. 20 atoms a binomial distribution of the two species is produced which is characterized by a 30
fold enrichment of Li/Na relative to the partial pressure ratio in the
feeding vapour. Assuming that the cluster is hit by Na or Li with a
probability proportional to the respective partial pressures and that
Li has a sticking coefficient of one, replacing a Na when it enters the
cluster, several thousand atomic condensation and reevaporation steps
are required to produce the observed Li-enrichment. This comes near to
the description of a boiling drop keeping its size in an equilibrium
vapour. The enrichment factor then naturally translates into the free
energy change of replacing Na for Li as estimated from the bond
energies of Li_2, Na_2, and LiNa. Note that this statement includes the
whole binomially distributed ensemble Li_xNa_{20-x}. This is confirmed by
the independent observation of the same phenomena, though smaller
enrichment of Na, with Na-K clusters [18]. We shall attempt to model
the history of cluster growth and the survival of the most stable
droplets from an extensive evaluation of these observations for all
cluster sizes. The "evaporative ensemble" described by Klots [19]
illustrates some of these processes. The "hot" clusters resulting have
thermal excitations amounting to a few hundred Kelvin. Even after
running out of collisions the larger clusters still are able to evapor-
ate one or two atoms before entering the detector. Those which had a
loss on transit will be 2-300 K colder than those which had not. The

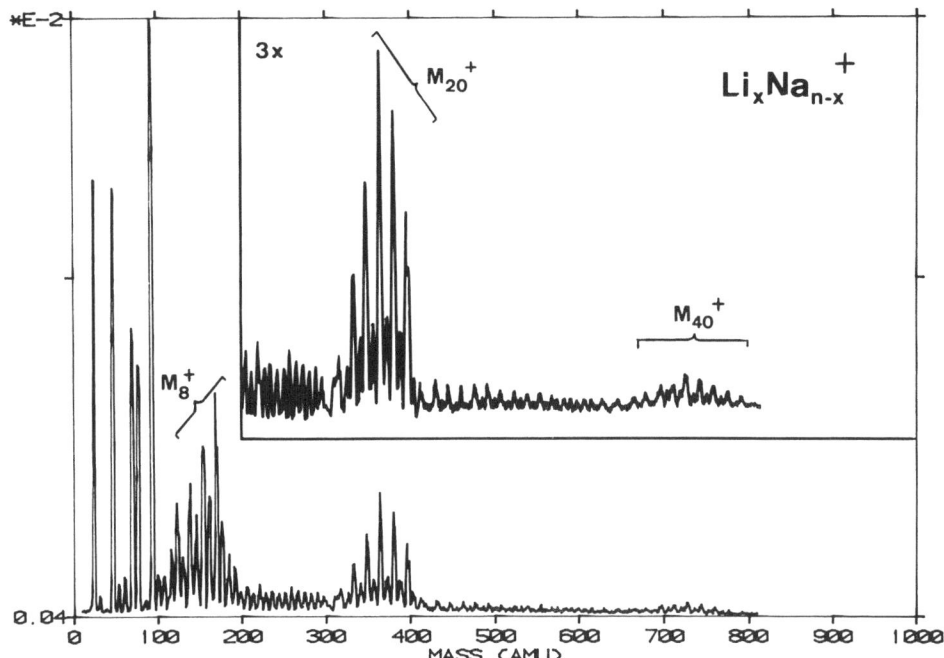

Fig.2 Photoionization mass spectrum obtained for a mixed sodium/lithium
vapour expansion. A 15 degree conical nozzle with 0.6 mm throat and an
oven temperature of 765°C were used. "Magic" numbers observed in the
pure Na expansion are now replaced by binomial peak underlines corresponding to the ensemble of Li_xNa_{n-x} species having correct n. Statistical analysis shows that the lithium content observed in mixed
clusters is significantly higher than that present in the vapour feed.

processes have to be modelled by the theory of unimolecular reactions,
a first attempt at it is in Klots' paper. If this is a plausible story
of what happens to produce a size-abundance distribution of clusters it
becomes evident that minute details of the experimental setup can drastically influence its outcome, e.g. the time window after formation
which the location of the detector defines in relation to the velocity
distribution of every cluster size in the beam (only supersonic beams
with a high Mach number have nearly zero velocity slip between cluster
sizes). The detected clusters will usually exhibit vastly different
"temperatures". The most stable ones are the hottest because they can
support a higher internal excitation or a longer lifetime before
evaporating an atom.
Because of the odds of forming nearly the same size-abundance distribution in two different experimental systems one could proclaim the term
"corresponding cluster-ensemble" if this happens to be true. An "identical cluster-ensemble" requires equal history of formation and passage
of the system through a heat bath at the same temperature. We usually
launch a thermometer with every expansion. It is the dimer whose vibra-

tional and rotational temperatures are probed optically. They are typically 60-80 K for the vibration, 20-40 K for the rotation (20 and 7 respectively for extremely cold beams [20]). These numbers do not differ
significantly between cluster sources 1) and 2). Since vibration and
rotation are not in equilibrium in the dimer population, it is very unlikely that this thermometer is of any value for the larger clusters.
As is now well known, Knight et al. [15] proposed for the first time an
electronic cause for the dominant species with 8, 20, 40, 58, ...
atoms. Closed shells of a spherical jellium-system have this number of
electrons and are distinguished from their neighbors with one more or
one less electron by a slightly enhanced cohesive energy. The spherical
symmetry - while never exactly realized in a finite system - is well
approximated in hot, molten droplets. Frozen clusters, if formed near
thermodynamic equilibrium, should exhibit isomeric forms with the most
stable structure. Iñiguez et al. [21] have computed cohesive energies
of crystalline Na-clusters (selfconsistent solution in the approximation of a non spin-polarized local density formalism with Coulomb repulsion of the cores at the occupied lattice sites) with bcc, fcc, and
hcp lattices. These showed a strong selectivity for closed coordination
shells, in fact much stronger than that of a spherical jellium system
for closed electronic shells. Among the numbers produced are 7 and 19
as found earlier. So it seems to be appropriate that we should have
found snow in Switzerland, while Knight in California had produced
rain. The 3. section will compare ionization potentials with the
jellium predictions in molten clusters.
Ionic metal clusters can be investigated with a richer and perhaps
simpler methodology than neutral ones. Several mass spectrometric
techniques exist which allow to establish parent-daughter-granddaughter
a.s.o. correlations in an ionization induced fragmentation tree [22].
Sometimes special experimental situations can be found which allow to
distinguish a genuinely formed cluster from a fragment of equal mass.
We were able to study e.g. the ionization induced unimolecular decay of

$$Na_8^+ \Rightarrow Na_7^+ + Na$$

and to give a threshold energy E_d of 0.58 eV for it [23]. More than 30%
of the Na_7^+ peak comes from fragmentation of Na_8^+ except if the ionizing radiation is filtered to be lower in energy than $IP_8 + E_d$. This was
repeated with several clusters up to Na_{22}. The fine structure of the
Na_x abundance distributions must not be correlated to any model of
cluster stability before every one of the peaks has been cleaned up for
fragmentation. We believe that the fit of ellipsoidal jellium deformation parameters using contaminated abundances is totally meaningless
[24].
Source 3) has become most popular for refractory metals through the
work of Smalley, Kaldor and others [25]. It has produced most interesting results for transition metal clusters. It is usually assumed, that
this source generates very cold clusters as proven by measuring the
dimer temperatures. While this is not sufficient proof it seems plausible that the noble gas seeded prolonged transit through the fast-flow
exit tube offers intense collisional energy transfer. Since V-T transfer is so inefficient, especially for the soft modes of larger
clusters, a still lacking theoretical model of this source would be

very welcome. An interesting observation has recently been made by Meiwes-Broer et al.: Cu_x^+ clusters show magic numbers 9, 21, 41.. if produced by Ar^+-sputtering. This produces very hot, molten clusters. If the Cu_x^+ cluster ensemble is generated by laser pulse evaporation into He or Ne and exit through a fast-flow tube no magic numbers can be detected in the mass spectrum [26]. Many other experiments have shown similar results: Cold clusters show lacking or different local abundance maxima compared to molten droplets.

Source type 4) has been in use in several distinct embodiments. Clusters are swept away with a noble gas flow which in some sources has been cooled to liquid N_2. Large clusters are obtained. K. Sattler has shown Sb_x cluster formation with strong selectivity for Sb_4 polymers [27], which was to be expected from the well known chemical stability of these particles for all group 15 elements. Schulze's group has recently found [28] jellium closed shell preferred stability for Ag_x^+ Ions, with perhaps some superimposed geometrical maxima as well.

As has become obvious a remarkable step forward in the interpretation of mass spectrometric abundance distributions can be done if mixed clusters are prepared. This allows to partially uncouple fragments from genuine clusters. But the main advantage is that one or several foreign atoms serve as probes into the electronic system. The lower symmetry thus obtained reveals otherwise undetected features [29]. We have by now incorporated Mg, Ca, Sr, Ba, Zn, Hg, Eu, and Yb into alkali clusters and see the beginnning of a metal-metal coordination chemistry which, of course, cannot be rationalized in terms of the simplistic paradigm of a spherical jellium.

Conclusion of this section: A proof for exceptional stability of dominant clusters has to be based on equilibrium partial pressures of the neutral species. This can be measured properly in a Knudsen cell mass spectrometer using a high pressure liquid-vapour two phase system. As Leo Brewer has shown many years ago, the species dominant in the vapour become more and more complex the higher the temperature (and concomitant saturation pressure) contrary to oversimplified entropy arguments. Thus large, molten clusters have to grow. However, the higher the pressure in a Knudsen cell the more difficult it becomes to meet Knudsen effusion conditions (orifice small compared to the mean free path) and thus measure an equilibrium distribution with a mass spectrometer. Hence high temperature techniques with proper confinement and optical observation of the system have to be developed. It is remarkable how close Hensel's group in Marburg has already come to this goal [30]. Of course, we will not abandon beams and mass spectrometry for a large number, in fact for almost all other types of experiments. Eventually a reliable theory of clusterbeams and proper experimental tools to monitor partial pressures and thermal excitations (e.g. Scoles' quantum bolometer with particle specificity [31]) of the clusters may be developed which allow to transform cluster abundances into meaningfull thermodynamic information.

3. IONIZATION POTENTIALS: PROBES INTO THE ELECTRONIC STRUCTURE

Relative ionization cross sections as a function of photon energy can be measured in a size specific way with a mass spectrometer using a monochromator/light source (plasma or synchrotron) combination, or a tunable laser. The latter offers a situation where (almost) no corrections have to be applied for an apparatus function. However, the high power density may induce multiphoton processes, or the population of narrow, intensely autoionizing Rydberg states may complicate the issue. The most accurate ionization potentials are obtained from the observation of Rydberg series above the ionization threshold whose terms can be precisely extrapolated to the groundstate of the singly charged ion [32]. This has only been done for metal dimers. For most of the larger clusters monochromator/lamp ionization efficiency curves have been measured. Various methods for the deconvolution of the data to yield a (vertical) ionization potential are used [33]. A precision of ± 0.02 eV and an accuracy of ± 0.05 to ±0.1 eV can be obtained depending on

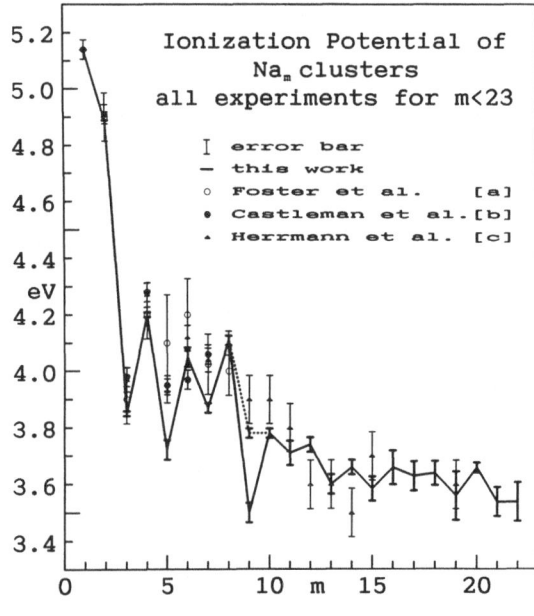

Fig.3 Collection of all IP's of Na_m, m<23, to date.
[a] see [34] goes to Na_8;
[b] see [35] to Na_7;
[c] see [36], [37] to Na_{14} with a few values < Na_{65}. Fat error bars represent new, higher precision determinations [38].

the width of the energy spread from the monochromator, the clusterion signal to noise ratio, and on the peculiarities of the ionization threshold behaviour. In fig.3 all the Na_x IP's do date are collected, and in fig 4. a set of more precise recent IP's up to Na_{22} is compared with two models.
In Fig. 5 the IP's of ten cluster ensembles are plotted against the reciprocal of the cluster radius. The ordinate is the difference between the extrapolated (polycrystalline) bulk workfunction W_{inf} of the respective metals and $W(R)$, the IP of the cluster with radius R. To assign a radius we assume the cluster has spherical shape and a volume proportional to the number of atoms with a size derived, e.g. from the

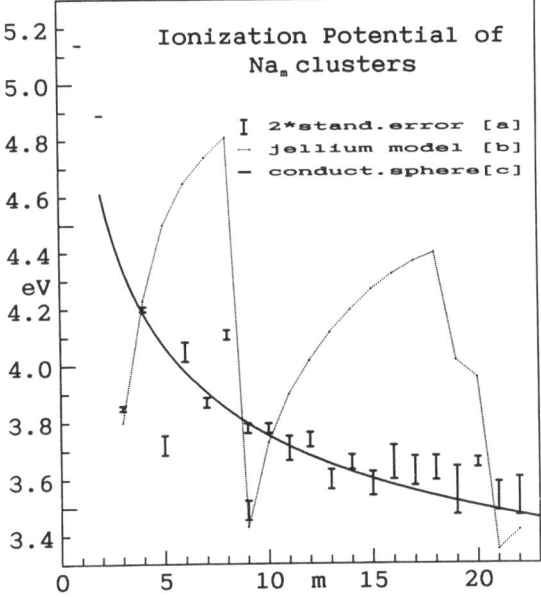

Fig.4. Ionization Potentials of Na_m, m<23. The error bars show twice the standard deviation. Overlays are the predictions of a jellium sphere calculation and the classical conductive sphere model. The former predicts large discontinuities in adiabatic ionization potentials which are not observed. The latter does not represent the discrete energyspectrum of the small molecules.

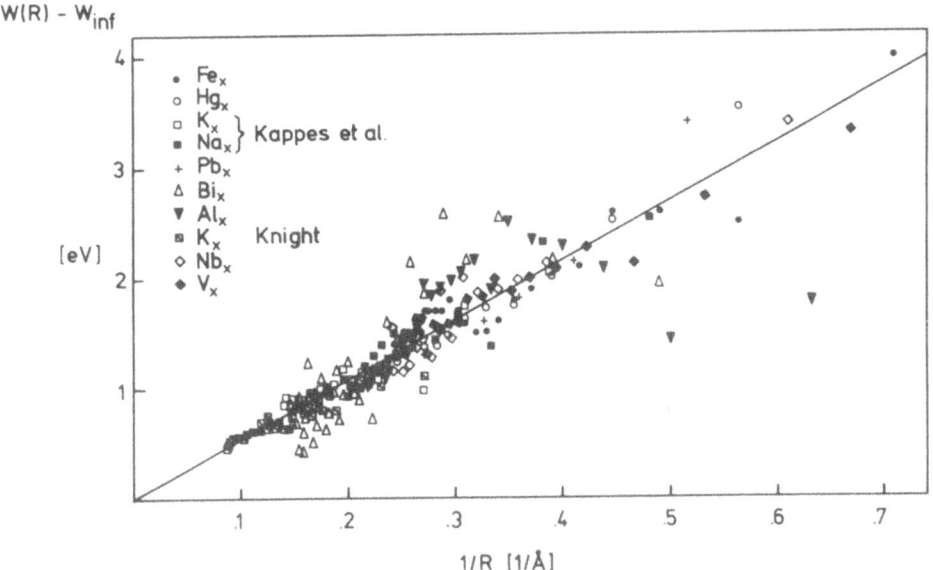

Fig.5 "Reduced" ionization potentials for several metal cluster data sets against 1/R. Plotted is the difference between measured ionization potential and extrapolated bulk workfunction based on the spherical drop model. The straight line corresponds to the absolute electrostatic correction for curvature as a function of particle size.

Wigner-Seitz radius of the crystalline bulk. The straight line is not a
fit, but the classical correction to the ionization energy due to the
difference in electrostatic image potential of the metallic half space
and a conductive sphere of Radius R as derived by Smith and Wood [39]:

$$W(R) = W_{inf} + \frac{3e^2}{8R} \qquad\qquad (1)$$

(1) (but IP versus R) has already been tried by Leckenby et al. [34],
who determined the first few IP's of Na- and K-clusters. Their limited
data set and precision gave, however, inconclusive proof of the
validity of (1). While the extrapolation for Na_x, K_x, and Pb_x to $1/R \Rightarrow$
0 leads to a reasonable prediction of W_{inf} (within 0.2, 0.07, and 0.25
eV resp.), Fe_x (0.69), and Hg_x (2.39 eV [40]) IP's only fit the slope
of (1) without extrapolating correctly. For the small radii down to the
atom the scatter suggests the inapplicability of a classical picture.
It is, however, remarkable that from x<66 (Na), x<35 (K), and x<8 (Pb)
good bulk workfunctions can be extrapolated. It is amusing to note,
that (1) allows to predict bulk workfunctions even from the first
ionization potential of the atom within 5 to 10% accuracy for 2/3 of
the metallic elements! They happen to be metals with half filled s-
bands or partially filled p- or d-bands. The extrapolation is worst for
group 2 (Mg,Ca,Sr,Ba) and 12 (Zn,Cd,Hg) full s-band "metals".- The
interpretation of the IP's of Na_x and Li_x (not shown) has been
attempted by quantum chemistry which has produced a good fit with the
data up to x=9 [41], the largest clusters treated so far. IP's have
also been predicted by the spherical jellium potential or, even more
simply, with the particle in a spherical well, whose finite depth has
been calibrated at Na_2 [36]. The latter was not able to describe IP's
up to x=15 and the refined, selfconsistent and correlated jellium
potential [42] does not do better as shown in fig. 4. The shell closing
effects are greatly exaggerated. It is doubtful, within the limits of
accuracy, whether the jellium shells leave any significant trace in the
sets of IP's. Knight's group [43] claims that their precision shows
small shell closing effects in the IP's of K-clusters to be significant. If that is true they are less than 1/3 of what has been predicted. This is strange if one regards the ±correct generation of the
magic numbers as a success of the spherical jellium calculation. Our
own feeling is, that the spherical jellium, or in fact any calculation
with spherical boundary conditions and a smeared out positive background, is only a number generator for the closed shells. In order to
adequately describe physical response properties, a much more realistic
potential is necessary than what even the self-consistent, correlation
corrected, jellium can give. All our hope rests on Koutecky [44].
The large spread in the results of IP measurements of different groups,
as shown in fig.3 is partly due to bad signal to noise ratios at the
beginning of cluster research. The systematic differences are, however,
caused by different methodology for the extraction of an IP from a
photoion yield curve as a function of photon energy. Apart from the
deconvolution of the apparatus function the main uncertainty pertains
to the assessment of the prethreshold ion current. Fig. 6 shows the
variable length of the "photoion tail" for the same set of data as in
fig. 4 . Many authors [35] assume that this is some sort of "Boltzmann

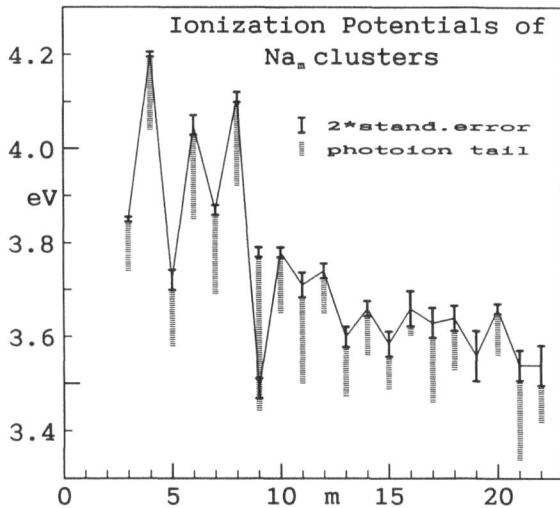

Fig.6. Recent series of Na$_x$ Ionization Potential measurements, identical to fig.4. Added are the onsets of the ion current in the prethreshold region. Note the variable length of this tail. Na$_9$ is an anomalous case with a distinct kink in the photoion yield curve, lying 0.28 eV higher than the lower threshold. Note the odd-even oscillation of the IP's which is now well documented in the contiguous series up to Na$_{41}$.

tail" reflecting prethreshold ionization from vibrationally excited clusters. They get rid of it by taking the logarithm of the ion current, fitting a straight line to the toe section of the curve and localize the ionization threshold where the ion yield curve breaks away from the linear part. Very often that break is not easily made out. Most of the argument we have with A.W. Castleman about the IP of Na$_6$ (which he has publizised for many years now [45]) stems from a difference in opinion where to cut the Boltzmann tail. He will perhaps be pleased to learn, that our new, much more precise IP falls between our older and his measurement. However, the latter is now clearly outside any defendable error limit. Furthermore a distinct odd-even oscillation in the IP's is now well established up to Na$_{22}$ and even up to Na$_{41}$ for the post-threshold behaviour (> 0.5 eV above the IP).- There is more serious business connected to ionization thresholds in polynuclear metal clusters. If the clusters are crystalline we expect isomers to form a non equilibrium mixture with distinguishable IP's. Nothing of this sort has been observed with certainty up to now. Knight's group has found anomalous behaviour of the threshold of K$_9$ which also exists in exactly the same form in Na$_9$. That is the reason why we give an upper and a lower IP, fig.4, whereas Knight has now dropped the former double assignment for the lower value which fits the jellium prediction better [43]. The anomalous threshold is an intrinsic property of the Na$_9$ population and not due to fragment contamination.

The term "corresponding cluster ensemble" can perhaps be illustrated by a large enough series of contiguous IP's of two experiments. We have an almost ideal correspondence between our Na_x- and Knight's K_x-IP values up to x=22 if the electrostatic term due to curvature is subtracted [38]. Another correspondence relation can be established for photon induced fragmentation. This is probably more sensitive to phononic excitation than ionization. Equal photofragmentation rates over every species of a cluster ensemble characterizes "correspondence" [38].

Conclusion: We made a few critical remarks based on recent experiments from different laboratories pertaining to cluster abundances, to ionization potentials, and their interpretations. They are meant to stimulate endeavours to better define cluster systems in order that we might agree, eventually, on the main features of what we observe.

Acknowledgments: We thank our collaborators M. Schär, U. Heiz, A. Vayloyan who have contributed their measurements for this paper. This work has been supported by the Swiss National Foundation Grant 2.348-0.86.

REFERENCES

[1] Reviews in Z. Physik D, **3**,86; Chem.Rev. **86**,4,1986
[2] See for example Proc.Sixth Int.Conf.Liquid Amorph.Met., Garmisch, August 1986 (in press)
[3] G. Delacretaz, E. Grant, R. Whetten, L. Wöste and J. Zwanziger, Phys.Rev.Lett., **56**,2598(1986); M. Broyer, G. Delacrètaz, P. Labastie, J. Wolf, and L. Wöste, Phys.Rev.Lett.,**57**,1851(1981)
[4] E. Rohlfing & J. Valentini, Chem.Phys.Lett., **126**, 113(1986).
[5] W. Knight, R. Monot, E.Dietz, and A. George, Phys.Rev.Lett., **40**,1324(1978); M. Hoffmann, Ph.D. thesis U. of Bern, Bern 1981
[6] D. Cox, D. Trevor, R. Whetten, E. Rohlfing, and A. Kaldor, Phys. Rev. B, **32**,7290(1985); J.Chem.Phys., **84**,4651(1986).
[7] K. Clemenger, Phys.Rev.B, **32**,1359(1985); Ph.D. thesis, U. of California, Berkeley, 1985.
[8] M. Kappes & E. Schumacher, to be published.
[9] S. Davis, K. Klabunde, Chem Rev., **82**,152(1982).
[10] A. Kaldor, D. Cox, D. Trevor, and M. Zakin, Z.Phys. D,**3**,195(1986).
[11] M. Geusic, M. Morse, and R. Smalley, J.Chem.Phys., **82**,590(1985).
[12] S. Richtsmaier, E. Parks, K. Liu, L. Pobo, and S. Riley, J.Chem. Phys., **82**, 3659(1985).
[13] M. Kappes & S. Leutwyler, in "Atomic & Molecular Beam Methods", G. Scoles, Ed., Oxford University Press, in press.
[14] M. Kappes, R. Kunz & E. Schumacher, Chem.Phys,Lett., **91**,413(1982).
[15] W. Knight, K. Clemenger, W. de Heer, W. Saunders, M. Chou, and M.Cohen, Phys.Rev.Lett., **52**,2141(1984).
[16] O. Hagena and W. Obert, J.Chem.Phys., **56**,1793(1972).
[17] M. Kappes, M. Schär & E. Schumacher, J.Phys.Chem., **91**,658(1987).
[18] C. Brechignac & Ph. Cahuzac, Z.Physik D,**3**,121(1986).
[19] C. Klots, J.Chem.Phys., **83**,5354(1985); Z.Physik D,**5**,83(1897).
[20] G. Delacretaz & L. Wöste, Surf. Sci., **156**,770(1985).
[21] M.P. Iñiguez, J.A. Alonso & L.C. Balbas, Solid State Comm.,

57,85(1986); C. Baladron, M.P. Iñiguez & J A. Alonso, ibid. **50**,549(1984).

[22] e.g. "The Mass Spectra of Organic Molecules", J. Beynon, R. Saunders & A. Williams, Elsevier, Amsterdam, 1968.

[23] M. Kappes, A. Vayloyan & E. Schumacher, Z. Physik D, in press.

[24] K. Clemenger, see ref. [7].

[25] e.g. M. Morse, M. Geusic, R. Heath & R. Smalley, J.Chem.Phys. **82**, 2293(1985)

[26] H. Siekmann, P. Jonk, G. Ganteför, K. Meiwes-Broer & H. Lutz, to be published (private comm.)

[27] J. Mühlbach, E. Recknagel & K. Sattler, Surf.Sci.,**106**,188(1981).

[28] W. Schulze et al., J.Chem.Phys., in press.

[29] M. Kappes, P. Radi, M. Schär & E. Schumacher, Chem.Phys.Lett., **119**,11(1985); Ch. Yeretzian, Diploma Thesis, U. of Bern,1986.

[30] F. Hensel et al., see this volume.

[31] G. Scoles, in "Atomic & Molecular Beam Methods", G. Scoles, Ed., Oxford University Press, in press.

[32] S. Leutwyler, T. Heinis, M. Jungen, H.-P. Härri & E. Schumacher, J.Chem.Phys. **76**, 4290(1982).

[33] P. Guyon & J. Berkowitz, J.Chem.Phys., **54**,1814(1972).

[34] R. Leckenby, E. Robbins, P. Trevalion, Proc.Roy.Soc.A, **280**,408 (1964); E. Robbins, R. Leckenby & P. Willis, Adv.Phys., **16**,739 (1967); P. Foster, R. Leckenby & E. Robbins, J.Phys.B, **2**,478(1969)

[35] K. Peterson, P. Dao, R. Farley & A.W. Castleman Jr., J.Chem.Phys., **80**,1780(1984).

[36] A. Herrmann, L. Wöste & E. Schumacher, J.Chem.Phys. **68**,2327(1978).

[37] M. Kappes, M. Schär, P. Radi & E.Schumacher, J.Chem.Phys., **84**,1863 (1986).

[38] M. Kappes, M. Schär, A. Vayloyan & E. Schumacher, to be published.

[39] J. Smith, J.Am.Inst.Aeronaut.Astronaut., **3**,648(1965); D. Wood, Phys.Rev.Lett. **46**,749(1981)

[40] Fe: E. Rohlfing, D. Cox, A. Kaldor, K. Johnson, J.Chem.Phys., **81**,3846(1984); Hg: A. Hoareau, B. Cabaud, & P. Melinon, Surf.Sci., **106**,195(1981); Pb: Y. Saito, K. Yamauchi, K. Mihama & T. Noda, Jpn.Appl.Phys., **21**, L396(1982).

[41] J. Field, H. Stoll, H.-W. Preuss, J.Chem.Phys., **71**,3042(1979); J. Martins, J. Buttet, R. Car, Phys.Rev.Lett., **53**,655(1984); D.R. Snider, R.S. Sorbello, Surf.Sci., **143**,204(1984); J. Koutecky et al. see Chem.Rev. **3**,3(1986);

[42] M.Y. Chou, A. Cleland & M. Cohen, Solid State Comm., **52**,645(1984).

[43] W.D. Knight, W.A. de Heer, W.A. Saunders, K. Clemenger & M. Cohen, Chem.Phys.Lett., **134**,1(1987).

[44] J. Koutecky, this volume.

[45] see ref.35 and A.W. Castleman Jr. and R. Keesee, Ann.Rev.Phys. Chem., **37**,525(1986); A.W. Castleman Jr. and R. Keesee, Y.Physik D, **3**,167(1986); A.W. Castleman Jr., and R. Keesee, Chem.Rev., **86**,584 (1986).

ELECTRONIC STRUCTURE AND BASIC PROPERTIES OF SMALL ALKALI METAL CLUSTERS

J. Koutecký, V. Bonačić-Koutecký, I. Boustani
Institut für Physikalische und Theoretische Chemie
Freie Universität Berlin, Takustr. 3, 1000 Berlin 33,
Federal Republic of Germany;

P. Fantucci
Dipartimento di Chimica Inorganica e Metallorganica
Centro CNR, Universita di Milano, I-20133 Milano,
Italy

and

W. Pewestorf
Institut für Physikalische und Theoretische Chemie
Freie Universität Berlin, Takustr. 3, 1000 Berlin 33,
Federal Republic of Germany

ABSTRACT

The number of electrons and the cluster geometry are the determining factors for the stability of differently charged alkali metal clusters. The interplay of the tendency to accomplish an optimal electronic shell structure and the possibility to achieve it with the appropriate geometrical shape for the given nuclearity is demonstrated on new examples (Li_3^-, Li_7^-, Li_9^-, Li_{20} and $Be-Li_6$). The anionic Li_n^- clusters have been systematically investigated. The similarities in properties of neutral Li_n and Na_n clusters demonstrate the predominant importance of the valence electrons for the properties of alkali metal clusters.

1. INTRODUCTION

Quantum chemists applying their methods for the investigation of cluster properties are confronted with many problems which remind of a typical dilemma of the quantum chemistry two or three decades ago: Indeed, the systems which are of any direct general interest for the chemistry are so larhe that their sufficiently accurate theoretical treatment meets large technical difficulties. Consequently, it was very difficult

303

J. Jortner et al. (eds.), Large Finite Systems, 303–317.

to obtain from the theory interesting, stimulating and challenging results which have a real predictive value. However,
the situation in the theory of the electronic structure of
clusters differs from the state of affairs in classical applications of the quantum chemistry at least in one point:
The most important facts concerning the chemical bond in the
usual molecules have been known and have been organized in
general and very ingenious logical systems, so that the role
of the theory was to justify them on a more sound theoretical
ground. On the contrary, the experimental cluster science is
a very young field of scientific activity and its notions are
just now in a stormy development. Consequently, any new im-
pulses coming from the theory are much more appreciated by
the cluster scientists than the quantum chemistry results
for customary molecules have been appreciated by the chemists.

Since an alkali metal atom has only one valence electron
the alkali metal clusters represent the simplest possible objects of investigation applying quantum chemistry methods to
clusters science. On the other hand, the investigation of
clusters formed from heavier alkali metal atoms with larger
number of core electrons can show to which extent the core
electrons influence the specific cluster properties.

Four leading factors governing electronic structure and
geometry of small neutral and cationic stable alkali metal
clusters have been found until now:[1,2]
1) Degeneracies, occupancies and nodal properties of one-electron functions (molecular orbitals in one-electron theory or
natural orbitals in many-electron theory) are of dominant importance. Their partial occupancy can give rise to the deformation of a highly compact (and therefore usually very symmetrical) geometry towards a less compact structure. Large
degeneracies of the one-electron functions are unfavorable
for clusters with small number of electrons. Consequently,
small planar clusters can be more stable than their three-
dimensional isomers.
2) Compactness of cluster geometry is a favorable feature for
stability. Therefore, small clusters exhibiting fivefold symmetry and clusters with large number of tetrahedrons as subsystems are stable.
3) Topologies properties of delocalized bonds are important.[3]
4) Inclusion of polarization functions in the AO basis describes adequately developing metallic features of clusters
and therefore is important for their stability.
All four listed factors are in mutual competition.

In this contribution the already published results of investigations on neutral and cationic Li_n clusters are summarized and analyzed.[1,2,4-9] New results concerning anionic Li_n^-
clusters, neutral Li_{18}-Li_{20} clusters, small neutral Na_n
clusters and Be-Li_k mixed clusters extend substantially our
knowledge about structure and electronic properties of alkali
metal clusters.

The geometry optimization[18] has been carried out with the conjugated gradient method[18] in the framework of the Hartree-Fock procedure. For the Hartree-Fock optimized cluster geometries the calculations with the multireference biexcited configuration interaction method (MRD-CI)[19] with subsequent extrapolation procedure and generalized Davison corrections have been carried out. In addition, for the SCF optimized geometries of small Na_n (n<6) clusters exhibiting highest stability after introducing the correlation effects by MRD-CI procedure, the search for the lowest CI energy and the best geometry has been performed. The reliability of the MRD-CI extrapolation procedure has been examined[20] by comparison with the results obtained from the direct CI[20] calculations for the most interesting Na_n clusters.

II. METHOD USED.

In this work the emphasis is not put on the surely very important methodical details which have been either published or will be published in separate papers. The AO basis sets are chosen so that they are simultaneously as small as possible, but they are capable of describing the leading features of the electronic cluster structure. This choice should make possible to carry out the investigation for relatively large systems. The aim of this study is not to obtain exact numbers but to extract the qualitative and semiquantitative features of the cluster structure. The reliability of the results are tested on smaller clusters employing larger basis sets and improving the description of correlation effects.

Only most relevant information will be given. The all-electron ab-initio investigation of Li_n and Li_n^+ clusters is carried out with a minimal AO basis set augmented by one p function: (6s,1p/2s,1p).[10,11] The quality of results is tested using the larger 6-31G AO basis set.[12] The (6s,1p/2s,1p) AO basis augmented by one more diffuse s Gaussian with the exponent $\alpha_s=0.03$ is employed in the calculations of Li_n^- clusters.[13] This AO basis set is also used for the calculation of the electron calculations of Na_n clusters is (9s,4p/3s, 1p)[14] augmented by one diffuse p function with exponent $\alpha=0.065$ whereas AO basis for effective core potential calculations of Na_n clusters is (4s,1p/3s,1p).[15,16,17] The AO basis for Be_n-clusters is 3-21G and 6-31G.[17]

III. COMPARISON OF SMALL NEUTRAL, CATIONIC AND ANIONIC LI CLUSTERS.[2,13]

The atomization energies per atom of small most stable neutral as well as of most stable charged Li_n clusters (E_b/n, E_b^+/n, E_b^-/n) have tendency to increase with the growing cluster

size (Fig. 1). These measures of the cluster stability are
defined as follows:

$$E_b/n = (nE_1 - E_n)/n$$

$$E_b^+/n = ((n-1)E_1 + E_1^+ - E_n)/n \qquad (1)$$

$$E_b^-/n = ((n-1)E_1 + E_1^- - E_n^-)/n$$

where E_n, E_n^+ and E_n^- is the energy of a neutral, cationic and
anionic moiety with n atoms, respectively. The topologies of
most stable clusters are schematically shown in Fig. 1.

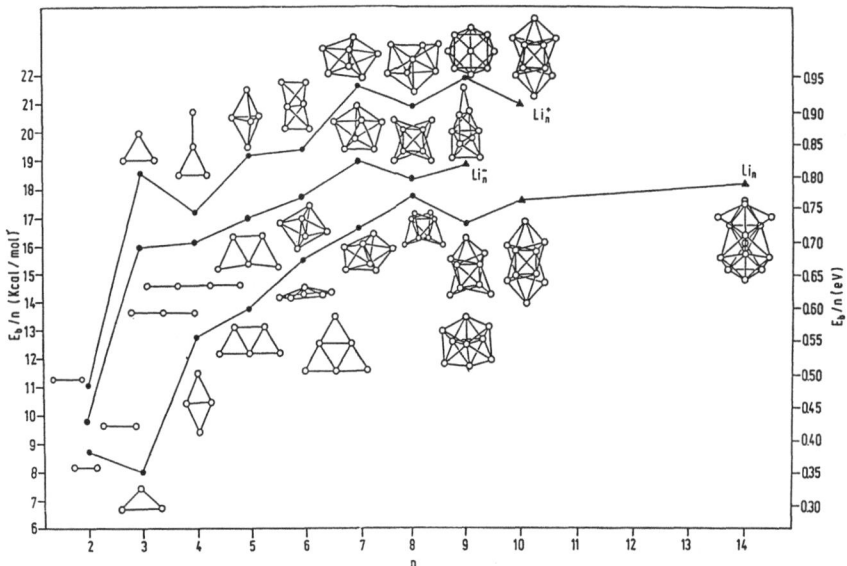

<u>Fig. 1</u>: Binding energy per atom (atomization energy per
atom, E_b/n) of Li_n, Li_n^+, and Li_n^- clusters as function of
number of atoms in cluster (n). MRD-CI extrapolated energies
with Davison corrections for cluster geometries found with
SCF energy optimization are plotted. Topologies of the SCF
energy optimized cluster are shown. The preliminary results
for which the normal frequencies have not been determined
are denoted with Δ. The scale on the left and on the right
is in kcal/mol and eV, respectively.

The curve E_b/n for neutral clusters is steeper in the intervals between smaller odd cluster sizes (n=2m-1) and larger
even cluster sizes (n=2m) than in the adjacent intervals.
Opposite is true for the curves corresponding to charged
clusters (cf. Fig. 1). This illustrates that even number of
electrons is favorable factor for stability of a cluster.

Nevertheless, there is no regular oscillation within E_b/n, E_b^+/n and E_b^-/n between the values of binding energy per atom for clusters with odd and even number of electrons. The E_b^+/n and E_b^-/n curves exhibit very similar behaviour.

The relative large stability of "even" clusters is better illustrated by maxima of the second differences for the clusters with even number of electrons (cf. Fig. 2). The second differences are defined as:

$$\Delta^2 E_n = E_{(n+1)} + E_{(n-1)} - 2E_n$$
$$\Delta^2 E_n^+ = E_{(n+1)}^+ + E_{(n-1)}^+ - 2E_n^+ \qquad (2)$$
$$\Delta^2 E_n^- = E_{(n+1)}^- + E_{(n-1)}^- - 2E_n^-$$

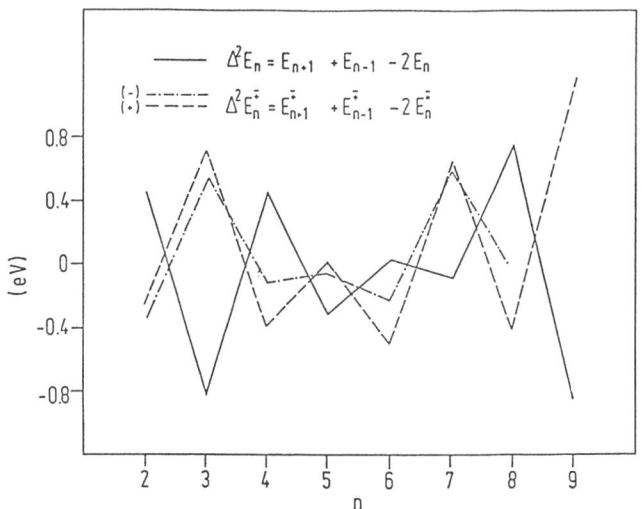

Fig. 2: Second differences $\Delta^2 E_n$, $\Delta^2 E_n^+$ and $\Delta^2 E_n^-$ for neutral, cationic and anionic Li_n clusters as functions of nuclearity n.

The minima in Figs. 1 and 2 for the neutral Li trimer are clearly noticeable as well as the maxima for $Li_8(T_d)$, for Li_9^+ with the shape of an antiprism having one central atom, and for Li_7^- with pentagonal bipyramid form. The maxima for neutral clusters can be connected with the so called "magic numbers" for the large abundances, in the mass spectroscopical detection of clusters.[21-23] Similarily, the maxima of E_b^+/n for n=2 and n=9 can be interpreted as symptoms of "magic numbers" in the case that the stability of cationic clusters is of decisive importance due to the fragmentation process after the photoionization.[24,25]

The changes in the general character of neutral and the

cationic cluster shape with increasing nuclearity is quite
well understood in the terms of the four leading factors mentioned in the Introduction. For example, the fact that most
stable neutral Li_n clusters are planar for $n<6$ confirms the
importance of the fully occupied highest molecular orbitals
for the stability consideration. The planar forms exhibit
lower degeneracy of the one-electron functions than the three-
dimensional shapes. Therefore, for small number of electrons
it is easier to fill up completely degenerate levels. Symmetrical three-dimensional structures with not fully occupied
degenerated MO's can be easily deformed due to Jahn-Teller,
pseudo Jahn-Teller effect and related phenomena. On the other
hand, the degeneracy of highest occupied molecular orbitals
and the existence of several empty low lying MO's can be favorable when the electron correlation effects are taken into
account. Consequently, the CI can lead to an exchange in the
energy sequence of cluster isomers corresponding to the different HF local energy minima in favour of the structures
with MO spectrum exhibiting high degeneracy or almost degeneracy.

The neutral, cationic and anionic Li_n clusters exhibit
very different optimal geometries.[13] This circumstance is
connected with different occupation of one-electron functions
and with the electrostatic repulsion which is mainly important for anionic clusters. Smaller Li_n^--clusters have linear

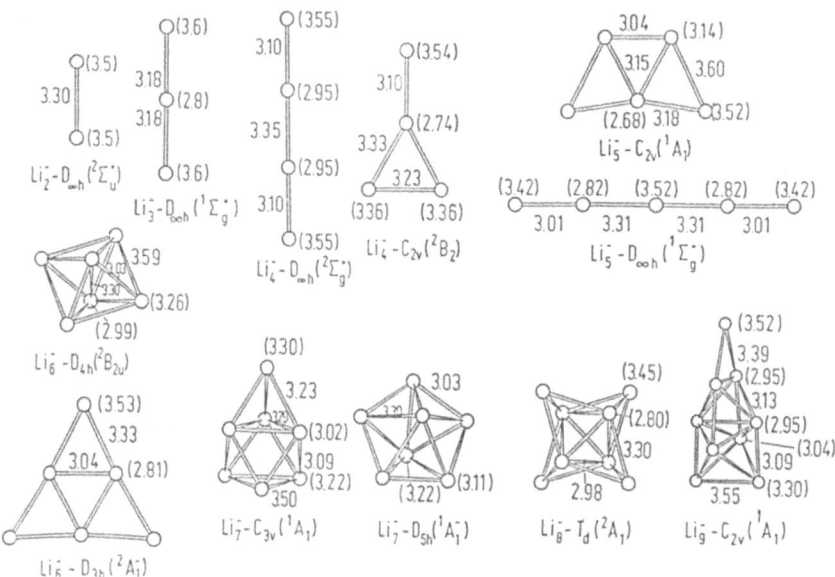

Fig. 3: Schematic geometries of anionic Li_n^--clusters. Interatomic distances (Å), the Mulliken electronic charges (in
brackets) as well as the symmetry and the label of the ground
state (in bracket) are denoted.

form. The linear arrangement makes possible a relatively
small Coulomb repulsion of the negative extra charges which
can be localized on the terminal atoms (cf. Fig. 3). The Li
atoms with smaller number of nearest-neighbours have generally larger negative charge in the Li_n^--clusters.

   The electron affinities (EA)[13] which have been computed
with the AO basis set augmented by one more diffuse s-type
Gaussian for both anionic and neutral Li_n-clusters are shown
in Fig. 4. The values of the EA's are small and positive. The
EA curve exhibits maxima and minima for the nuclearities with
minimal and maximal IP values, respectively (Fig. 4). This
feature shows again a special stability of clusters with
even number of electrons.

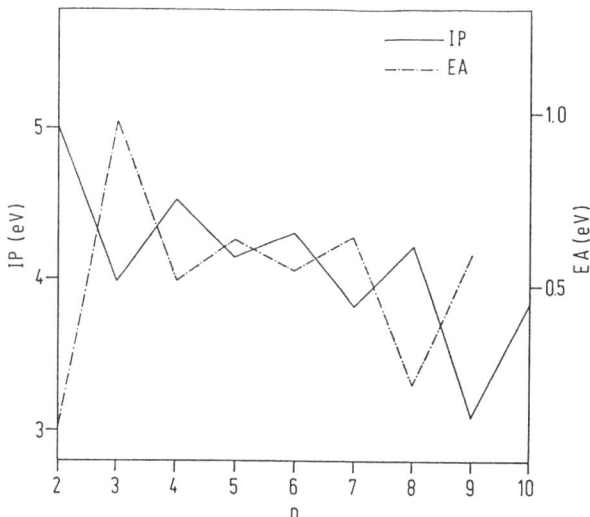

Fig. 4: Adiabatic ionization potentials and electron affinities of energy optimized Li_n clusters. The left and right
scale (eV) is valid for ionization potentials and electron
affinities, respectively.

IV. LI_n CLUSTERS WITH n NEAR TO 20.[26]

Experimentally found "magic" number twenty is in the literature interpreted as an indication of the special stability
of the neutral Li_{20} cluster.[23] For this reason, the pilot
study of larger Li_n cluster with n=18-21 has been carried
out. Indeed, for the Li_{20} with nearly spheric shape (T_d - SS)
(cf. Fig. 5) a large binding energy per atom has been found.
The other studied shape of Li_{20} (T_d-fcc) which can be taken
as a tetrahedral section from the fcc crystal lattice has a

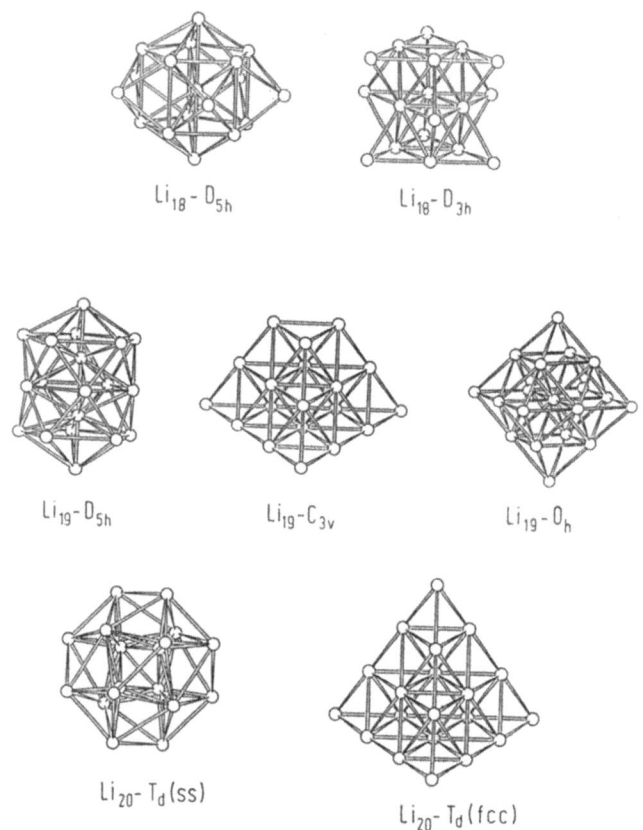

$Li_{18}-D_{5h}$ $Li_{18}-D_{3h}$

$Li_{19}-D_{5h}$ $Li_{19}-C_{3v}$ $Li_{19}-O_h$

$Li_{20}-T_d(ss)$

$Li_{20}-T_d(fcc)$

Fig. 5: Schematic geometries of larger studied Li_n clusters
(n=18,19,20).

slightly lower Hartree-Fock energy than Li_{20} (T_d-SS) but the
MRD-CI method can recover much larger correlation energy for
the latter isomer. Therefore, the T_d-SS geometry of Li_{20} is
found to be more stable when MRD-CI procedure is employed.
The MRD-CI E_b/n for Li_{20} (T_d-SS) and for Li_{20} (T_d-fcc) is
19.4 kcal/mol and 18.5 kcal/mol, respectively, whereas the
Hartree-Fock E_b/n for Li_{20} (T_d-SS) and for the Li_{20} (T_d-fcc)
is 9.3 kcal/mol and 9.6 kcal/mol, respectively.

The nearly fivefold degeneracy of the highest valence molecular orbital resembles strongly to the fivefold degeneracy
of the d atomic orbitals. Furthermore, one nondegenerate level of a molecular orbital with a nodal character reminding
of 2s AO lies very close to the energies of those MO's which
look like d AO's (cf. Fig. 6). As already mentioned the described effective degeneracy is probably the reason for the
large correlation energy contribution to the binding energy

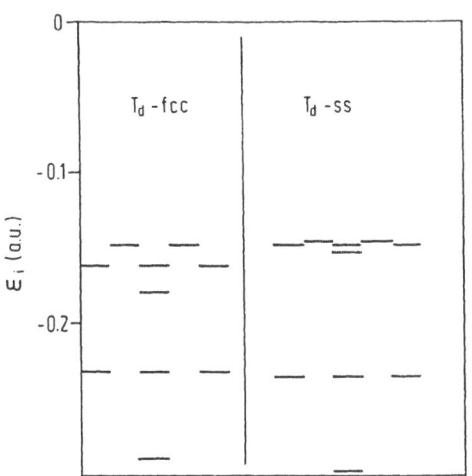

Fig. 6: MO energies (a.u.) of Li_{20} isomers.

for L_{20} (T_d-SS).
   Presently only some preliminary results concerning Li_{18} and Li_{19} clusters can be reported: Two isomers of Li_{18} have been studied as possible candidates for very stable cluster shapes: $Li_{18}(D_{5h})$ and $Li_{18}(D_{3h})$ (cf. Fig. 5). $Li_{19}(D_{5h})$ having a form of double-icosahedron exhibits higher SCF energy than C_{3v} and mainly than O_h isomer (cf. Fig. 5). The C_{3v} form which is a deformed section of a fcc lattice has the lowest Hartree-Fock energy among all three Li_{19} isomers but again the correlation effects cause the inverse of the energy sequence in favour of the double-icosahedron.
   The Hartree-Fock E_b/n for all studied Li_{18} and Li_{19} clusters is lower than the value of E_b/n for Li_{20}. Important conclusion which can be drawn from the preliminary results is that also for larger alkali metal clusters the differences in the total electronic energy of various isomeric clusters forms can be quite large. One can observe two general tendencies: i) the geometry of a cluster with given n should have smallest possible number of atoms with small number of nearest-neighbours and, ii) the degeneracy of the fully occupied frontier one-electron functions should be as large as possible.

V. Na-CLUSTERS.[16]

The optimized geometries of small Na_n clusters (n=2-9) have been studied with four procedures: a) the optimal geometry of Na_n clusters has been searched with conjugated gradient method in the framework of the HF-all-electron procedure. Subsequently, the MRD-CI energies have been calculated for the optimal HF geometries. b) the optimal Na cluster geometries have been found with Powel's[22] method on the basis of the HF-effective core potential (ECP) approximation. For ECP optimized geometries the ECP MRD-CI procedure have been applied. c) The optimal geometries of the Na_n cluster (n<6) have been determined with Powel's procedure for the optimal ECP MRD-CI energy. d) The energies of selected Na_n cluster (n<6) have been determined with the direct configuration interaction method.

The atomization energy per atom as a function of nuclearity n for Na_n-clusters has very similar features as the corresponding curve for the Li clusters. E_b/n for Na_n calculated with all electron procedure as well as with the pseudopotential approach (ECP) are shown in Fig. 7. The E_b/n values calculated with the direct CI method cannot be distinguished from corresponding AE MRD-CI data in the scale of the Fig. 7. Similarily, the data for the ECP procedure b) and c) are indistinguishable on the Fig. 7.

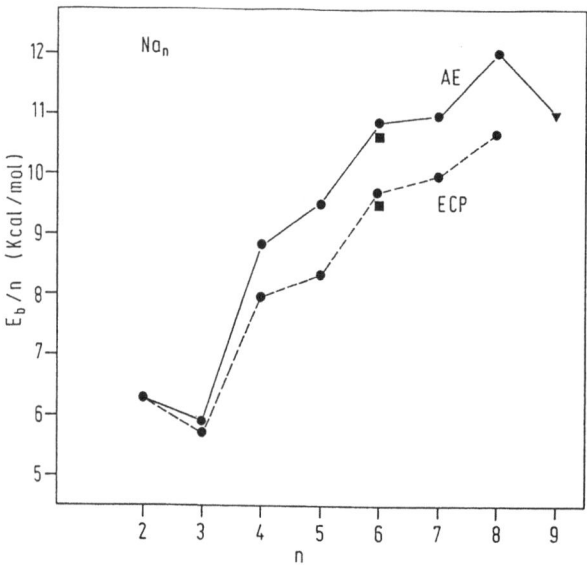

Fig. 7: MRD-CI atomization energies (E_b/n) for the SCF energy optimized neutral Na_n clusters calculated with all-electron procedure (AE) and effective core potential approach (ECP). The AE preliminary result for Na_9 is denoted as ∇. The E_b/n values for $Na_6(C_{5v})$ ⊙ and $Na_6(D_{3h})$ are shown.

The "optimal" geometries of small Na clusters exhibit the

same topologies as the optimal neutral Li clusters. Therefore, these geometries are not shown in Fig. 7. Also, the Na
isomers having comparable energies with the "best" Na cluster
forms (e.g. Na_5, Na_6) have the same topologies as the analo-
geous Li_n clusters (e.g. Li_5, Li_6). The only difference occurs for $^nD_{5h}$-Na_7 which has larger distance between the apical
atoms of the pentagonal bipyramid (4.0 Å) than between the
Na atoms forming the pentagon (3.8 Å). In contrast these two
distances are nearly the same in the Li_7-D_{5h} (3.1 Å). The relatively small increase of the atomization energy per atom
with increase of the cluster size from the Na hexamer to the
Na heptamer can be due to this less optimally compact geometry of Na_7-D_{5h}. However, the shortening of the distance d
between the apical Na atoms changes the MRD-CI energy only
very little. The total MRD-CI energy is higher by <5 kcal/
mol when d is shortened from 4.0 Å to 3.1 Å. This shows that
the energy minimum is very shallow.

The large stability of Na_4 and Na_6 clusters shows a far
reaching analogy between the electronic properties of Li and
Na clusters. The preliminary MRD-CI E_b/n for Na_9 indicates
that E_b/n for Na_9 is smaller than for Na_8 (cf. Fig. 7).

It is worth of mentioning that the agreement among predictions obtained from various quantum chemistry HF-CI methods is extremely good. Therefore, ECP-CI procedure seems
to be very promising for predicting the properties of larger
Na_n clusters as well as of clusters built from heavier alkali
metal atoms. In contrast, the binding energies per atom obtained from local spin density approach (LSD) exhibit an
overal overestimation of the binding and also larger increase
in E_b/n with increase of nuclearity than the corresponding
quantities obtained from both AE and ECP HF-CI procedures.

VII. Be-Li_k MIXED CLUSTERS[28].

In varying the chemical composition of mixed clusters one
can change the number of valence electrons in the cluster
with constant number of atomic nuclei. Therefore, in the study of the mixed clusters the electronic and geometrical effects determining the cluster stability can be clearly distinguished. The simplest possible case is the investigation
of the structure of the Be-Li_k clusters. The comparison of
Be-$Li_{(n-2)}$ mixed clusters with Li_n cluster can serve as a
very clear test of the validity of the simple electron shell
model.

The dependence of the atomization energy per atom

$$E_b/k' = | -E_k + kE_{Li} + E_{Be} | /(k+1)$$

where E_k, E_{Li} and E_{Be} is the energy of the BeLi_k cluster, Li
atom and Be atom, respectively is shown in Fig. 8. Both the
Hartree-Fock and the MRD-CI binding energies per atom in-

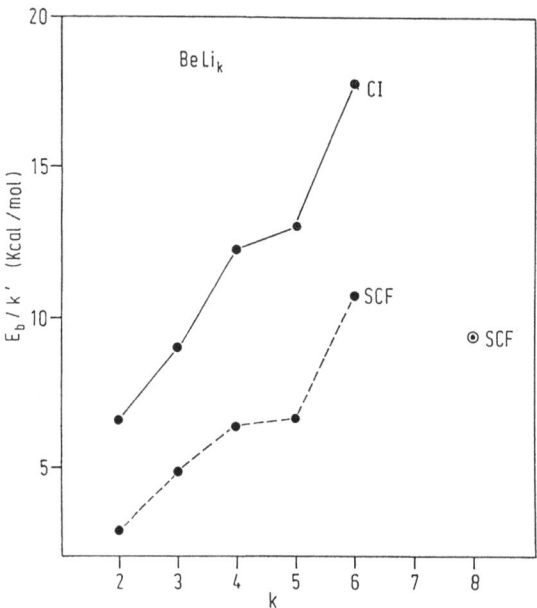

Fig. 8: CI and SCF atomization energies per atom (E_b/k') of Be-Li$_k$ mixed clusters for SCF energy optimized geometries. The preliminary SCF E_b/k' value for Be-Li$_8$ is denoted as ⊙.

crease with the cluster size. The Be-Li$_6$ with eight valence electrons is found to be very stable in agreement with the ideas of the electron shell theory. The investigation of Be-Li$_8$ is not yet conclusive. Relatively low E_b/k' is estimated (cf. Fig. 8). A very small increase of the average atomization energy with increasing k from four to five is remarkable.

In the optimal geometries of Be-Li$_k$ the Be atom is surrounded by k Li atoms. An exception is the Be Li$_3$ mixed cluster which has the form of a trapezium with the longest interatomic distances between the Be atom and two Li atoms. Be-Li$_4$ is a square formed by Li atoms with the Be atom in its middle. Two forms of Be-Li$_5$ have comparable SCF energies: the planar pentagon of Li atoms with the Be atom in its center (D_{5h}) and Li trigonal bipyramid with Be again centrally located (D_{3h}). The Be-Li$_5$(D_{3h}) is more stable according to the results obtained from the MRD-CI procedure. In the very stable Be-Li$_6$ the Be atom lies in the center of a regular octahedron. The triple degeneracy of the highest occupied MO in Be-Li$_6$ corresponds fully to the assumption of the electron shell theory[23]. This triple degeneracy is a consequence of the symmetry of Be-Li$_6$ of course. The Be-Li$_8$ which seems to be less stable has the form of a cube with the Be atom in its middle.

VII. CONCLUSIONS

1) The very details of the electronic and geometric properties of small neutral, cationic and anionic alkali metal clusters can be understood and described with the appropriate quantum chemistry methods. If the stability of the clusters with different charges should be determined, the geometry of the nuclei framework seems to play an important role. This cluster geometry decides also to which extent and in which way the electron shell structure influences cluster stability. Therefore, the fifth important factor should be added to the four factors listed in the Introduction all of which in mutual concurrence determine electronic structure and geometry of stable small neutral and charged alkali metal clusters: Optimal cluster geometry minimizes Coulomb repulsion in the charge density of charged clusters.

2) The quantum chemical investigation supports strongly the idea of especialyy large stability for nearly spherical Li_{20} and consequently the finding of the "magic number" 20.

3) The similarities in the propeties of small Na_n and Li_n clusters are confirmed with a variety of quantum chemistry methods.

4) The calculations of Be-Li_k clusters seem to support the general ideas of the shell model because Be-Li_6 with eight valence electrons is found to be a very stable mixed cluster.

5) The convergence of various quantum chemistry methods used in this work to nearly identical results for Na_n clusters shows clearly the reliability of the predictions of the cluster geometries and stabilities although the involved energy differences between various cluster isomers are in some cases relatively small.

In general, quantum mechanical methods which are capable to account for the given factors mentioned in the Introduction and in the Section III yield similar predictions for electronic properties of small alkali metal clusters if the factors respected play dominant role for determining these properties.

6) According Brechignac[25] the experimentally determined abundances in the mass spectroscopical detection of clusters depend prevalently on the energy of the ionizing photon (or electron). If this energy is not sufficient to cause the dissociation of clusters the relative stability of neutral clusters is predominantly important for the occurence of abundances. Under these experimental conditions the "magic numbers" 4,8 and 20 are observed for Na and K clusters. If the ionization energy can also cause fragmentation, than the stability of cationic clusters is important and high peaks for K_3^+; K_9^+ and K_{21}^+ are observed. These experimental findings correspond precisely to the maxima of theoretical curves Δ^2E_n (4,8) and $\Delta^2E_n^+$ (3,9) as well as to the theoretical prediction of the high stability of the neutral Li_{20} cluster.

ACKNOWLEDGMENT:

This work was supported by the Deutsche Forschungsgemeinschaft (Sfb 6 "Structure and Dynamics of Interfaces", Sfb
337 "Energy and Charge Transfer in Molecular Aggregates")
and Consiglio Nazionale delle Recerche (CNR)). The authors
thank to Dr. C. Brechignac (Laboratoire Aime-Cotton; Dr. W.
Schulze, (Fritz-Haber-Institut der Max-Planck-Gesellschaft)
und Prof. E. Schumacher (University of Bern) for stimulating
discussions.

REFERENCES:

1.   J. Koutecký and P. Fantucci, Chem. Rev. 86, 539 (1986)
2.   I. Boustani, W. Pewestorf, P. Fantucci, V. Bonačić-
     Koutecký and J. Koutecký, Phys. Rev. B35, (15 June 1987)
3.   Y. Wang, T. F. George, D. M. Lindsay and A. C. Bori,
     J. Chem. Phys. 86, 3493 (1987)
4.   H.-O. Beckmann, J. Koutecký and V. Bonačić-Koutecký,
     J. Chem. Phys. 73, 5182 (1980)
5.   G. Pacchioni, D. Plavšić and J. Koutecký, Ber. Bunsen-
     gesellschaft Phys. Chem. 87, 503 (1983)
6.   D. Plavšić, J. Koutecký, G. Pacchioni and V. Bonacić-
     Koutecký, J. Phys. Chem. 87, 1096 (1983)
7.   J. Koutecký and G. Pacchioni, Ber. Bunsenges. Phys. Chem.
     88, 233 (1984)
8.   P. Fantucci, J. Koutecký and G. Pacchioni, J. Chem. Phys.
     80, 3251 (1984)
9.   G. Pacchioni and J. Koutecký, J. Chem. Phys. 81, 3588
     (1984)
10. S. Huzinaga, J. Andselm, M. Klobukowski, E. Radzio-
     Andselm, A. Sakai and H. Tatewaki, Gaussian Basis Sets
     for Molecular Calculations (Elsevier, Amsterdam, 1984)
11. H.-O. Beckmann and J. Koutecký, Surf. Sci., 120, 127
     (1982)
12. J. D. Dilland and J. A. Pople, J. Chem. Phys. 62, 2921
     (1975)
13. I. Boustani and J. Koutecký, unpublished results
14. L. Gianolio, R. Pavani and R. Clementi, Gazz. Chim.
     Ital., 108, 181 (1978)
15. W. J. Stevens, H. Bark and M. Krauss, J. Chem. Phys.
     82, 6026 (1984)
16. V. Bonačić-Koutecký, P. Fantucci and J. Koutecký, un-
     published results
17. J. Binkley,J. A. Pople, and W. J. Hehre, J. Am. Chem.
     Soc. 102, 939 (1980)
18. M. F. Guest and J. Kendrick, Daresbury Laboratory Report
     No CCP1/86, unpublished
19. R.J. Buenker and S. D. Peyerimhoff, Theor. Chim. Acta,
     35, 33 (1974); R. J. Buenker, S. D. Peyerimhoff and W.
     Butscher, Mol. Phys. 35, 771 (1978); R. J. Buenker and

       R. A. Phillips, J. Mol. Struct. Theochem. 123, 291 (1985).
20.   P. E. Siegbahn, J. Chem. Phys. 72, 1647 (1980)
21.   O. Echt, K. Sattler and E. Recknagel, Phys. Rev. Lett.
       47, 1121 (1981)
22.   M. M. Kappes, R. W. Kunz and E. Schumacher, Chem. Phys.
       Lett. 91, 413 (1982).
23.   W. D. Knight, K. Clemenger, W. A. De Heer, W. A. Saun-
       ders, M. Chou and M. L. Cohen, Phys. Rev. Lett. 52, 2141
       (1984)
24.   C. Brechignac and Ph. Cahudzac, Z. Phys. D3, 121 (1986)
25.   C. Brechignac, private communications
26.   I. Boustani and J. Koutecký, unpublished results
27.   M. Y. D. Powell, Computer J., 7, 155 (1964)
28.   W. Pewestorf and J. Koutecký, unpublished results

Laser Spectroscopy of Matrix-Isolated Clusters: Diatomic and Triatomic Manganese

K.D. Bier, T.L. Haslett, A.D. Kirkwood and M. Moskovits
Department of Chemistry,
University of Toronto,
Toronto, Canada M5S 1A1

The properties of metal clusters both with molecules "adsorbed" upon them or naked has been a subject of intense research for some time, exploding recently, apparently uncontrollably. The questions that one wishes to answer, in broad terms, relate to the manner in which the physical and chemical attributes of a material depends upon its state of aggregation. In particular one might wish to determine how rapidly the infrared spectrum of a molecule bonded to an ever increasing metal cluster approaches that of the same molecule bonded to the surface of the bulk metal. Likewise one might wish to determine the manner in which the eigen-energy spectrum of a cluster approaches the density of states characteristic of the bulk material. These sorts of questions are best answered spectroscopically. While properties such as ionization potential [1] and reactivity towards H_2 and other molecules [2] have been determined for a number of metals up to cluster sizes involving tens of atoms, precise spectroscopic data are available for far fewer systems and for much smaller clusters. These come from a variety of techniques such as two-photon ionization mass-spectroscopy [3], laser-induced fluorescence spectroscopy [4] of jet-cooled clusters, and from absorption spectroscopy [5], laser-induced resonance fluorescence [6], ESR [7], Mössbauer [8], magnetic-circular dichroism [9], studies, among others, of matrix-isolated clusters.

To date, data are available on about three dozen metal diatomics, half a dozen triatomics, and a handful of higher clusters. The field has been complicated somewhat by a number of misassignments, most egregious among which is the revelation [10] that the fluorescence spectrum attributed to Ni_2 [11] was in reality that of Se_2 presumably left over in the experimental system from a previous run. Likewise the resonance fluorescence spectrum that we ascribed to Mn_2 [12] was found in subsequent studies with more refined instrumentation to be that of a higher Mn cluster, possibly Mn_3. The original error due to

J. Jortner et al. (eds.), Large Finite Systems, 319–331.

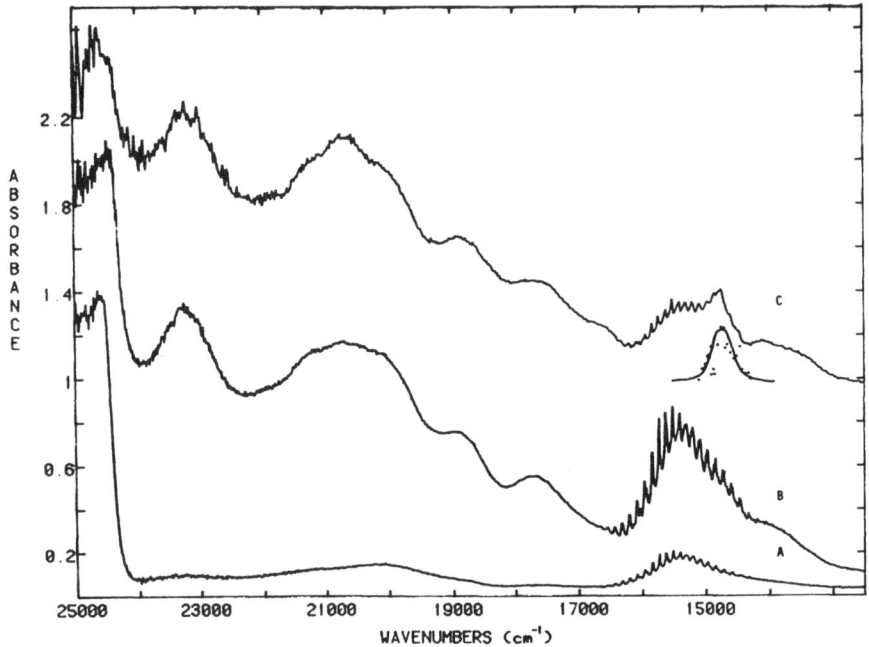

Figure 1:

Visible absorption spectrum of manganese isolated in
solid Ar: (A) (B) (C) progressively more concentrated
in Mn. Dotted spectrum in C is the excitation spectrum
of resonance Raman signal attributed to Mn_3.

the coincidental overlap of the bands of the two species was first
pointed out by Morse [13]. There abounds, in addition,
interpretational controversies such as that regarding the symmetry of
the excited state of Cu_3 [14]. Likewise differences between
structures and assignments of several clusters proposed on the basis
of experimental data and those arising from quantum chemical
calculation have not yet been resolved [15].

In this article we present new resonance fluorescence data for Mn_2
and confirm the assignment by means of laser depopulation
spectroscopy and temperature effects. We then discuss the bonding
trends in the first row transition metal dimers. A spectrum due to
Mn_3 is presented. The molecule is shown to be a D_{3h} triatomic subject
to a weak dynamic Jahn-Teller effect.

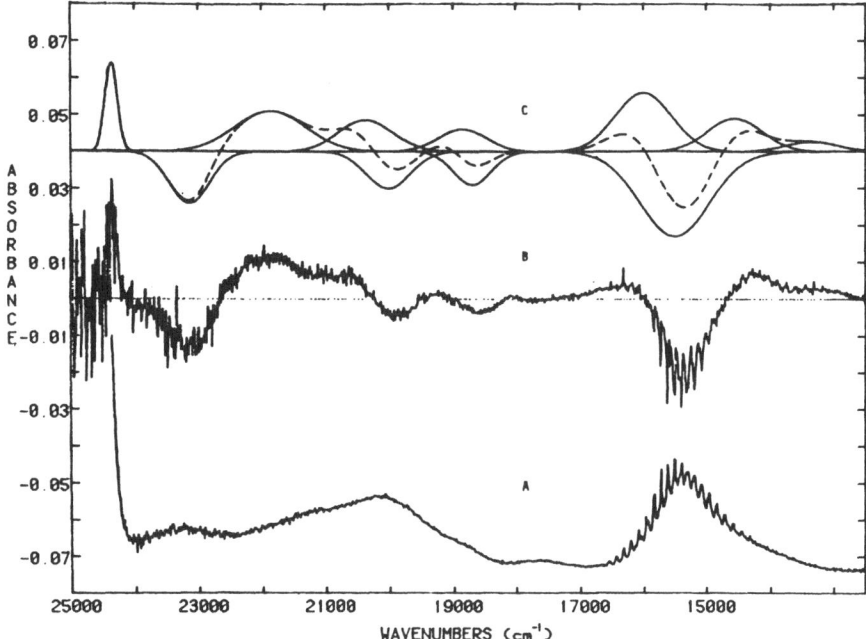

Figure 2:

Manganese isolated in solid argon:
(A)   Absorption spectrum, (B) Raman difference spectrum
(C) data of 2B fitted with Gaussians. 15516 cm^{-1} laser
used to generate (B).

Experimental

Matrices were prepared under ultrahigh vacuum conditions (low 10^- $_{10}$ torr). Manganese was sublimed from a zone-refined nugget resistively heated in a tungsten basket. Metal fluxes were determined by means of a quartz crystal microbalance which received the metal flux when the matrix substrate was raised by means of a bellows-sealed manipulator. The polished aluminum substrate was cooled to 11K by means of a DISPLEX closed cycle refrigerator. Argon and krypton were leaked into the vacuum chamber through a precision metering valve attached to a dosing tube directed at the cold surface. Visible absorption spectra of the transparent matrix were recorded by reflection using a quartz-halogen lamp source and a SPEX 1400 series double monochromator followed by photon counting. Resonance Raman and resonance fluorescence spectra were excited by means of a Spectra Physics Ar$^+$ laser alone or pumping a Coherent model CR-599 dye laser using DCM dye. The monochromator control, signal collection and all other instrumental control were carried out on a Tektronix 4052

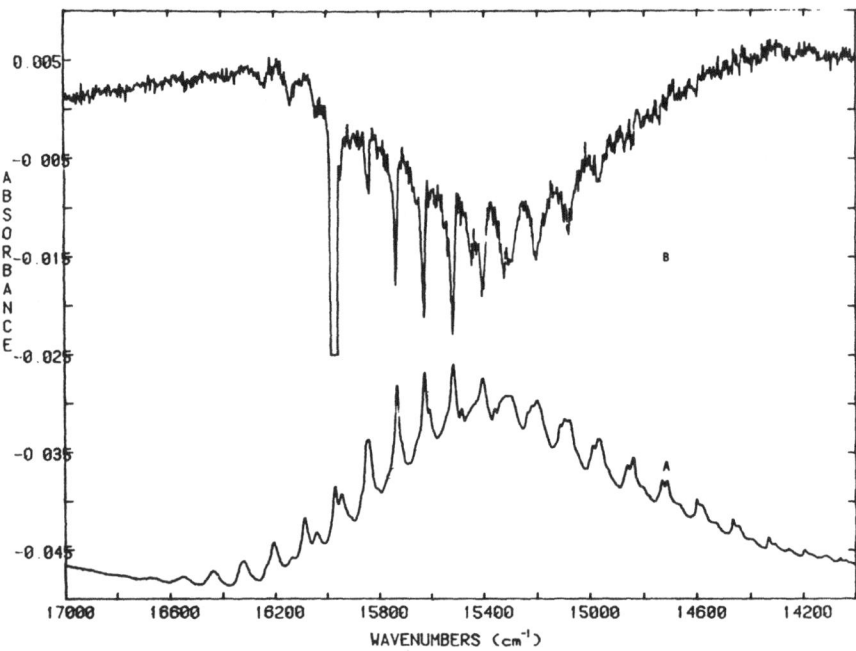

Figure 3:

(A) (B) high resolution position of figure 2A and B

computer interfaced appropriately. Laser depopulation experiments
were carried out by simultaneously illuminating the matrix with
chopped laser light while measuring its absorption spectrum. The
computer controlled a toothed-wheel chopper and directed the
absorbance measured when the light laser was on or off to the
appropriate array in memory. The difference spectrum was then
generated by taking the log of the ratio of the two arrays when the
experiment was terminated. This corresponds to a subtraction of the
absorbance spectra.

Results and Discussion

Fig. 1 shows the visible absorption spectra obtained from three
manganese-containing argon matrices each progressively more
concentrated in metal. Except for the band centered at 15500 cm^{-1},
the spectra consist of broad, featureless, overlapping bands
belonging to at least three different species. The band at 15500 cm^{-1}
was first reported by DeVore et al. [16] and assigned to Mn_2. At high

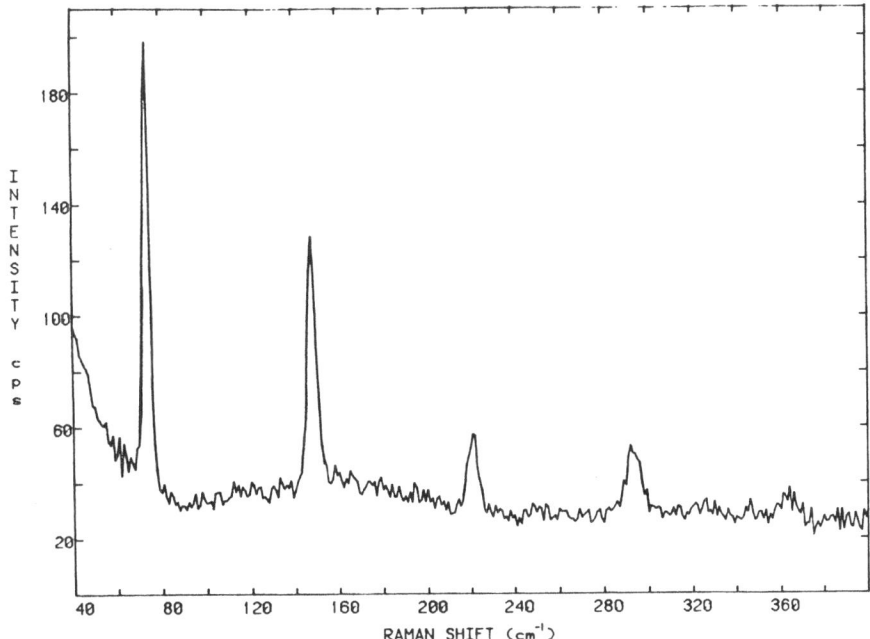

Figure 4:

Resonance Raman spectrum of Mn_2 isolated in solid Kr.

Mn concentration, a feature is clearly seen growing under this band (fig 1c) centered at 14750 cm^{-1}. The resonance Raman spectrum reported by our group in ref. 12 and assigned to Mn_2 is actually due to the excitation of the carrier of this sharper absorption band centered at 14750 cm^{-1} and not of Mn_2 that is in turn responsible for the highly featured absorptions centered at 15500 cm^{-1}. The resonance Raman excitation spectrum (fig 1c) proves this. Hence the carrier of the resonance Raman spectrum reported in ref. 12 as due to Mn_2 is actually that of a higher cluster of manganese.

Untangling the various contributions to the absorption spectra shown in fig 1 is made easier by using laser depopulation spectroscopy. By tuning the laser into an absorption of a known species one depopulates its ground state slightly and all absorptions originating from that ground state are a little reduced. Displaying the difference in absorbance obtained with the laser on and off ($\Delta A = A_{on} - A_{off}$) identifies bands belonging to the species excited as negative differences. In addition, if the lifetime in the excited state populated by the laser is sufficiently long or if the molecule relaxes to another state in which it is relatively long-lived, one will see <u>positive</u> features in the ΔA spectrum corresponding to

absorptions connecting the excited state populated to higher excited
states. An example of such a difference spectrum is shown in fig 2
obtained by exciting into one of the components of the 15500 cm^{-1}
absorptions (15516 cm^{-1}). The four negative features correspond to
four bands belonging to Mn$_2$ buried among the other absorptions
visible in fig 2a. One can see, moreover, seven positive bands in the
laser depopulation spectrum. Fitting Gaussians to the differences
(fig 2c) allows one to determine the frequencies of the band centers
of the observed difference bands more accurately; and while one
cannot guarantee that the positive features originate from the
excited state pumped by the laser, nor for that matter that all of
the positive features originate from the same excited state, by
making these assumptions one can estimate the frequencies of seven
higher excited states of Mn$_2$.

Table 1 summarizes the results. There is remarkable concord
between the frequencies (or wavelengths) so determined and those
reported by other authors [17] [18] on the basis of uv-visible matrix
absorption spectroscopy. In fact, spectral features previously seen
but left unassigned because of experimental difficulties [18] have
been assigned to Mn$_2$ on the basis of this spectroscopy.

The portion of the absorption spectrum and laser depopulation
spectrum of Mn$_2$ showing the region of the 15500 cm^{-1} band with higher
resolution is shown in fig 3. From it, it is clear that the band is
actually composed of several progressions (we count five) not all of
which originate from the same lower state. This is obvious from the
laser depopulation spectrum (fig 3b) which shows that excitation with
15970 cm^{-1} dye laser radiation results in a difference spectrum
dominated by two of the six progressions. In particular the
difference spectrum is almost devoid of features at frequencies lower
than 15000 cm^{-1} where the absorption spectrum is still rich in bands.

The extensive groundwork laid by Rivoal at al [17], Baumann et al
[19] and Nesbet [20] makes the interpretation of these spectra rather
easy. It is accepted that Mn$_2$ is an antiferromagnetically coupled Van
Der Waals dimer whose lowest electronic states are described more or
less adequately by the Heisenberg exchange Hamiltonian

$$H = - \lambda \vec{S}_a \cdot \vec{S}_b$$

whose eigenenergies are given to first order by an expression similar
to the Landé interval rule

$$E(S) = -(\lambda/2)[S(S+1) - 35/2] \qquad (1)$$

with S=0 to 5. (The quantity 35/2 arises from two spin 5/2 (^{6}S) Mn
atoms interacting antiferromagnetically. Nesbet assumed that there
was a single sσ molecular orbital in Mn$_2$ created from d^{6}s^1 atomically
prepared states and only the six d electrons are localized on each
atom and disposed antiferromagnetically. In that case the
eigenenergies would be given by E(S)=-($\lambda/2$)[S(S+1)-24] with S=0 to 6.

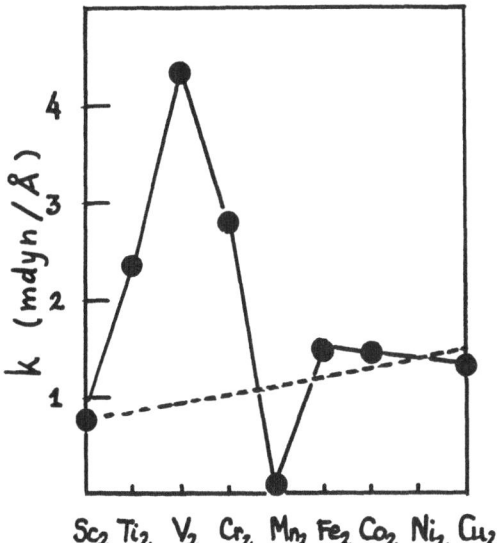

Figure 5:

Force constants for the first row transition metal
diatomics

Since we (and previous workers) have only been able to access the
lower three spin states we cannot distinguish between these two
alternatives.)

Table 1 UV-Visible Transitions of Mn$_2$

Fitted Band[a] Position (cm^{-1})	λ (nm)	Previously Observed Band Position (nm)
39871	250.8	253[b]
37376	267.6	275[b]
35891	278.6	275[b]
34366	291.0	288[b]
31516	317.3	312[b]
30066	332.6	329[b] 331.5[c]
28876	346.3	345[b] 347[c]

range from 357 to 400 nm   400[b]
not observable by this technique

23140	432.2	435[b]
20020	499.5	-
18700	534.8	-
15500	645.2	650[d]

a) Bands with λ below 357 nm are fit from Depopulation Difference
spectra by adding the laser frequency to the positive band
frequencies; bands with λ above 400 nm are observed in the visible
spectra. Frequencies are ±100 cm^{-1}.

b) From [18] - assigned to various unknown Mn species except for 275 nm which was a strong atomic transition. Note that our method allows Mn_2 bands under this transition to be discerned.

c) From [17] - assigned to Mn_2 from MCD measurements.

d) From [16] - assigned to Mn_2

Each of the states given by eq 1 are 2S+1-fold degenerate. This fact together with the smallness of the value of λ (10 ± 2 cm⁻¹) [17][19] implies that even at matrix temperatures (10-15K) there would be significant population in the lower three spin states belonging to the representations $^1\Sigma_g^+$, $^3\Sigma_u^+$ and $^5\Sigma_g^+$, in ascending order of energy.

With manganese isolated in krypton a similar situation prevails except that the structured absorption is now centered at 15000 cm⁻¹ and it is found to be composed of only three progressions. Varying the temperature of the matrix changes the relative intensities of the components of these absorption bands with both argon and krypton matrices. In the krypton matrices reversible intensity changes were observed. In argon matrices both reversible and irreversible intensity changes were observed indicating that some of the observed progressions were due to transitions originating from different spin states of Mn_2 while others were due to the replications of these transitions for Mn_2 in other matrix sites.

Table 2 Spectroscopic Constants of Mn_2 isolated in Ar and Kr

		T_e (cm⁻¹)	ω_e (cm⁻¹)	$\omega_e x_e$ (cm⁻¹)	D_e (eV)	k (mdyne/Å)
$X^1\Sigma_g^+$	in Ar	-	68	0.8	0.18	0.07
$X^1\Sigma_g^+$ (new site)	in Ar	-	59	0.5	0.21	0.05
$X^3\Sigma_u^+$	in Ar	-	71	0.3	0.52	0.08
$X^1\Sigma_g^+$	in Kr	-	76.3	0.3	0.62	0.09
$A^1\Sigma_u^+$?	in Ar	14510	119	0.6	0.73	0.23
$A^3\Sigma_g^+$?	in Ar	13650	140	0.6	1.01	0.32
$A^5\Sigma_u^+$?	in Ar	13740	150	1.3	0.54	0.36
$A^1\Sigma_u^+$?	in Kr	12730	122	0.2	2.32	0.24
$A^3\Sigma_g^+$?	in Kr	12750	146	0.5	1.32	0.34
$A^5\Sigma_u^+$?	in Kr	12650	155	0.1	6.75	0.39

? signifies tentative assignment

When excited with appropriate dye laser radiation, resonance Raman (or resonance fluorescence) progressions were detected (fig 4) from which the spectroscopic constants of the ground state and low lying spin states of Mn_2 were determined. These data are summarized in table 2.

Bonding in Transition Metal Diatomics

The vibrational frequencies of all first row transition metal diatomics with the exception of that of Ni_2 are now known. Fig 5 shows the trend in the metal-metal force constant as a function of the metal's position in the periodic table. The dashed line indicates an estimate of the expected value of the force constant if the s-electron alone were responsible for the metal-metal bond (i. e. if the bond were only a single $s\sigma$ bond). This is based on Siebert's formula [21].

Three main features are obvious in this figure. To the left of Mn_2 the measured force constants exceed substantially the contribution of the $s\sigma$ bond. For these metal diatomics the bond order is high and d-electron involvement substantial. To the right of Mn_2 the metal-metal force constants differ only slightly from those due to $s\sigma$ bonding alone. Finally the value of the force constant of Mn_2 indicates that it does not possess even a single bond, conforming with the view that it is best described as a Van Der Waals dimer.

These observed trends may be understood as an interplay among three contributions. First one must assume that the molecular orbitals derived from atomic d-orbitals are progressively more stabilized as one proceeds leftward from copper to scandium. That is, the energy interval between bonding and antibonding orbitals derived from atomic d-orbitals increases as one traverses the first transition row from right to left. This is a manifestation of the fact that the spatial extent of the atomic d-orbitals is greatest for Sc and least for copper. Hence on bringing together two atoms from the right hand side of the row to form a diatomic, $s\sigma$ electron interactions will be highly repulsive before the small internuclear separation is reached at which d-electron attraction is maximized. Hence d-electron contribution to the bonding will be minimal and the bond length will be relatively large. In contrast, for atoms on the left side of the row the $s\sigma$ electron repulsion will not be too great at the internuclear separation for which d-orbital overlap is optimum; d-electron involvement in the bonding will therefore be high and the bond length relatively short. A possible consequence of this s versus d interplay is that near the center of the row one might have a diatomic with two minima in its ground state potential; an outer minimum due mainly to s-electron bonding and an inner minimum due to d-electron overlap. This was indeed proposed to be the case for Cr_2 [22] [23].

The total d-orbital contribution to the metal-metal bonding is a
function both of the aforementioned d-orbital stabilization which
decreases from left to right and the d-electron population that
<u>increases</u> in that direction. The increase in d-electron population is
beneficial only from Sc to Cr. Beyond chromium the d-electrons begin
to fill antibonding orbitals thereby reducing the bond order. Based
on these two effects alone one would expect the metal-metal bond
strength to become a maximum at Cr_2 and decrease monotonically
thereafter. The amonotonic behavior observed in fig 5 is due to the
fact that the d-electron contribution to the bonding is offset by an
atomic preparation energy that maximizes at Mn_2. This is the energy
required to promote the metal atom from its ground state to the
center of the states derived from the $d^{n+1}s^1$ configuration [24]. A
more detailed explanation of this contribution is given in refs [12]
[24].

Trimanganese

Exciting the manganese in argon matrix with dye laser radiation in
the 14275 to 15050 cm^{-1} range produces the spectra shown in fig 6.
With the lowest energy excitation the spectrum consists of two
progressions with approximately 124 cm^{-1} spacing, one based on the
vibrational ground state, the other on a state 196 cm^{-1} above it. On
exciting with progressively higher frequency laser light it was found
that all but the 196 cm^{-1} line are in fact the first members of a
series of (v_a+1) equally spaced multiplets, v_a being the quantum
number associated with the 124 cm^{-1} "vibration".

Multiplets of this sort are suggestive of a number of effects. It
is known [25], for example, that degenerate vibrational states are
split in the presence of anharmonicity into $v/2+1$ or $(v+1)/2$
multiplets for v even and odd respectively. The spacing results from
a correction of the form gl^2 (l=v, v-2, v-4, ...) to the
eigenenergies on the degenerate vibration. The dependence upon the
square of the quantum number l results in unequal spacing between
adjacent members of the multiplets, unlike what we observe.

Likewise a molecule of the form X_3 which is subject to the
dynamical Jahn-Teller effect would, in the limit of large Jahn-Teller
coupling but with a low barrier to pseudorotation, possess vibronic
eigen-energies that depend on two "vibrational" quantum numbers, v_s
and v_a and on a pseudorotation quantum number j, introduced by
Longuet-Higgins et al [26], that takes half integer values. Here
again j enters the expression for the eigen-energies as its square,
hence once again, multiplets due to this quantum number are not
expected to be equally spaced. Indeed, this is what is reported by
Delacrètaz et al [27] for an excited state of Na_3 in which the
molecule appears to be an example of this particular limiting case of
the dynamic Jahn-Teller effect.

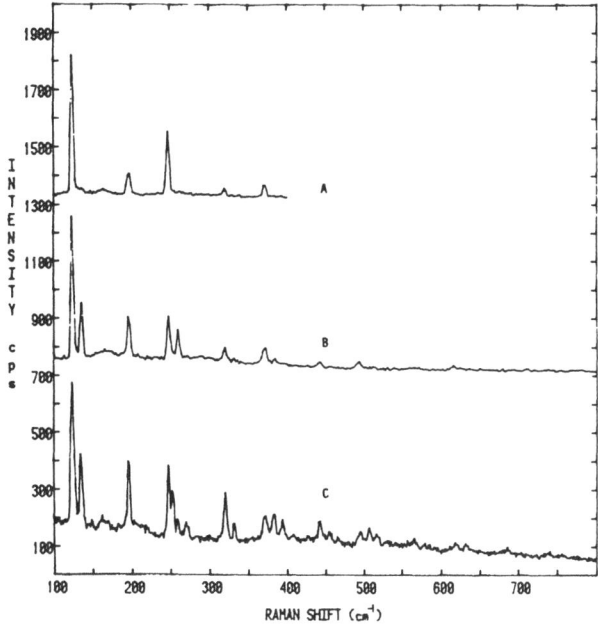

Figure 6:

Portions of the resonance Raman spectra of Mn_3 excited
with (A) 14550, (B) 14750 and (C) 15000 cm^{-1} dye laser
radiation

There is another limit of the dynamic Jahn-Teller effect for which
an expression for the vibronic energy eigenvalues is obtainable in
closed form. This is the case of very small Jahn-Teller coupling. A
triatomic molecule of D_{3h} parentage subject to the JT effect in this
limit would possess vibronic states whose energies, in the harmonic
approximation, are given by

$$G(v_s, v_a, l) = \omega_s(v_s + 1/2) + \omega_a(v_a + 1) \mp 2D\omega_a(l \pm 1) - x_s(v_s + 1/2)^2$$

$$- x_a(v_a + 1)^2 - x_{as}(v_a + 1)(v_s + 1/2) \qquad (2)$$

where $l = v_a, v_a - 2 \ldots 1$ or 0.
Two levels corresponding to the upper and lower signs are obtained
for each value of l, except for $l = 0$ when only the upper sign is used.
This expression results in $v_a + 1$ equally spaced multiplets for each
value of v_a, as is observed; the energy interval between members of
the multiplets must, moreover, be small in order to meet the
condition that the Jahn-Teller coupling and hence the parameter $D\omega_a$
be small. Eq 2 fits the data shown in fig 6c quite well yielding the
parameters $(\omega_s - 2x_s) = 196.79$ cm^{-1}, $\omega_a = 130.15$ cm^{-1}, $D\omega_a = 2.95$ cm^{-1},
$x_a = 0.06$ cm^{-1} and $x_{as} = 0.79$ cm^{-1}. (The quantities ω_s and x_s can not be
obtained independently from the available data). There are still

hints of the need for a somewhat higher level of approximation in order to explain the fine details of spectrum 6c. For example, one of the members of the triplet near 250 cm^{-1} is split into two components. This may be due to the fact that the approximation used to generate eq 2 involved only a linear coupling term in the Hamiltonian. This causes A$_1$ and A$_2$ vibronic states to remain degenerate. Upon including a quadratic coupling term [28] this degeneracy is lifted.

Based on this analysis we propose that the progressions shown in fig 6 and the absorption centered at 14750 cm^{-1} (678 nm) are due to Mn$_3$ whose shape in zeroth order is D$_{3h}$ and subject to a slight Jahn-Teller effect. For this geometry, the normal modes of the triatomic were described as linear combinations of the three bond stretching internal coordinates and the frequencies ω_a and ω_s were fit by means of the Wilson FG method [29] using a program obtained from the QCPE [30]. The resulting parameters were a metal-metal force constant and a stretch-stretch interaction constant of 0.38 mdyne/Å and 0.02 mdyne/Å respectively. The force constant calculated for the trimer is thus four times larger than that of diatomic manganese. Clearly Mn$_3$ is much more strongly bound than Mn$_2$.

The above discussion implies that the ground state of Mn$_3$ is either of E' or E" symmetry with unknown (but even) spin multiplicity; likely a doublet.

Acknowledgements

We are grateful to NSERC, the Connaught Fund (University of Toronto) and the Government of Ontario (URIF program) for financial support. TLH wishes to thank the Ontario Government for an OGS scholarship. KDB is grateful to the Deutche Forschungs- Gemeinschaft for a fellowship. MM wishes to thank the Guggenheim Foundation for a fellowship.

References

[1] A.Herrmann, E.Schumacher and L.Wöste; J.Chem.Phys., 68, 2327 (1978); E.A.Rohlfing, D.M.Cox, and A.Kaldor; J.Phys.Chem., 88, 4497 (1984).
[2] M.E.Geusic, M.D.Morse and R.E.Smalley; J.Chem.Phys., 82, 590 (1985); S.C.Richtsmeier, E.K.Parks, K.Liu, L.G.Poles and S.J.Riley, ibid., 82, 5470 (1985); R.L.Whetten, D.M.Cox, D.J.Trevor and A.Kaldor; Phys.Rev.Lett., 54, 1494 (1985).
[3] A.Herrmann, M.Hofmann, S.Leutwyler, E.Schumacher and L.Wöste; Chem.Phys.Lett., 62, 216, (1979); D.L.Michalopoulos, M.E.Geusic, S.G.Hansen, D.E.Powers and R.E.Smalley; J.Phys.Chem., 86, 3914 (1982).
[4] J.L.Gole, G.J.Green, S.A.Pace and D.R.Preuss; J.Chem.Phys., 76, 2247 (1982); V.E.Bondybey, Chem.Phys.Lett., 109, 436 (1984).

[5] W.Schulze, H.V.Becker and H.Abe; Chem.Phys., 35, 177 (1978);
     G.A.Ozin; Appl.Spectrosc., 30, 573 (1976); Cryochemistry,
     M.Moskovits and G.A.Ozin eds, (Wiley, New York, 1976).

[6] M.Moskovits and D.P.DiLella; ACS Symp., Series 179, 153
     (1982); V.E.Bondybey and J.H.English; J.Chem.Phys., 67, 3405
     (1974); H.Sonntag, B.Eberle and R.Weber; Chem.Phys., 80, 279
     (1983); U.Kettler, P.S.Bechthold and W.Krasser; Surf.Sci.,
     156, 867 (1985).

[7] G.A.Thompson and D.M.Lindsay; J.Chem.Phys., 74, 959 (1981);
     J.A.Howard, R.Sutcliffe and B.Mile; Chem.Phys.Lett., 112, 84
     (1984); L.B.Knight Jr., R.J.VanZee and W.Weltner, ibid., 94,
     296 (1983).

[8] P.A.Montano, G.K.Shenoy and T.Morrison; Phys.Rev., B25, 4412
     (1982).

[9] A.Grinter and D.Stern; J.Chem.Soc. Farad.II, 79, 1011
     (1983); J.C.Miller, R.L.Mowery, E.R.Krausz, S.M.Jacobs,
     H.W.Kim, P.N.Schatz and L.Andrews; J.Chem.Phys., 74, 6349
     (1981); and ref [17].

[10] V.E.Bondybey (personal communication).

[11] F.Ahmed and E.R.Nixon; J.Chem.Phys., 71, 3547 (1979).

[12] M.Moskovits, D.P.DiLella and W.Limm; J.Chem.Phys., 80, 626
     (1984).

[13] M.D.Morse; Chem.Rev., 86, 1049 (1986).

[14] M.D.Morse; Chem.Phys.Lett., 133(1), 8 (1987).

[15] S.P.Walch and B.C.Laskowski; J.Chem.Phys., 84, 2734 (1986).

[16] T.C.DeVore, A.Ewing, H.F.Franzen and V.Calder;
     Chem.Phys.Lett., 35, 78 (1975).

[17] J.C.Rivoal, J.Shakhs Emampour, K.J.Zerinque and M.Vala;
     Chem.Phys.Lett., 92, 313 (1982).

[18] W.E.Klotzbücher and G.A.Ozin; Inorg.Chem., 19, 3776 (1980).

[19] C.A.Baumann, R.J.VanZee, S.V.Bhat and W.Weltner Jr.;
     J.Chem.Phys., 78, 190 (1983).

[20] R.K.Nesbet; Phys.Rev., 135, A460 (1964).

[21] H.Siebert; Anwendungen der Schwingungspektroskopie in der
     Anorganischen Chemie., (Springer, Berlin, 1966).

[22] M.Moskovits, W.Limm and T.Mejean; J.Chem.Phys., 82, 4875
     (1985).

[23] M.M.Goodgame and W.A.Goddard III (personal communication).

[24] C.D.Gelatt Jr., H.Ehrenreich and R.E.Watson, Phys.Rev., B15,
     1613 (1977).

[25] G.Herzberg; Molecular Spectra and Molecular Structure, (Van
     Nostrand, New York, 1945), Vol 1,2 and 3.

[26] H.C.Longuet-Higgins, Adv.Spectrosc., 2, 461 (1961).

[27] G.Delacrètaz, E.R.Grant, R.L.Whetten, L.Wöste and
     J.W.Zwanzigen; Phys.Rev.Lett., 56, 2598 (1986).

[28] W.A.Gerber and E.Schumacher; J.Chem.Phys., 69, 1692 (1978).

[29] E.B.Wilson Jr., J.C.Decius and P.C.Cross; Molecular
     Vibrations, (McGraw-Hill, New York, 1971).

[30] D.F.McIntosh and M.R.Peterson, General Vibrational Analysis
     Programs Utilizing the Wilson GF Matrix Method for a General
     Unsymmetrical Molecule., Programs No.342, Quantum Chemistry
     Program Exchange, room 204, Dept. of Chem., Indiana
     University, Bloomington Indiana, USA 47401.

SMALL DIELECTRIC CLUSTERS: SIZE AND SHAPE DEPENDENCE OF PHOTOPHYSICAL AND PHOTOCHEMICAL PROPERTIES

Abraham Nitzan
School of Chemistry, the Sackler Faculty of Science,
Tel Aviv University, Tel Aviv 69978, Israel

ABSTRACT. Electromagnetic phenomena involving small metal and dielectric particles and molecules adsorbed on such particles are strongly dependent on particle size, shape and geometrical arrangement. This results from the combined effects of the particle on the local electromagnetic field in its vicinity and on energy transfer processes between adsorbed molecules and the particles. Absorption, light scattering, lifetimes and yields are therefore modified in a way strongly dependent on geometry. Such phenomena are discussed. Also, a conflict between two different classical approaches to the calculation of ionization potentials of small metal particles is resolved. The commonly cited result based on the curvature dependence of the image potential is shown to be wrong. The way based on solvation energy arguments gives the correct classical result. Small deviations of experimental results from the classical prediction are shown to result from quantum mechanical effcts.

I. INTRODUCTION

This paper discusses some phenomena associated with the electrostatic and electromagnetic response of small dielectric particles. We focus on optical properties of such particles and on their ionization potentials. Optical properties of small particles as well as the photophysical and photochemical processes involving molecules adsorbed on such particles are governed by the behavior of the electromagnetic field in and near the particles. We shall review effects on light absorption, light scattering, molecular lifetimes and photochemical yields. These phenomena will be shown to be sensitive to cluster size and shape and in turn to the morphology of rough surfaces and phorous media which may be modeled as random collection of small particles.

Another manifestation of size dependence of the electrostatic response of small clusters is the size dependent ionization potential of such clusters. We show that the commonly used result $W_I(R) = W_I(\infty) + \propto q^2/R$ for the ionization potential of a metal sphere of radius R in

J. Jortner et al. (eds.), Large Finite Systems, 333–343.
© *1987 by D. Reidel Publishing Company.*

terms of the electron charge q, the bulk work function $W_I(\infty)$ and the Coulombic interaction $\alpha q^2/R$ with $\alpha = 3/8$ is wrong, resulting from incorrectly using the classical image potential too close to the particle surface. The correct classical result is as written above with $\alpha = 1/2$. Deviations of experimental results from this behavior are mainly due to quantum mechanical effects.

II. LOCAL FIELDS AND OPTICAL PROPERTIES

Many observed optical properties of small dielectric particles and of molecules adsorbed on such particles are manifestations of the behavior of the local electromagnetic field in and near the particles [1]. Underlying this physics are various effects which often act in concert to determine the local field. These are (1). The high polarizability of small particles when the frequency of the incident (or emitted) radiation coincides with electronic or ionic morphology dependent resonance (MDR) of the particle; (2). the ability of structures of high curvature to concentrate electric field lines into a confined space (lightening rod effect); (3). image charges induced when a molecule is in proximity to a conductor or a dielectric and (4). the ability for small particles to efficiently radiate photons. All these factors play important roles in determining the nature optical phenomena associated not only with small particles but also with rough surfaces, porous dielectric materials and metallic colloid solutions.

A simple example demonstrating the effects listed above is provided by a small dielectric sphere of radius R and dielectric function $\varepsilon(\omega)$ affected by an incident uniform electric field $\vec{E} = E_0 e^{i\omega t}$ (Fig.1a) or by a nearby point dipole $\mu = \mu_0 e^{i\omega t}$ representing an excited molecule (Fig.1b). The sphere radius and the sphere dipole distance are assumed to be much smaller than the radiation wavelength. From classical electrostatics, the field $\vec{E}_{in}$ inside the sphere for the situation depicted in Fig.(1a) is given by $\vec{E}_{in} = g_1(\omega)\vec{E}$ and the field outside the sphere is the field of a point dipole $\vec{\mu} = g_2(\omega)\vec{E}$ seated in the sphere center, where the response functions $g_1(\omega)$ and $g_2(\omega)$ are [2]

$$g_1(\omega) = \frac{3}{\varepsilon + 2} \qquad (1)$$

$$g_2(\omega) = \frac{(\varepsilon(\omega)-1)R^3}{\varepsilon(\omega)+2} \qquad (2)$$

E_0

ϵ

(a)

ϵ

μ_0

(b)

For the system shown in Fig.(1b) we are interested in the field induced by the sphere at the position of the dipole μ, $\vec{E}_d = \overset{\leftrightarrow}{g_3}(\omega)$, and in the total dipole $\vec{\mu}_{tot} = \overset{\leftrightarrow}{g_4}(\omega)\vec{\mu}$ induced in the system. Classical electrostatics provide explicit expressions also for the tensors $\overset{\leftrightarrow}{g_3}(\omega)$ and $\overset{\leftrightarrow}{g_4}(\omega)$ in terms of $\epsilon(\omega)$, R, and the distance d of the dipole from the sphere surface.

    If the particle is not spherical, linear response functions (generally tensors) $g_i(\omega)$ still exist though usually will be difficult to calculate. Explicit expressions are available for spheroids. These response functions incorporate in their structures all the effects mentioned above. In particular the sphere dipolar plasmon resonance appears as a minimum of the denominator in Eqs.(1) and (2). Similarly for a sphere the response functions g_3 and g_4 show higher multiple resonances at $\epsilon = -(1 + 1/\ell)$, $\ell = 1,2...$. For spheroids the response functions depend also on the aspect ratio a/b (a and b are the long and short axes of the spheroid), showing an explicit particle shape dependence.

    All relevant optical properties of small particles and of molecules adsorbed on such particles may be expressed by the response tensors $\overset{\leftrightarrow}{g_i}(\omega)$ or their generalizations. In what follows I consider a few examples.

Lifetime of the particle plasmon resonance. [3]. The response function $g_2(\omega)$ (Eq.(2)) is nothing but the dynamic polarizability of the sphere at frequency ω. Using it, we may obtain the absorption cross-section associated with the sphere

$$\sigma_a = - \frac{12\pi\omega a^3}{c} \, Im\left(\epsilon(\omega) + 2\right)^{-1} \tag{4}$$

and fitting this to a simple Lorentzian shape near the peak center we may obtain the total decay width

$$\gamma = 2\left(\frac{\varepsilon_2(\omega)}{\varepsilon_1'(\omega)}\right)_{\omega = \omega_r} \tag{5}$$

(ε_1 and ε_2 are the real and imaginary part of ε, $\varepsilon' = d\varepsilon/d\omega$ and r is the resonance frequency) and the radiative decay rate

$$\gamma_r = 4\left(\frac{\omega_r R}{c}\right)^3 \frac{1}{\varepsilon_1'(\omega_r)} \tag{6}$$

It is seen that the non radiative damping does not depend on the sphere size but, as expected, γ_r is proportional to the sphere volume.

Lifetime of adsorbed molecules. Consider now an excited molecule, represented by an oscillatory point dipole μ_M, located at a distance d from the surface of a small dielectric particle. We may write the equation of motion for such a molecule in the form (see, e.g. ref.3)

$$\ddot{\mu}_M + \left(\omega_M^\circ\right)^2 \mu_M + \gamma_M^\circ \dot{\mu}_M = \left(\omega_M^\circ\right)^2 \alpha_M \vec{E}(\vec{R}_M, t) \cdot \hat{u}_M \tag{7}$$

Here ω_M° and γ_M° are the frequency and decay width of the excited free molecule, $\hat{u}_M$ is a unit vector in the direction of the molecular transition dipole, α_M is the component of the molecular polarizability tensor in that direction and $\vec{E}(\vec{R}_M, t)$ is the field, at the position of the molecule, caused by the polarization induced on the particle by the molecular dipole; $\vec{E}(R_M, t) = \overleftrightarrow{g}_s(\omega, R_M) \vec{\mu}(t)$. We can follow a procedure first introduced for a molecule absorbed on plane surface [14] to obtain the surface modified frequency and damping rate:

$$\omega_M^2 = \left(\omega_M^\circ + \Delta\omega_M\right)^2 \; ; \; \gamma_M = \gamma_M^\circ + \Delta\gamma_M \tag{8}$$

$$\Delta\omega_M = \omega_M^{o}\left[1 - \alpha_M\,\hat{u}\cdot Re\,\overleftrightarrow{g}_3\cdot\hat{u}\right]^{1/2} - \omega_M^{o} \tag{9}$$

$$\Delta\gamma_M = \left(\alpha_M/\omega_M\right)\left(\omega_M^{o}\right)^2\,\hat{u}\cdot Im\,\overleftrightarrow{g}_3\cdot\hat{u} \tag{10}$$

The change in damping rate $\Delta\gamma_M$ is due to non-radiative relaxation of the molecule caused by energy transfer to the particle. The radiative relaxation rate is modified as well: It is due now to the total induced dipole $\vec{\mu}_{tot} = \overleftrightarrow{g}_4(\omega,R_M)\vec{\mu}_M$. A closer examination of the different components of the tensors $\overleftrightarrow{g}_3$ and $\overleftrightarrow{g}_4$ shows that when the molecule lies very close to the particle surface ($|R - R_M| \ll R$) $\overleftrightarrow{g}_3$ is not very different from the corresponding quantity for a plane surface. Therefore the surface induced change in the non-radiative relaxation rate is sensitive to the molecule orientation and to its distance from the surface ($\Delta\gamma_M \sim d^{-3}$) but not to the particle size and shape. In contrast, $\overleftrightarrow{g}_4$ is sensitive also to latter parameters.

Absorption by adsorbed molecules. [3,5] Consider the molecule-small particle system subjected to an incident radiation field represented by its electric component $E_o e^{i\omega t}$. The local field at the position of the molecule is given by $(1+\vec{A}(\omega,R_M))E_o e^{i\omega t}$, where $\vec{A}.\vec{E}_o e^{i\omega t}$ is the field at the position of the molecule of the dipole $g_2(\omega)\vec{E}_o e^{i\omega t}$ induced on the particle. The later may be much larger than the incident field, in particular near plasmon resonances of the particle when $g_2(\omega)$ becomses large. There are also situations of destructive interferences when g_2 is negative. Since absorption by the molecule is associated with the local field felt by the molecule we may observe strong effects due to the proximity to the dielectric particles. This effect is relatively long range, for a sphere becoming smaller like R_M^{-3} with increasing molecule-sphere center distance.

Surface enhanced light scattering. [1,6] The potentially strong effect of the particle on the absorption by an adsorbed molecule is one component of the strong enhancement of Raman scattering processes observed for molecules adsorbed on rough dielectric surfaces and dielectric particles. Another such factor arises from the emitted radiation. Indeed, it may be shown that the Raman scattering signal is modified near a dielectric particle by the factor

$$R_{RS} = |(1 + A(\omega))\,(1 + A(\omega'))|^2 \tag{11}$$

where ω and ω' are the frequencies of the incident and the scattered radiation. Resonant processes (resonance Raman scattering, fluorescence) are affected in the same way by the local field enhancement, however an additional reducing factor is caused by the particle induced damping discussed above, which usually acts to reduce the yield of the light scattering/emission processes. We therefore find, as a rule, that resonant processes are enhanced considerably less by the presence of the dielectric particle than non resonant processes (reduction of resonant processes may also occur) [7]. In addition, because the opposing effect of local field enhancement and particle induced damping have very different distance dependence, their combined effect on resonant optical processes give rise to a complex dependence on the molecule-particle distance: A maximum at a typical distance of 10-20 A is predicted and observed [8].

Photochemistry. [5,9] The enhanced local fields near small dielectric particles may lead to enhanced photochemical processes in much the same way that other optical processes are modified as described above. Photochemistry, initiated by a resonant absorption process shows dependence on geometrical parameters similar to resonant radiative processes.

Other phenomena. The effect of the local dielectric inhomogeneity associated with the small particle on the local electromagnetic field has many other implications that are only mentioned here. The plasmon resonance induced enhancement of the field inside small particles (expressed by $g_1(\omega)$, Eq.(1)) may give rise to enhanced photoemission [10]. Energy transfer between molecules adsorbed on such particles may be strongly affected both because the presence of the dielectric particle opens another dissipation channel and because of (particle induced) local field effect [11]. Non linear optical phenomena are expected to be even more sensitive to the presence of the dielectric particles than linear optical processes as discussed above [12]. Lasing and superfluorescence associated with molecules distributed inside small dielectric droplets are discussed elsewhere in this volume [13]. Finally, I mention some very pronounced effects associated with local fields in other microstructures such as porous dielectric media, cavity sites and clusters of dielectric particles [14].

The same formalism as described above can be used to discuss systems of an entirely different kind. Recent experiments indicate that the radiative lifetimes of molecules are affected by clusutering with rare gas atoms. In a classical approach we may regard the atoms as point dielectric particles, interacting with the oscillating molecular dipole and with each other via their polarizabilities. For any given geometrical arrangement the total dipole induced in such a cluster for the given molecular dipole may be calculated and this leads to the clustering effect on the radiative lifetime [15]

$$\frac{\Gamma_R}{\Gamma_R^{(f)}} = \left(\frac{\omega}{\omega_0}\right)^3 \left|\frac{\mu_{tot}}{\mu}\right|^2 \tag{12}$$

where ω_0 and ω are respectively the original and the shifted molecules transition frequencies. The ratio $|\mu_{tot}/\mu_M|$ between the total dipole and the original molecular transition dipole is very sensitive to the cluster geometry. It is found, for example, that an atom positioned along a direction parallel to the molecular transition dipole leads to shortening of the radiative lifetime while an atom located along an axis perpendicular to $\vec{\mu}_M$ causes a lengthening of this lifetime. This provides a potentially useful way of studying cluster structure [15].

III. IONIZATION POTENTIALS

In addition to the modification of electromagnetic interactions and optical processes near small particles, it is to be expected that charge transfer phenomena involving such particles will also be affected. A well known example is the particle size dependence of the ionization potential [16]. By comparing the work needed to bring an electron from the surface of a small conducting sphere of radius R to infinity to the same work associated with a planar surface one gets [17]

$$W_I'(R) = W_I'(\infty) + \frac{3}{8}\frac{q^2}{R} \tag{13}$$

where $W_I(R)$ is the ionization potential of the sphere, $W_I(\infty)$ is the bulk work function and q - the electron charge. While Eq.(13) which has been widely used is a mathematically correct result of classical electrostatics, it has to be contrasted with another such result: Born's theory of solvation may be used to calculate the difference in solvation energies of a charge q in an infinite dielectric and in a sphere of the same dielectric. This leads to [18]

$$W_I'(R) = W_I'(\infty) + \frac{1}{2}\frac{q^2}{R} \tag{14}$$

which, for $\varepsilon \to \infty$ leads to

$$W_I(R) = W_I'(\infty) + q^2/2R \tag{15}$$

in contrast to Eq.(13). It might be added that Eq.(13) is found to be in a somewhat better agreement with experiments on small Na and K clusters, while Eq.(15) seems to account slightly better for the results on bigger Ag particles [19].

The resolution of the conflict between Eq.(13) and (15) comes from realizing that Eq.(13) is obtained from a limiting process in which the difference between the classical image potential for an electron near a conducting sphere and near a conducting plane is evaluated in the limit when the electron surface distance vanishes. However, it is well known that using the classical image potential too close to a surface is unjustified and there is no guarantee that even the difference between two such potentials behaves properly. If a proper cutoff in the image potential is introduced, in a way consistent with the requirement that the work needed to move a charge through the interface between two different dielectric media is equal to the difference between the solvation energies of the charge in the bulks of these media, Eqs.(14) and (15) are recovered.

It remains to understand the apparent success of Eq.(13) in fitting the experimental result for sodium and potassium. To this end we have carried density functional calculations for a jellium sphere based on the functional used by Smith [20] for studying the work function associated with planar metal surfaces. The electron density is taken to be

$$
\rho_-(r) =
\begin{cases}
\dfrac{k^2}{4\pi q} \dfrac{B_1}{r} \left(e^{kr} + e^{-kr} \right) + B_0 & ; r < R \\[3mm]
\dfrac{\lambda^2}{4\pi q} \dfrac{C_1}{r} e^{-\lambda r} & ; r > R
\end{cases}
\tag{16}
$$

with a positive background $\rho_+(r) = \rho_0 \Theta(r-R)$ (Θ is the step function). Considerations of continuity of the potential and the electron density and of charge conservation lead to explicit expressions for all parameters in terms of k, λ and the (given) total charge Q on the sphere. The parameters k and λ are varied in order to minimize the energy expressed as a functional of $\rho_-(r)$:

$$
E = E_{kin} + E_x + E_c + E_{coul}
\tag{17}
$$

where E_{kin}, E_x, E_c and E_{coul} are the kinetic, exchange, correlation and Coulomb energies given explicitly as functionals of $\rho_-(r)$. The calculations are done for Q=0 and for Q=1 and the ionization potentials are obtained from the corresponding differences. In this way we can obtain the contributions of the different terms in E to the ionization potential:

$$\Delta E \cong W_I = \Delta E_{kin} + \Delta E_{xc} + \Delta E_{coul} \tag{18}$$

$(\Delta E_{xc} = \Delta E_x + \Delta E_c)$. The results, using an electron density appropriate for sodium $r = 4$) are summarized in the table. There we present, for the different energies ΔE, the values of the parameters A, B and C which provide the best fits for ΔE_{kin}, ΔE_{xc}, ΔE_{coul} and W_I written as

$$\Delta E = A + \frac{B}{R} + \frac{C}{R^2} \tag{19}$$

	A	B	C
ΔE_{kin}	-0.0402	-0.0773	-0.227
ΔE_{xc}	0.128	0.0239	0.221
ΔE_{coul}	0.0103	0.457	0.111
W_I	0.0980	0.403	0.104

It is seen that while the B coefficient in W_I is very close to 3/8, the corresponding term in ΔE_{coul} is closer to 1/2; the difference results from the quantum mechanical contributions ΔE_{kin} and ΔE_{xc}. It is found that the deviation of ΔE_{coul} from 0.5 is due to another quantum mechanical effect - that of the spilloff of electronic charge into the $r > R$ region. The success of Eq.(13) in fitting the experimental result on alkali clusters was thus fortuitous, and the deviations of these results from the $\varepsilon \to \infty$ limit of Eq.(14) is entirely due to quantum mechanical effects.

CONCLUSION

I have discussed several aspects by which size and shape effects appear in optical and electron transfer phenomena involving small particles. Optical phenomena are affected strongly by plasmon resonances in the particles and by energy transfer between adsorbed molecules

and the particles. Size dependence of ionization potentials are dominated by electrostatic interactions though small quantuum mechanical
effects are evident.

ACKNOWLEDGEMENT

This research is supported by the U.S.-Israel Binational Science
Foundation.

REFERENCES

1.   a). J.I. Gersten and A. Nitzan, Surf. Sci $\underline{158}$, 165 (1985).
      b). H. Metiu, Progress in Surf. Sci. $\underline{17}$, 153 (1984).
      c). R.K. Chang and T.E. Furtak, eds., $\underline{Surface}$ $\underline{Enhanced}$ $\underline{Raman}$
          $\underline{Scattering}$ (Plenum, N.Y., 1982).
2.   See, e.g. J.D. Jackson, $\underline{Classical}$ $\underline{Electromagnetic}$ $\underline{Theory}$ (McGraw-
      Hill, N.Y., 194).
3.   J.I. Gersten and A. Nitzan, J. Chem. Phys. $\underline{75}$, 1139 (1981).
4.   R.R. Chance, A. Prock and R. Silbey, Adv. Chem. Phys. $\underline{37}$, 1.
      (I. Prigogine and S.A. Rice, Eds. Wiley, N.Y., 1978).
5.   A. Nitzan and L.E. Brus, J. Chem. Phys. $\underline{74}$, 5321 (1981); $\underline{75}$,
      2205 (1981).
6.   a). J.I. Gersten and A. Nitzan, J. Chem. Phys. $\underline{73}$, 3023 (1980),
      b). D.S. Wang, M. Kerker and H. Chew, Appl. Opt. $\underline{19}$, 2135, 2256,
          4159 (1980).
7.   D.A. Weitz, S. Garoff, J.I. Gersten and A. Nitzan, J. Chem. Phys.
      $\underline{78}$, 5324 (1983).
8.   A. Wokaun, H.P. Lutz, A.P. King, U.P. Wild and R.R. Ernst,
      J. Chem. Phys. $\underline{79}$, 509 (1983).
9.   a). G.M. Gonscher and C.B. Harris, J. Chem. Phys. $\underline{77}$, 3767 (1982)
      b). C.J. Chen and R.M. Osgood, Phys. Rev. Letters $\underline{50}$, 1705 (1983).
10.   A. Schmidt-Ott, P. Schurtenberger and H.C. Siegmann, Phys. Rev.
      Letters $\underline{45}$, 1284 (1980).
11.   X.M. Hua, J.I. Gersten and A. Nitzan, J. Chem. Phys. $\underline{83}$, 3650
      (1985).
12.   a). A.M. Glass, A. Wokaun, J.P. Heritage, J.G. Bergman, P.F. Liao
          and D.H. Olson, Phys. Rev. $\underline{B24}$, 4906 (1981).
      b). P. Ye and Y.R. Chen, Phys. Rev. $\underline{B28}$, 4288 (1983) and
          references therein.
13.   G. Kurizki and A. Nitzan, this volume.
14.   a). N. Liver, A. Nitzan and K.F. Freed, J. Chem. Phys. $\underline{82}$, 3831
          (1985).
      b). N. Liver, A. Nitzan and J.I. Gersten, Chem. Phys. Letters $\underline{111}$,
          449 (1984).
15.   N. Liver, A. Nitzan, A. Amirav and J. Jortner, to be published.

16. a). A. Herrmann, E. Schumacher and L. Woste, J. Chem. Phys. $\underline{68}$,
         2327 (1978).
     b). E. Schumacher, M. Kappes, K. Marti, P. Radi, M. Schar and
         B. Schumacher, M. Kappes, K. Marti, P. Radi, M. Schar and B.
         Schmidhalter, Ber. Bunseng. Phys. Chem. $\underline{88}$, 220 (1984).
     c). M.M. Kappes, M. Schar, P. Radi and E. Shumacher, J. Chem.
         Phys. $\underline{84}$, 1863 (1986).
     d). A.W. Castleman, Jr. and R.G. Keesee, Ann. Rev. Phys. Chem.
         $\underline{37}$, 525 (1986).
17. D.M. Wood, Phys. Rev. Letters $\underline{46}$, 749 (1981); J.M. Smith, AIAA J.
     $\underline{3}$, 648 (1965).
18. L.E. Brus, J. Chem. Phys. $\underline{79}$, 5566 (1983); $\underline{80}$, 4403 (1984).
19. G. Makov, A. Nitzan and L.E. Brus, to be published.
20. J.R. Smith, Phys. Rev. $\underline{181}$, 522 (1969); D.R. Snider and R.S.
     Sorbello, Sol. St. Comm. $\underline{47}$, 845 (1983).

FLUIDS THAT EXHIBIT CHANGES IN ELECTRONIC STRUCTURE

F. Hensel
Institute of Physical Chemistry
Philipps-University of Marburg
D-3550 Marburg, FRG

ABSTRACT. Experiments on fluid metals near the liquid-vapour critical points have shown that there are profound changes in their electronic structure in that region. A metal-nonmetal transition occurs with decreasing density. In the present paper the problem of interest is the interplay between the liquid-vapour critical point density fluctuations and the electronic transition.

Selected recent experimental results including the equation of state, electrical, dielectric and optical properties are used to demonstrate that the strong variation in the electronic structure in course of the metal-nonmetal transition which manifests itself in a correspondingly strong thermodynamic-state-dependence of the effective interparticle interaction noticeably influences the thermodynamic and kinetic features of the vapour-liquid phase transition of metals.

1. INTRODUCTION

Whilst considerable progress has been made over the last decades in understanding the liquid-vapour critical phenomena of molecular fluids, the nature of the critical point phase transition of metallic fluids, like Hg, i.e. fluids in which the electronic structure changes substantially with nuclear rearrangement, are relatively less well understood. As is well known, under ordinary conditions mercury is a metallic liquid, whereas the dilute vapour is an insulator. Consequently, going from liquid-like densities to vapour-like densities involves a metal-insulator transition. One of the oldest questions, that is still unresolved is, what is the relation of the metal-insulator transition to the liquid-vapour phase separation? The fundamental theoretical difficulty of dealing with this question is that the interatomic interactions are very different in the two states. In the dilute vapour the mercury atoms interact through weak van der Waals interactions, whereas the cohesion in the liquid is well described by the nearly-free-electron approximation. By contrast, for molecular fluids, like Ar, the interatomic potential energy function may be considered independent of density to a good approximation. This contrast was first discussed by

345

J. Jortner et al. (eds.), Large Finite Systems, 345–359.
© *1987 by D. Reidel Publishing Company.*

Landau and Zeldovitch [1] who suggested the possibility of separate
first-order electronic and liquid-vapour transitions in fluid metals.
Subsequent theoretical attempts to model the statistical mechanics of
the metal-insulator transition in fluids reach similar conclusions
but are still insufficient to provide a clear-cut answer to this
question from theory [2], [3], [4], [6], [7]. At present no theory
exists which incorporates both the fluid aspects and the variation in
the electronic structure from extended to localized states.

Several attempts have been made to explore this problem experimentally, but the subject has remained elusive until recently. As the
experimental difficulties in the critical region are rather severe this
is not surprising. Corresponding to the theorist's intractable many-body
effects is the experimental difficulty that the high cohesive energies
of metals place the critical region at temperatures and pressures too
high for easy experimental investigation. The resulting problems of
temperature and pressure measurement and control together with the
highly reactive nature of fluid metals have chiefly limited the accuracy
with which properties have been measured in the past. The main problem
was that the analysis of electrical, optical, magnetic and thermophysi-
cal data close to the critical point of metals was hampered by the
presence of spurious effects due to temperature gradients.

This situation is now changing. Improvements and extensions of
experimental technique have made it possible to measure the equation of
state, the electrical transport properties and partly optical properties close to the critical points of Hg, Cs and Rb with quite high precision and comparatively accurate temperature control and optimal elimination of temperature gradients. This new experimental information
seems to be accurate enough to allow one to determine the exact location of the critical points of Hg (T_c=1478°C, p_c=1673 bar, ρ_c=5.8 g/cm^3)
[8], Cs (T_c=1651°C, p_c=92.5 bar, ρ_c=0.38 g/cm^3) [9], and Rb (T_c=1744°C,
p_c=124.5 bar, ρ_c=0.29 g/cm^3) [9] and to study the asymptotic behaviour
of the physical properties of metals near the vapour-liquid critical
point [10], [11], [12], [13].

While from the experimental standpoint it is now unquestionable
that there is no sharp (first-order) electronic transition in the
homogeneous fluids Cs, Rb and Hg, it is characteristic that the liquid-
vapour phase separation tends to separate the nonmetallic and metallic
fluids [13], i.e. the gross change in the electronic properties in
course of the metal-insulator transition which manifest itself in a
correspondingly strong thermodynamic-state-dependence of the effective
interparticle interaction occurs in the vicinity of the critical point.
To evaluate the significance of this, however, one must first understand
the underlying mechanism of the metal-insulator transition and in
particular the interplay between the liquid vapour critical phenomena
and the electronic transition.

The problem of the metal-nonmetal transition in expanded fluid
metals has received considerable attention in the past. The pioneering
studies by Mott [14], [15], [16] attributed the transition to Wilson
and Hubbard band crossing but modified in an important way by structural disorder which can cause localization of states by the Anderson
process [17] when the structural disorder is great enough. Other theo-

ries of the experimentally observed phenomena have made use of classical
percolation ideas [18], or have postulated specifically in relation to
expanded mercury, the existence of a Frenkel excitonic insulator transition [19], [20]. Still other approaches invoke the role played by
thermally generated positive and negative ions for the metal-nonmetal
transition [21], [22], [23].

    Studies of the interplay between the liquid-vapour critical point
density fluctuations and the electronic properties have received less
attention and are rather new [8], [12], [13] but nonetheless might be
especially important for the understanding of the metal-nonmetal transition and the thermodynamics of the liquid-vapour condensation of
fluid metals.

    The purpose of the present article is to discuss very recent ex-
perimental results which clearly demonstrate that the character of the
liquid-vapour critical point phenomena of fluid metals is noticeably
different from that of insulating molecular fluids and that density
fluctuations are intimately related to the metal nonmetal transition.

2. METAL-INSULATOR TRANSITION IN MERCURY

The large amount of work devoted over the last 15 years to the experimental and theoretical investigation of the metal-nonmetal transition
in fluid mercury has been fully reviewed, [13], [24], [25], [26], [21],
[27], [7] and need not detain us here. It is sufficient to summarize a
few noteworthy observations. Experimental measurements such as those
of the conductivity [8], the thermoelectric power [8], the Knight shift
[28], [29], the optical reflectivity [30], [31] and the Hall effect
[32] show that as the density is lowered, a gap in the density of
states appears to develop at the relatively high density of 9 g/cm^3.
The opening of this gap means that mercury changes macroscopically to
a nonmetallic, effectively "semiconducting" state at the same density,
i.e. before the critical point density ρ_c=5.8 g/cm^3 is approached. On
the other hand, most theoretical calculations [33], [34], [35] have
yielded values for the gap closing density that are very close to ρ_c.
It should be noted, however, that these calculations did not take into
account the situation close to the critical point where critical density fluctuations may strongly affect the electronic properties. The
importance of density fluctuations for fluid mercury densities smaller
than 9 g/cm^3 and larger than 4 g/cm^3 in the temperature region around
T_c is clearly demonstrated by figure 1 which shows the isothermal com-
pressibility $\chi_T = 1/\rho \cdot (\partial \rho / \partial p)_T$ at constant supercritical temperatures
as a function of density. χ_T begins to rise quite rapidly as the
density falls below 9 g/cm^3 at temperature close to T_c=1478°C.

    Direct evidence for a strong interplay between the liquid-vapour
critical point transition and the electrical transport properties of
mercury stems from recent very accurate simultaneous measurements of
the electrical conductivity σ, the thermoelectric power S and the
density ρ by Götzlaff [8], [36]. In figure 2 the electrical conductivity σ is plotted in the vicinity of the critical point as a function of
density ρ at constant temperature T for a few selected isotherms. At a

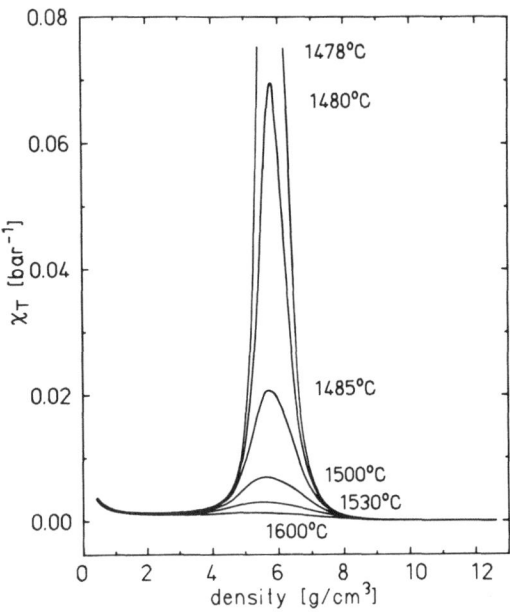

Figure 1. Isothermal compressibility $\chi_T = 1/\rho \cdot (\partial\rho/\partial p)_T$ of fluid mercury at constant temperatures T (T_c = 1478°C) as a function of density (ρ_c = 5.8 g/cm^3).

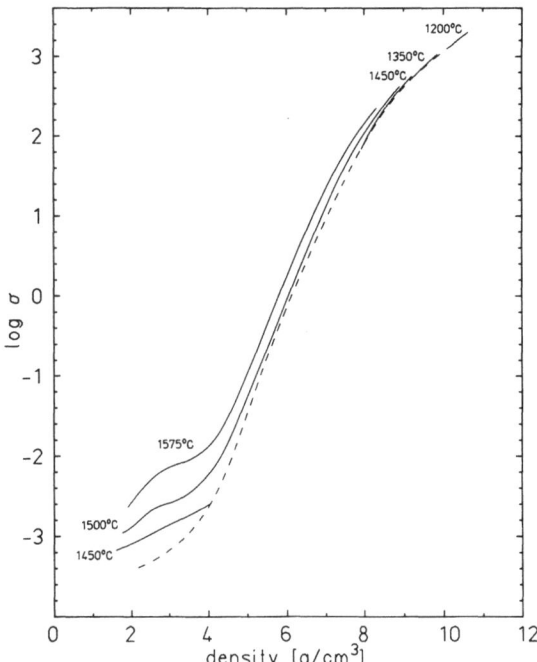

Figure 2. Electrical conductivity σ of fluid mercury at constant temperatures close to T_c = 1478°C as a function of density ρ.

density of 9 g/cm^3 the conductivity is about the "minimum metallic conductivity" (i.e. about 200 ohm^{-1}cm^{-1}). For densities smaller than 9 g/cm^3, σ falls more rapidly and approaches a value of about 1 ohm^{-1}cm^{-1} at the critical density ρ_c=5.8 g/cm^3. The more rapid fall of σ for densities smaller than 9 g/cm^3 has been considered as a strong indication for the onset of the transition to a nonmetallic state. It is clear from the results of figures 2 and 3 that there is a close correlation between the slope of the σ-ρ-curves and the critical point.

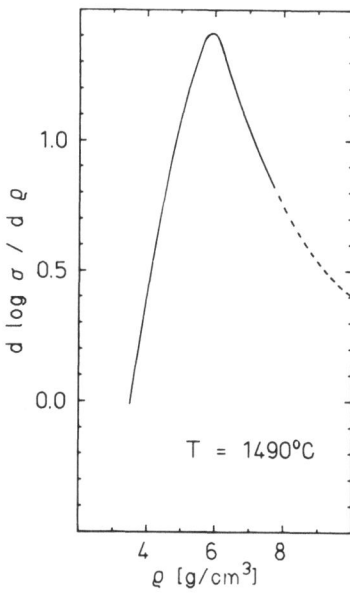

Figure 3.   The density coefficient $(\partial \ln\sigma/\partial\rho)_T$ of fluid mercury at constant temperatures as a function of density (T_c=1478°C, ρ_c=5.8 g/cm^3).

There is no doubt that the steepest fall in the conductivity of Hg is observed at the critical point. $(\partial \ln\sigma/\partial\rho)_T$, which is positive and nearly constant for ρ larger than 9 g/cm^3, has a strong maximum accurately at the critical density. The pattern of the σ-ρ-curve is especially interesting for densities below the critical density where fluid mercury forms a dense, partially ionized plasma consisting of neutral species, ionic species and electrons. In this case, in addition to screened Coulomb interaction among the charges, electron-neutral interaction plays an important role for the transport [37], [38], [39]. If the density of neutrals is high enough the electron interacts with many neutral atoms at the same time. We believe that in the region of small degree of ionization, i.e. for densities well below the metal-nonmetal transition range, and for temperatures close to the critical temperature T_c=1478°C the thermally ionized quasi-free electron can be trapped by density fluctuations. Perhaps the term "enhanced scattering" [39] may be preferable to "trapped", because the traps are shallow.

       Both the density dependence of σ for densities smaller than ρ_c=5.8 g/cm^3 and the strong positive temperature dependence of σ for

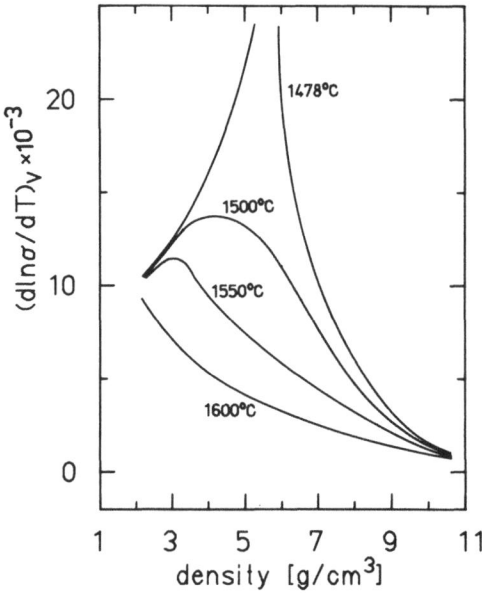

Figure 4. The temperature coefficient $(\partial\ln\sigma/\partial T)_\rho$ of fluid mercury
at constant density as a function of density at different temperatures
close to the critical point (T_c=1478°C, ρ_c=5.8 g/cm³)

temperatures close to the critical temperature T_c=1478°C (figure 4) are
completely consistent with this assumption. Heating fluid mercury to
remove it from the vapour-liquid critical region decreases the magnitude
of the microscopic density fluctuations and thereby diminishes the
number and quality of trapping centers. The large temperature coefficient of the coductivity near the critical region indicates that the
entropy change associated with the trapping of an electron is large,
i.e. that the fast electrons interact with the fluid less than the
slow ones. As is well known, it is valid and often helpful, to think
of the thermoelectric power S of a conductor as a transport entropy of
electric charge [40], [41]. Consequently, thermoelectric measurements
are especially suited to study the interplay between the liquid vapour
critical point density fluctuations and the electrical transport [42],
[39] in fluid mercury. Several experiments have suggested that in Hg at
pressures and temperatures near and above critical values the thermoelectric power vanishes [43], [44], [45], [46], [8]. From simultaneous
measurement of S and density ρ Götzlaff [36] was able to evaluate the
density dependence of the thermoelectric power at a constant temperature
near T_c (figure 5). A remarkably strong increase of S up to large positive values is observed in the density region where electron-neutral
interaction becomes important (cp. also figure 2). The energy transport
by neutrals induced by the electron current gives a large contribution
(a drag effect) to the thermoelectric power in the region of small
degree of ionization. This strongly supports the suggestion that the

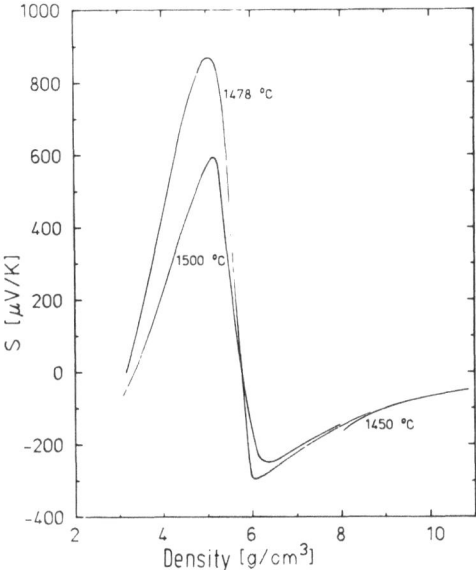

Figure 5. Thermopower S of fluid mercury at constant temperatures as a function of density ρ (T_c=1478°C, ρ_c=5.8 g/cm³).

electrons can be trapped by density fluctuations, and that there is a negative heat of transport which is present only if the density of electrons is small enough so that they do not compete unduly for the surrounding atoms. A negative heat of transport can be envisaged if the trapping process involves entropy and if the fast electrons interact with the fluid less than slow ones, and carry less heat with them [47].

For densities above the ionization catastrophe density the thermopower tends to small negative values (fully ionized degenerate plasma). From the results in figure 5 it is clear that the change in the electrical transport behaviour of expanded Hg is intimately related to the vapour-liquid critical point.

The interplay between the critical point phase transition and the rapid changes in the electronic structure, when the ionization catastrophe is traversed, becomes especially evident when the critical point is approached from the insulating vapour side. The approach has recently been extensively studied by measurements of the density- and temperature dependence of the optical properties of the vapour. At very low densities a linespectrum is observed with the main absorption lines at 4.89 eV and 6.7 eV corresponding to transitions between the 6s ground state and the 6p triplet and singlet state of the Hg-atom. As the density is increased the sharp lines broaden due to interactions with neighboring atoms resulting in a relatively steep absorption edge which moves rapidly to lower energies with increasing density. [12], [37], [30], [31], [35], [48]. Bhatt and Rice [49] and Uchtmann et al. [50] have shown that a uniform density increase is insufficient to

explain this line broadening and that one must take density fluctuations
into account. The absorption edge is then lowered by the environment
of the atom being excited, and the edge is thus explained in terms of
absorption by excitonic states of large randomly distributed clusters.
From the large values of the absorption coefficient it can be concluded
that the singlet exciton 6^1p_1) with large oscillator strength broadens
faster than the triplet exciton (6^3p_1) with small oscillator strength.
A detailed analysis [50] of the density dependence of the absorption
edge shows that the singlet contribution dominates for densities larger
than 1 g/cm^3 whereas for $\rho<1$ g/cm^3 the shape of the edge is dominated
by the triplet transition.

The shift of the effective gap in the excitation spectrum can also
be viewed in terms of a nonlinear enhancement of the real part of the
dielectric constant ε_1 [10] with increasing density as demonstrated in
figure 6 which shows results for ε_1 at the constant photon energy
1.27 eV in the form of isotherms plotted versus pressure. As the

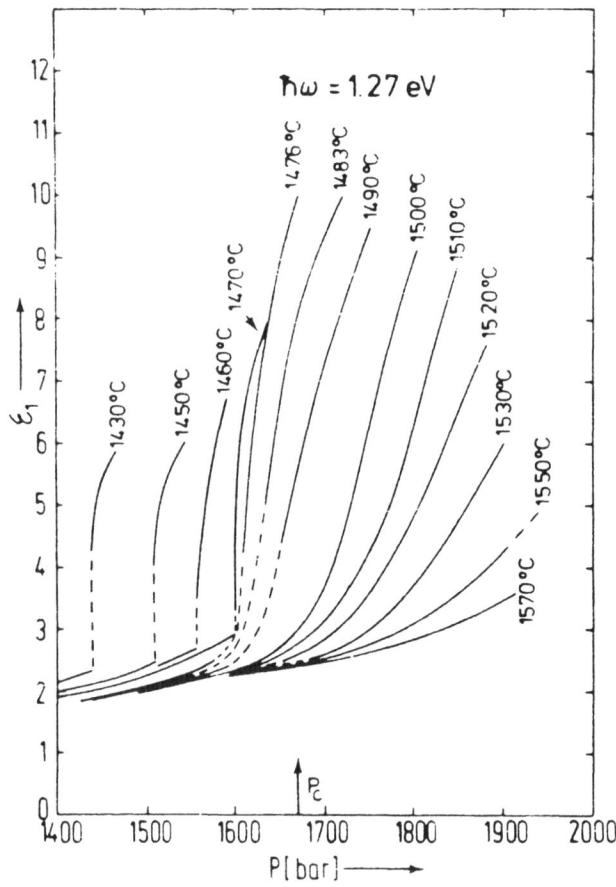

Figure 6. The real part of the dielectric constant ε_1 at the constant
photon energy 1.27 eV as a function of pressure p for different sub- and
supercritical temperatures (p=1673 bar).

pressure is increased at a constant temperature T larger than about
0.96 T_c (i.e. 1400°C) ε_1 initially follows the Clausius Mosotti model
of the polarisibility of induced dipoles before it shows a strong upward deviation from Clausius-Mosotti behaviour. It is obvious from the
pattern of the ε_1-curves in figure 6 that the strong dielectric anomaly
is inextricably related to the critical point phase transition of
mercury.

This becomes especially obvious when ε_1 is studied in the critical
region as a function of temperature at constant density, as shown in
figure 7 for a number of selected isochores. The data show close to the
critical isochore clearly the presence of a large anomalous contribution
in the dielectric constant reaching a magnitude of about 70% of the
background at $\Delta T/T_c = 10^{-3}$. This is in contrast to the behaviour of nonmetallic fluids for which a comparatively weak dielectric anomaly
(smaller than 0.1% for CO at $\Delta T/T_c < 10^{-4}$ [51]) occurs [52]. We consider
the finding of a large amplitude of the dielectric constant anomaly
for mercury as evidence for a strong interplay between the vapour-liquid
critical point phase transition and the large changes in the electronic
structure as the metal-nonmetal transition is approached.

Different models for the strong upward deviation of ε_1 from
Clausius-Mosotti behaviour have been put forward [37], [21], [53], [54]
[55], [56], [57]. Consistent with the behaviour of the conductivity σ
and the thermopower S, self-trapped negative ion states have been proposed [37], [55]. The basic argument was that thermally generated electrons could stabilize high-density fluctuations if the temperature and the

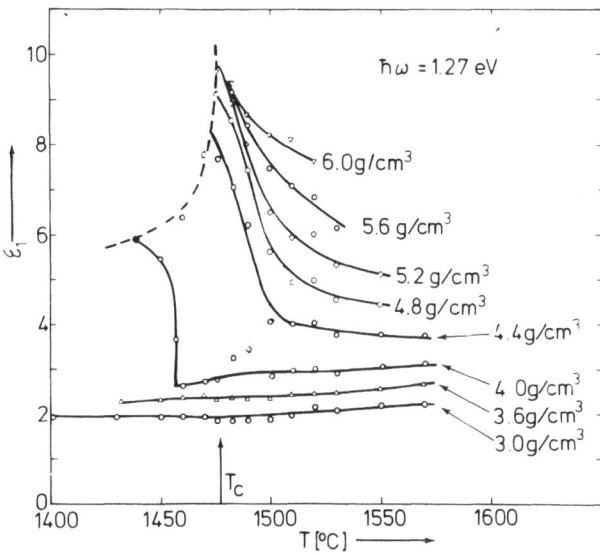

Figure 7.   The real part of dielectric constant ε_1 at constant density
versus temperature ($T_c=1478°C$, $\rho_c=5.8$ gc/cm^3).

average atomic density had suitable values. Alternatively, Turkevich and Cohen [57], [58] have proposed that dense fluid mercury close to the critical point constitutes a disordered inhomogeneous excitonic insulator phase. Such phases in which excitons condense into a macroscopically occupied state were first postulated by Kohn et al. [59] as intermediate phases in the metal-insulator transition. Turkevich and Cohen have explained the behaviour of ε_1 of fluid mercury by suggesting that improved coordination, i.e. clustering, can reduce the bottom of the Frenkel exciton band, leading ultimately to an exciton condensation instability.

3. THERMOPHYSICAL PROPERTIES OF METALS IN THE CRITICAL REGION

The existing experimental observations for Hg and the alkali metals Cs and Rb [24], [60] indicate that there exists a clear link between the liquid-vapour and the metal-nonmetal transition. The occurrence of the latter implies that the nature of the interparticle interaction must change dramatically, from metallic to a van der Waals-type interaction. By contrast, for most insulating molecular substances like Xe the intermolecular interaction may be considered independent of density to a good approximation. This contrast has been discussed by Goldstein and Ashcroft [61] who argued that the strong dependence on density, near ρ_c, of the electronic structure may considerably influence the thermodynamic features of fluid metals in the critical region.

Since the available equation of state data for fluid Hg, Cs and Rb approach the critical point close enough we are able to test experimentally the validity of this hypothesis. For that purpose we compare in fig. 8 as an example the reduced densities of gaseous and liquid Xenon (inner curves) and rubidium (outer curves) as a function of $\Delta T/T_c = |T_c-T|/T_c$. Pure reduced correlation between Rb and Xe is observed. Thus the experimental evidence shows that metals and nonmetals cannot be included together in a group obeying a principle of corresponding

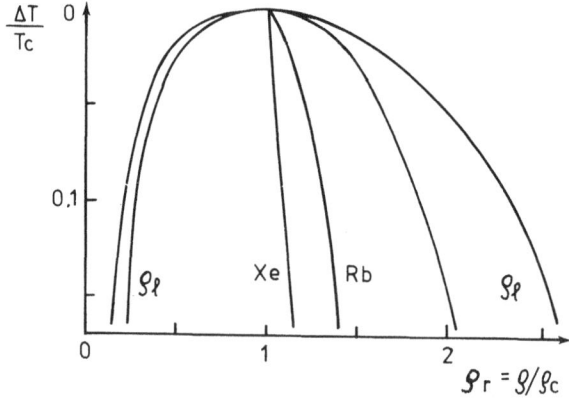

Figure 8. The reduced densities of gaseous and liquid Xenon (inner curves) and rubidium (outer curves).

states. The coexistence curve of Rb is remarkably asymmetric compared to
that of the simple nonmetallic fluid Xe. In addition, it is obvious from
figure 8 that Rb violates the law of rectilinear diameter over a surprisingly large temperature range. By contrast, the deviations from this
law are extremely small for the coexistence curves of nearly all simple
nonmetallic one-component fluids [62].

As mentioned above, it has been suggested [61] that the contrast
between the diameter data of metallic and nonmetallic one-component
fluids arises from many-body effects whose magnitudes distinguish
the particle interactions in metallic fluids from those in nonmetallic
fluids. In particular, it is argued that the strong thermodynamic state
dependence of the effective interactions in a metal, especially as the
metal-nonmetal transition is traversed, corresponds to the mixing of
thermodynamic fields present in certain solvable lattice models [63].
These models, thermodynamic arguments [64], and renormalization-group
studies [65] predict that the average value of the density, i.e. the
diameter, will have the asymptotic form

$$\rho_d = (\rho_L + \rho_V)/2 = \rho_c + D \ (\Delta T/T_c)^{1-\alpha} \tag{1}$$

where $\Delta T/T_c = |T_c-T|/T_c$ and the exponent α is the same as that of the
divergence of the constant volume specific heat for a pure fluid.

The predicted singularity in the diameters is difficult to verify
experimentally for several reasons. Firstly, it has been shown [66]
that if one particular function, e.g. ρ, has a $|\Delta T/T_c|^{1-\alpha}$ singularity,
then any less symmetric function ρ' is an analytic function of ρ (e.g.
$\rho = V^{-1}$), behaves as $|\Delta T/T_c|^{2\beta}$. Thus the sought-for effect will be
missed unless the correct function is chosen. The mass density has long
been known empirically to give a more symmetric coexistence curve than
volume. However, this is only a strong, but not a conclusive argument
for supposing that ρ is the appropriate function for the order parameter.
Secondly, the size of the asymptotic range depends on the choice of
the order parameter.

Renormalization-group theory has been used to produce series expansions for representing data over a wide range of thermodynamic space.
The expansions provide the following correction terms for the diameter

$$\rho_d = \rho_L + \rho_V/2\rho_c = 1 + D_0|\Delta T/T_c|^{1-\alpha} + D_1|\Delta T/T_c| + \ldots \tag{2}$$

Since $(1-\alpha)$ is not very different from unity, the true singularity is
difficult to separate out from the analytic temperature terms. The coefficient D_1 does not even have to be much larger than D_0 for the analytic term to dominate over the entire range accessible to experimenta-
tion. The latter certainly causes the invisibility of the $|\Delta T/T_c|^{1-\alpha}$
anomaly in most nonmetallic fluids.

Up to now the only convincing experimental evidence for the
existence of a $(1-\alpha)$ term for one-component systems is the analysis of
the diameters of Cs, Rb [9] and Hg [36]. The apparent experimentally
determined $(1-\alpha)$ values are very close to the value 0.89 predicted by
the renormalization-group theory. This finding strongly supports the
suggestion [61] that the strong state-dependence of the effective inter-

particle interactions, and especially the changes in such forces in
course of the metal-nonmetal transition, lead to very large amplitudes
of the $(1-\alpha)$ anomaly in the diameters of liquid-vapour coexistence
curves of metals.

4.   HOMOGENEOUS NUCLEATION OF SUPERSATURATED METAL VAPOURS

The study of the condensation of supersaturated metal vapours to the
liquid state by homogeneous nucleation has received only little attention. This is essentially due to the fact that according to the classical nucleation theory [67], [68] it is extremely difficult to get metals
to nucleate. The nucleation expression of the classical theory is

$$J = A(p_\infty/kT)^2 (2\sigma/\pi m)^{1/2} \, v_c \, \exp(- \Delta G^*/kT) \qquad (3)$$

where J is the number of critical size clusters formed per sec and cm^3,
and ΔG^* is the Gibbs free energy for formation of the critical size
cluster $\Delta G^* = 4\pi r^{*2} \cdot \sigma/3$. The critical radius r^* is given by the Gibbs-
Thompson-Helmholtz equation $r^* = 2\sigma v_c/kT\ln(S)$. The saturation ratio S is
defined as p_v/p_∞, p_v is the partial pressure of the condensable vapour,
T is the temperature, p_∞ is the vapour-liquid equilibrium pressure at
the same T, k is Boltzmann's constant, σ is the surface tension, m is
the mass of a single vapour molecule, v_c the volume of the condensable
molecule in the condensed phase, and A is a correction factor used to
match theory to experiment. The magnitude of the nucleation rate is
very sensitive to surface tension (cubic in the exponential expression).
Because metals have surface tensions tens to hundreds of times higher
than most molecular fluids, the classical theory requires sometimes
extremely high supersaturations. Due to the extremely high supersaturations the starting or critical radius r^* for formation of a stable
nucleation site in a supersaturated metal vapour (e.g. Hg at temperatures between 260-400 K) is unphysically small, i.e. the critical
cluster sizes are in the size range N=1 to N=10. Various investigations
have criticized the classical theory because it has a conceptual difficulty for metals like Hg because surface tension is meaningless when
clusters are too small (N=1 to 10) to contain interior atoms. In addition such small clusters of Hg must be expected to have a dominant
nonmetallic character [69]. Experimental support for the existence
of a size-dependent metal-insulator transition in small Hg-clusters
comes from susceptibility measurements [70] of small particles of Hg
obtained by forcing the metal into cavities of the mineral NaX-zeolite.
After releasing the pressure some mercury leaves the porous material;
the remaining metal takes the form of spherical drops containing 7-10
atoms. From the susceptibility it can be concluded that these particles
are nonmetallic.
    A first experimental indication that the size-dependent metal-non-
metal transition which may noticeably influence all physical properties
including density [69] and surface tension [71] plays an important role
for the homogeneous nucleation process in supersaturated mercury vapour
stems from a recent reliable experimental determination of the critical

supersaturation for the homogeneous condensation of mercury in the temperature range 260 to 400 K. The measurements were made using an upward thermal diffusion cloud chamber. A few selected results are shown in figure 9 where we have plotted the measured nucleation as a function of the mercury vapour supersaturation at temperatures between 258 and 398 K. The critical supersaturations $S = p_V/p_\infty$ giving rise to a nuclea- tion rate of about one drop $cm^{-3}sec^{-1}$ are plotted as function of tempe- rature in the inset of fig. 9. The corresponding values calculated in

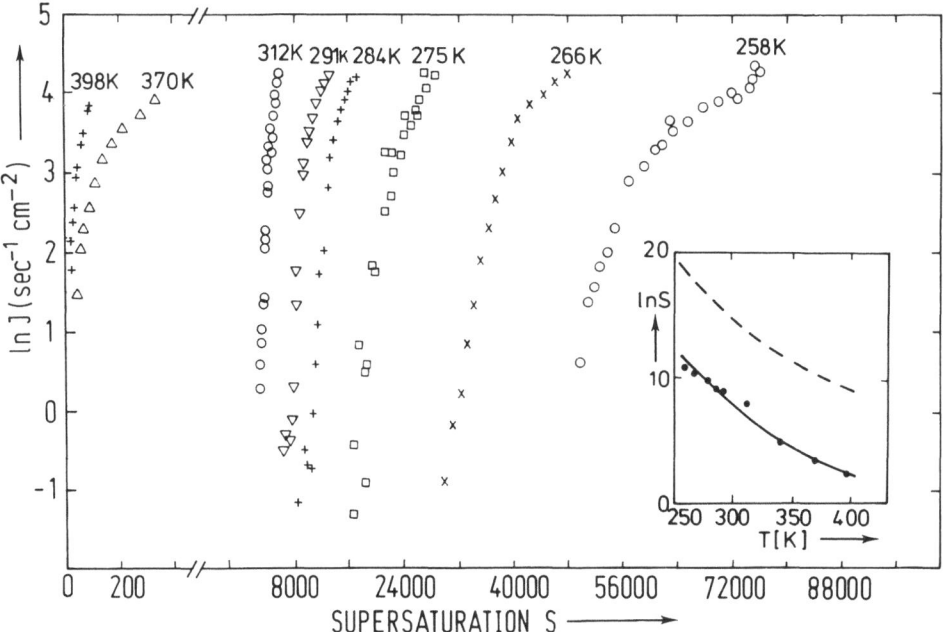

Figure 9. Homogeneous nucleation droplet flux for mercury vapour as a function of supersaturation in the temperature range between 258-398 K. The inset shows the measured critical supersaturation S giving rise to a nucleation rate of about one drop $cm^{-3}sec^{-1}$ in comparison with those (dashed curve) calculated with the Becker-Döring-Zeldovitch theory. See text for details.

the framework of the conventional Becker-Döring-Zeldovitch-theory are presented by the dashed line in the inset of fig. 9. We obtain these values using the macroscopic surface free energy, without making any correction for the dependence of surface tension on radius. The experi- mental and theoretical values differ significantly by more than 3 orders of magnitude. It is noteworthy that the change in the value of the bulk liquid surface tension necessary to bring the classical nucleation theory in agreement with the experimental observation is about 40%. In contrast most molecular liquids require only a very small adjustment of the bulk liquid surface tension to bring nucleation theory and experiment into agreement.

REFERENCES

[1] L.Landau and G.Zeldovitch, Acta Phys.Chem.USSR 18, 1940 (1943).
[2] N.F.Mott, Phil.Mag. 37, 377 (1978).
[3] S.Nara, T.Ogawa and T.Matsubara, Prog.Theo.Phys. 57, 1474 (1977).
[4] W.Ebeling and R.Sändig, Ann.Phys.(Leipzig) 28, 289 (1973).
[5] W.Ebeling, W.D.Kraeft and D.Kremp, Theory of Bound States in
     Plasmas and Solids. Berlin: Akademie-Verlag, 1976.
[6] J.A.Krumhansl in: Physics of Solids at High Pressures, C.T.Tomizuka
     and R.M.Emrick (eds.), New York, Academic Press, 1965.
[7] F.Yonezawa and T.Ogawa, Prog.Theo.Phys. (Japan) Suppl.72, 1 (1982).
[8] W.Götzlaff, G.Schönherr and F.Hensel, Z.Phys.Chem.N.F. 1987,
     to be published.
[9] S.Jüngst, B.Knuth and F.Hensel, Phys.Rev.Letters 55, 2160 (1985).
[10] F.Hensel, S.Jüngst, B.Knuth, H.Uchtmann and M.Yao, Physica 139B
     and 140B, 90 (1986).
[11] F.Hensel,M.Yao and H.Uchtmann, Phil.Mag. B52, 499 (1985)
[12] H.Uchtmann, U.Brusius, M.Yao and F.Hensel, Z.Phys.Chem.N.F. 1987
     to be published.
[13] F.Hensel in: High Pressure Chemistry and Biochemistry, Nato ASI
     Series, ed. by R.Van Eldik and J.Jonas, p.137 (1987) by D.Reidel
     Publishing Company.
[14] N.F.Mott, Phil.Mag. 6, 287 (1961).
[15] N.F.Mott, Metal-Insulator Transition, Taylor and Francis, London
     1974.
[16] N.F.Mott, Metal Non-Metal Transitions in Disordered Systems,
     Edit. L.R.Friedman and D.P.Tunstall. Scottish Universities, Summer
     School in Physics, Ediburgh, 1978.
[17] P.W.Anderson, Phys.Rev., 109, 1492 (1958).
[18] M.H.Cohen and J.Jortner, Phys.Rev.Letters, 30, 699 (1973).
[19] L.A.Turkevich and M.H.Cohen, Ber.Bunsenges.Phys.Chem., 88, 292 (1984)
[20] L.A.Turkevich and M.H.Cohen, Phys.Rev.Letters, 53, 2323 (1984).
[21] V.A.Alexeev and I.T.Yakubov, Phys.Rep., 96, 1 (1983).
[22] J.P.Hernandez, G.Schönherr, W.Götzlaff and F.Hensel, J.Phys.C,
     17, 4421 (1984).
[23] J.P.Hernandez, Phys.Rev.Letters, 57, 3183 (1986).
[24] W.Freyland and F.Hensel in: The Metallic and the Nonmetallic States
     of Matter: An Important Facet of Chemistry and Physics of Condensed
     Matter, P.P.Edwards and C.N.R.Rao (eds.), London, Taylor and
     Francis, 1985.
[25] F.Hensel, angew.Chem., 92, 598 (1980); Angew.Chem.Int.Ed.Engl.,
     19, 593 (1980).
[26] N.E.Cusack in: Metal Non-Metal Transitions in Disordered Systems,
     L.R.Friedman and D.P.Tunstall (eds.), Edinburgh, 1978.
[27] H.Endo, Prog.Theo.Phys. (Japan), Suppl. 72, 100 (1982).
[28] U.El-Hanany and W.W.Warren, Phys.Rev.Letters, 34, 1276 (1975).
[29] W.W.Warren and F.Hensel, Phys.Rev.B, 26, 5980 (1982).
[30] H.Ikezi, K.Schwarzenegger,A.L.Passner and S.L.McCall, Phys.Rev.,
     18, 2494 (1978).
[31] W.Hefner, R.W.Schmutzler and F.Hensel, J.Phys.(Paris)Colloq.,
     41, C8-62 (1980).

[32] U.Even and J.Jortner, Phil.Mag., 30, 325 (1974).
[33] L.R.Matheiss and W.W.Warren, Phys.Rev.B, 16, 624 (1977).
[34] F.Yonezawa and T.Ogawa, Supp.Prog.Theor.Phys., 72, 1 (1982).
[35] H.Overhof, H.Uchtmann and F.Hensel, J.Phys.F: Metal Phys. 6, 523 (1976).
[36] W.Götzlaff, Doctoral Thesis, University of Marburg, W.-Germany 1987.
[37] W.Hefner and F.Hensel, Phys.Rev.Letters, 48, 1026 (1982).
[38] J.P.Hernandez, Phys.Rev., 53, 2320 (1984).
[39] F.E.Höhne, R.Render, G.Röpke and H.Wegner, Physica, 128A, 643, (1984)
[40] S.R.De Groot, Thermodynamics of irreversible processes, Inter-science, New York (1952).
[41] C.Wagner, Progress in Solid State Chemistry, 7, 1 (1972).
[42] D.K.C.MacDonald, Thermoelectricity: an introduction to the principles, John Wiley, New York (1962).
[43] L.J.Duckers and R.G.Ross, Phys.Letters A, 38, 291 (1972).
[44] V.A.A.Alexeev, A.A.Vedenov, V.G.Orcharankov and Yu F.Ryzhkov, Sov.Phys.JETP Lett., 16, 49 (1972).
[45] M.Yao and H.Endo, J.Phys.Soc.Japan, 51, 1504 (1982).
[46] F.E.Neale and N.E.Cusack, J.Phys.F: Metal Phys., 9, 85 (1978).
[47] N.F.Mott, Phil.Mag., 31, 217 (1975).
[48] U.Brusius, Doctoral Thesis, University of Marburg, W.-Germany 1986.
[49] R.N.Bhatt and T.M.Rice, Phys.Rev., B20, 466 (1979).
[50] H.Uchtmann, J.Popielawski and F.Hensel, Ber.Bunsenges.Phys.Chem., 85, 555 (1982).
[51] M.W.Pestak and M.H.W.Chang, Phys.Rev.Letters, 46, 939 (1981).
[52] J.V.Sengers, D.Bedeaux, P.Mazur and S.C.Greer, Physica, 104A, 573 (1980).
[53] I.M.Lifshitz and S.A.Gredescul, Sov.Phys.JETP, 30, 1197 (1970).
[54] A.N.Lagarkov and A.K.Sarychev, Teplofiz Vys.Temp. 16, 903 (1978).
[55] J.P.Hernandez, Phys.Rev.Letters, 48, 1682 (1982).
[56] W.Hefner, B.Sonneborn-Schmick and F.Hensel, Ber.Bunsenges.Phys. Chem., 86, 844 (1982).
[57] L.A.Turkevich and M.H.Cohen, Phys.Rev.Letters, 53, 2323 (1984).
[58] L.A.Turkevich and M.H.Cohen, Ber.Bunsenges.Phys.Chem., 88, 292 (1984)
[59] D.Jerome, T.M.Rice and W.Kohn, Phys.Rev., 158, 462 (1967).
[60] W.Freyland, Comm. Solid State Phys., 10, 1 (1981).
[61] R.E.Goldstein and N.W.Ashcroft, Phys.Rev.Letters, 55, 2164 (1985).
[62] R.E.Goldstein, A.Parola, N.W.Ashcroft, M.W.Pestak, M.H.W.Chan, J.R.de Bruyn and D.A.Balzarin, Phys.Rev.Letters, 58, 41 (1987).
[63] J.S.Rowlinson, Adv.Chem.Phys., 41, 1 (1980).
[64] N.D.Mermin and J.J.Rehr, Phys.Rev.Letters, 26, 1155 (1971).
[65] J.F.Nicoll, Phys.Rev., A24, 2203 (1981).
[66] M.J.Buckingham in: Phase Transition and Critical Phenomena, C.Domb and M.S.Green (eds.), Academic Press, London (1972) Vol. 2.
[67] M.Volmer, Kinetik der Phasenbildung, Theodor Steinkopff Verlag, Dresden, 1939.
[68] J.Frenkel, Kinetik Theory of Liquids, Dover Publ., New York 1955.
[69] A.R.Miedema and J.W.F.Dorleijn, Phil.Mag.B, 43, 251 (1981).
[70] V.N.Bogonolov, T.I.Volkonskaya, A.I.Zadorozhnii, A.A.Kapanadze and E.L.Lutsenko Soviet Phys.Solid St., 17, 1110 (1975).
[71] M.F.O'Evelyn and S.A.Rice, J.Chem.Phys., 78, 5081 (1983).

COHERENT CONTROL AND LASER CATALYSIS OF CHEMICAL REACTIONS*

Moshe Shapiro and Tamar Seideman
Dept. of Chemical Physics
The Weizmann Institute
Rehovot, Israel 76100

ABSTRACT. The feasibility of laser control over branching unimolecular reactions, and laser catalysis of bimolecular reactions is investigated. Laser control in unimolecular reactions is being discussed using a two-colour excitation scheme. It is shown that specific products are obtained in preference to others as a result of interference between final states produced in two different excitation (and dissociation) routes. Computational results on the I*/I branching ratios in the photodissociation of CH_3I, illustrating the degree of control achievable are presented.

For bimolecular reactions, we discuss laser catalysis by resonant light scattering via a bound upper electronic state. The laser acts as a catalyst since no net photons are absorbed or emitted. We present computations for the $H + H_2$ reaction in which we show how optically induced reaction-barrier crossings can be achieved.

Common features of the two-colour unimolecular control and laser catalysis in the bimolecular case are discussed.

*Supported in part by the Minerva foundation, Munich, Germany and by the U.S. Israel Binational Science Foundation grant no. 84-00141

I. INTRODUCTION

Since the advent of tunable and high power laser technology, reaction dynamicists have sought a means of altering reaction pathways via laser irradiation[1]. In particular, the possibility of controlling dissociation (and ionization) via laser "pulse shaping" has excited the imagination of many investigators. Recently[2],[3], it has been demonstrated, however, that regardless of the pulse shape a single pulse is insufficient for controlling the selectivity of reactions. It was however shown[3] that this goal can be attained using multiple pulse configurations.

J. Jortner et al. (eds.), Large Finite Systems, 361–377.
© 1987 by D. Reidel Publishing Company.

The complementary field in which lasers are meant to assist and cata-
lyze, otherwise forbidden, bimolecular reactions, has been extensively
discussed in the theoretical literature[4]-[16].   So far, most of the
theoretical expectations have not been realized, although the experimental investigations[17]-[26], have been progressing steadily.

Experimental follow up of the theoretical ideas have been hindered
mainly by the projected need for extremely high laser powers. Such high
powers are required in many of the schemes involving free-free type
transitions of one form or another, suggested in the
past[6d],[10],[12],[14].   Reliance on such transitions severely limits
the use of narrow-band light sources, such as lasers, because the oscillator strengths are spread over a wide continuous range, i.e., the
strength per unit wavenumber tends to be small.

In contrast, free-bound transitions, which involve relatively sharp
lines are more suitable to excitation by narrow band sources[8].   A reactive free-bound scheme that may overcome the need for very high laser
powers was recently suggested[16].   It involves using stimulated light
scattering via an excited adduct made up of the reaction partners. Thus,
reactants whose energy is insufficient to surmount a reaction barrier,
would undergo an optical excitation to a bound barrier-free excited
state.   While in this state, the system may shuttle freely between the
reactants' and products' configurations. Due to the bound nature of the
excited state, the reaction can only come to conclusion when the products return to the ground electronic state by emitting a photon. The
photon emission is most effective when stimulated, preferably at the
same wavelength initially absorbed. In this way the laser merely serves
as a catalyst, since the number of laser photons remains unchanged.

The $H+H_2$ exchange reaction has all the necessary ingredients for the
successful application of the above scheme[16], save for the fact that
the H_3 transition state would only absorb light in the VUV. Since powerful VUV lasers are, as yet, non-existent, this reaction is not ideal for
demonstrating the effect.   Nevertheless because of its fundamental importance, (in a sense it is the "Hydrogen atom" of reaction dynamics),
we use the $H + H_2$ reaction as our primary example for laser catalysis.

In this paper we discuss both unimolecular control and bimolecular
laser catalysis and bring several computational examples. We then argue
that the unifying theme of both fields is (laser and material) coherence.

II. THEORY OF UNIMOLECULAR CONTROL

The basis of our approach is qualitatively straightforward.   Consider
a molecule at total energy E which can decay (fragment, ionize or emit a
photon) into several different products (channels).   The fact that a
number of channels are accessible at a fixed energy means that the sys-

tem possesses a set of degenerate continuum eigenstates, each of which correlates with a particular asymptotic channel[27],[28]. The key question is how to gain control over the relative branching amongst product channels.

As an explicit example we consider a case, in which we wish to control the branching ratio to two product channels denoted $|E,a\rangle$ and $|E,b\rangle$, accessible at energy E. $|E,q\rangle$ are free translational states, with q=a,b signifying the chemical identity of the photofragments. In ordinary photodissociation we subject the molecule to a radiation pulse whose electric-field is in general of the form,

$$\vec{\varepsilon}(t) = \int d\omega\ \vec{\varepsilon}(\omega)\ \cos(\omega t + \theta_\omega) \tag{1}$$

Here $\vec{\varepsilon}(\omega)$ denotes the electric field at frequency ω . The total (matter+radiation) Hamiltonian is:

$$H = H_m + H_{rad} + H_{int} , \tag{2a}$$

where H_m is the molecular Hamiltonian H_{rad} the Hamiltonian of the radiation field of Eq. (1) and H_{int} the radiation-matter interaction in the dipole approximation,

$$H_{int} = -\mu \int d\omega\ \varepsilon(\omega)\ \cos(\omega t + \theta_\omega) . \tag{2b}$$

where μ is the component of the transition dipole in the direction of the electric field.

Assuming the molecule to be given initially (t=0) in a single bound molecular eigenstate $|E_1\rangle$, the absorption of a photon results[27] in the creation of a <u>superposition</u> of two continuum states,

$$|\Psi(t)\rangle = (\pi i/\hbar)\int d\omega\ \varepsilon(\omega)\ \exp(-i\theta_\omega)\ [\ |E,a^-\rangle\mu_1{}^a + |E,b^-\rangle\mu_1{}^b\ ]$$
$$\exp(-iEt/\hbar), \tag{3}$$

$|E,a^-\rangle$ and $|E,b^-\rangle$ are <u>incoming</u> scattering states whose energy E, is given as,

$$E = \hbar\omega + E_1 . \tag{4}$$

An incoming state is a solutions of the full Schroedinger equation correlating in the distant future with a <u>single</u> free state. Thus,

$$\langle E',p|E,q^-\rangle\ \exp(-iEt/\hbar) \underset{t\to\infty}{=} \delta(E-E')\ \delta_{p,q}\ \exp(-iEt/\hbar) , \quad p,q = a,b. \tag{5}$$

The optical preparation of these states, given in first order perturbation theory by,

$$\mu_1{}^q = \langle E,q^-|\mu|E_1\rangle , \quad q=a,b. \tag{6}$$

completely determines the desired $t \rightarrow \infty$ limit. Thus, from Eqs. (3) and (5),

$$|\Psi(t)\rangle \underset{t \rightarrow \infty}{=} (\pi i/\hbar)\int d\omega \; \varepsilon(\omega) \; \exp(-i\theta_\omega) \; [\; |E,a\rangle\mu_1{}^a + |E,b\rangle\mu_1{}^b \;]$$

$$\exp(-iEt/\hbar). \tag{7}$$

The probability of observing a given reaction product at a given energy E' is therefore,

$$P_{a,1}(E') = \lim_{t \rightarrow \infty} |\langle E',a|\Psi(t)\rangle|^2 = [\; \pi \; \varepsilon(\omega_1) \; \mu_1{}^a/\hbar \;]^2, \tag{8}$$

where $\omega_1 = (E' - E_1)/\hbar$.

We see that "natural" photodissociation probabilities are independent of the phase θ_ω. The coherence of the light source therefore plays no role in such experiments, since the natural branching ratio is determined solely by the ratio of the field intensities $|\varepsilon(\omega_1)|^2/|\varepsilon(\omega_2)|^2$,

and the ratio of the molecular matrix elements squared $|\mu_1{}^a|^2/|\mu_1{}^b|^2$.

Consider now the dynamics which results when the molecule is first prepared in a $\underline{\text{superposition}}$ state, given by,

$$|\chi(t=0)\rangle = c_1|E_1\rangle + c_2|E_2\rangle . \tag{9}$$

Suppose further that we irradiate the sample by two frequencies,

$$\omega_1 = (E - E_1)/\hbar , \tag{10a}$$

and

$$\omega_2 = (E - E_2)/\hbar . \tag{10b}$$

If level $|E_1\rangle$ absorbs the ω_1 photon and level $|E_2\rangle$ absorbs the ω_2 photon, the final states produced in the continuum will have exactly the same energy E. Under these circumstances the system will have been excited to a superposition of degenerate $\underline{\text{scattering}}$ states,

$$|\Psi(t)\rangle = (\pi i/\hbar) \exp(-iEt/\hbar)\cdot$$

$$\{c_1 \; \varepsilon(\omega_1) \; \exp(-i\theta_1) \; [\; |E,a^-\rangle \; \mu_1{}^a + |E,b^-\rangle \; \mu_1{}^b \;] +$$

$$c_2 \; \varepsilon(\omega_2) \; \exp(-i\theta_2) \; [\; |E,a^-\rangle \; \mu_2{}^a + |E,b^-\rangle \; \mu_2{}^b \;]\} . \tag{11}$$

The probability of observing in the distant future the product state $|E,a\rangle$, is now a result of a $\underline{\text{coherent}}$ sum of the two routes of photo-producing $|E,a^-\rangle$. From Eqs. (11),(4),

$$P_a(E) = (\pi/\hbar)^2 \{ \; c_1 \; \varepsilon(\omega_1) \; \mu_1{}^a + c_2 \; \varepsilon(\omega_2) \; \mu_2{}^a \; \exp[i(\theta_1-\theta_2)] \; \}^2 . \tag{12}$$

We see that contrary to ordinary photodissociation, the probabilities, hence the branching ratios, are influenced by the <u>relative</u> time-independent phase,

$$\theta_{1,2} \equiv \theta_1 - \theta_2 , \tag{13a}$$

and by the relative strength,

$$x \equiv |c_1 \, \varepsilon(\omega_1)/c_2 \, \varepsilon(\omega_2)| \, . \tag{13b}$$

Specifically, the a:b branching ratio is given as,

$$R(a{:}b|E) \equiv P_a(E) \, / \, P_b(E) =$$

$$\frac{\{ \ |\mu_{1,1}^a| + x^2|\mu_{2,2}^a| + 2x \, \cos[\theta_{1,2} + \alpha_{1,2}^a]|\mu_{1,2}^a| \ \}}{\{ \ |\mu_{1,1}^b| + x^2|\mu_{2,2}^b| + 2x \, \cos[\theta_{1,2} + \alpha_{1,2}^b]|\mu_{2,2}^b| \ \}} \tag{14}$$

where $\mu_{i,j}^q$ and $\alpha_{i,j}^q$ are defined by the following expression,

$$\mu_{i,j}^q \equiv \mu_i^q \cdot \mu_j^q \equiv |\mu_{i,j}^q| \, \exp(i\alpha_{i,j}^q), \quad i,j{=}1,2 \, . \tag{15}$$

It follows immediately from Eq. (15) that by tuning the external strength ratio and relative phase to,

$$x = |\mu_{1,2}^a|/|\mu_{2,2}^a| \tag{16a}$$

and,

$$\theta_{1,2} = \pi - \alpha_{1,2}^a \tag{16b}$$

we shut–off channel a. Hence the reaction products are funneled exclusively to channel b. In an identical manner we can tune in the field intensities and relative phase to direct all the molecules to channel a.

For polyatomic photodissociation there are however more than just two final states. The molecule breaks apart to two (or more) composite fragments, each having a set of internal (vibrational, rotational, etc...) quantum numbers, denoted collectively as n. The expression for the branching ratio (Eq. (14)) still holds, but the molecular matrix elements must be replaced by cummulative sums over the internal quantum numbers[3],

$$\mu_{i,j}^q = \sum_n \langle E_i|\mu|E,n,q^-\rangle\langle E,n,q^-|\mu|E_j\rangle \, . \tag{17}$$

It follows from Eq. (17) that $\mu_{i,j}^q$ is no longer guaranteed to be factorizable as in the simple two channel case, (Eq. (15)). Hence by the Schwarz inequality,

$$|\mu_{i,j}^q|^2 \leq |\mu_{i,i}^q| \ |\mu_{j,j}^q| \tag{18}$$

with the equality being the exception rather than the rule. If the strict inequality holds, there are no zeros to either numerator or denominator in the branching ratio expression, (Eq.(14), and 100% control cannot be achieved. Nevertheless, as demonstrated elsewhere[3], and in the following example, it is still possible to substantially enhance (deplete) the branching fraction $R(a{:}b|E)$.

III. UNIMOLECULAR CONTROL - A COMPUTATIONAL EXAMPLE

We consider the photodissociation of CH_3I through the $^3Q_0$ and $^1Q_1$ electronic states to form $I^*(^2P_{1/2})$, (a=1), or $I(^2P_{3/2})$, (a=2). This process has been the subject of many experimental[29] and theoretical[30] studies.

Converged multichannel calculations previously performed[31] for CH_3I were utilized, in conjunction with Eqs. (17),(14) to compute the $I/(I{+}I^*)$, yield at $E = \hbar \omega_1 = 411265$ cm^{-1} for excitation from several different pairwise combinations of the five lowest bound states. A sample of the results[32], shown in Fig. 1 as a function of the relative strength parameter,

$$S = |x_2|^2/(1 + |x_2|^2), \text{ and relative phase } \theta_{1,2} = \theta_1 - \theta_2,$$

clearly demonstrates the broad range of control afforded over the I to I* product ratio.

As shown in Fig. 1, excitation from a superposition of the v=3_0,J=2, and v=3_1,J=2, states allows for a change in the yield of I* production from 70%, the "natural" value, to 30%, (at S=0.7 and $\theta_{1,2}$ = 100°). Thus, a change of 8° in phase difference, while maintaining S at 0.7 can increase the I*/I branching ratio by a factor of 4. Preliminary studies show that the extent of variation attainable correlates with the quantity

$$\left| \mu^a_{1,1} \mu^a_{2,2} - [\mu^a_{1,2}]^2 \right|^{-1} .$$

Some control over the probability of observing a single (q,n) quantum state can be attained too. This is done by not performing the sum over n in Eq. (17). We can then make use of the fact that,

$$|\mu_{1,2}(q,n)|^2 = |\mu_{1,1}(q,n)||\mu_{2,2}(q,n)|. \tag{19}$$

Tuning to $x = |\mu_{1,2}(q,n)|/|\mu_{2,2}(q,n)|$ and $\theta_{1,2} = \pi - \alpha_{1,2}(q,n)$

results in the elimination of all product in channel (q,n).

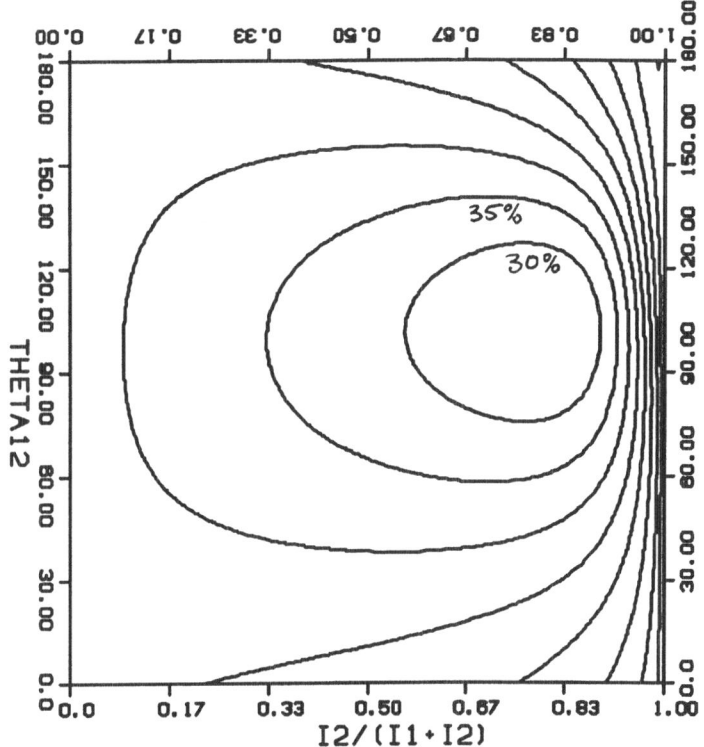

Figure 1: Contour plot of the yield of I* (i.e. percent of I* as product) in the photodissociation of CH_3I from a linear superposition of the $(3_0, J=2)$; $(3_1, J=2)$ states where 3_n denotes the nth level in the C-I stretch mode. The abscissa is labelled by the relative amplitude parameter $S = |x_2|^2/(1 + |x_2|^2)$, and the ordinate by the relative phase parameter $\theta_{1,2} = \theta_1 - \theta_2$

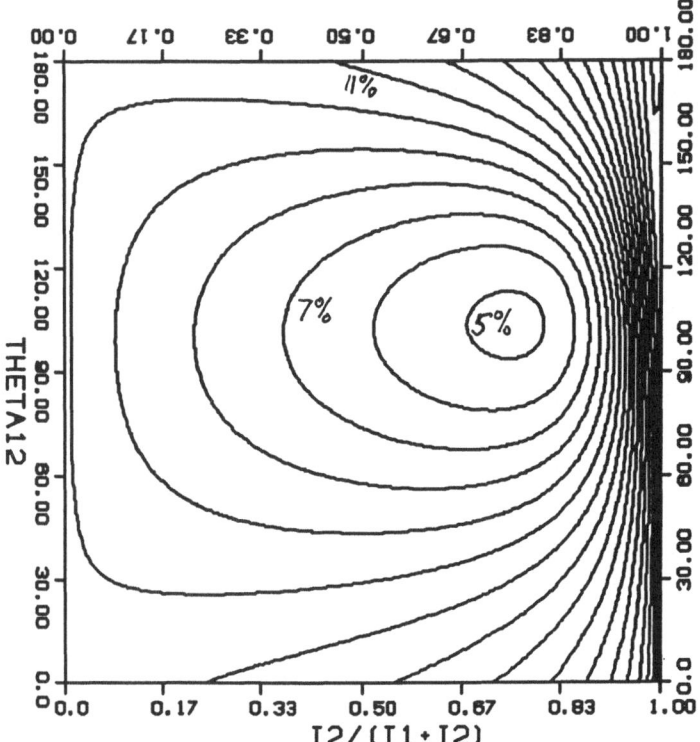

Figure 2: Contour plot of the yield of a quantum state of the products in the photodissociation of CH_3I from a linear superposition of $3_0, J=2$ and $3_1, J=2$ to yield $I^* + CH_3(v=3)$
The labelling of the abscissa and ordinate is as in Fig. 1

Figure 2 shows the yield of the $(I^*, v=1)$ quantum states, where $v=1$ denotes the second excited vibrational state of the CH_3 radical, obtained under coherent dissociation of the $3_0, J=2 + 3_1, J=2$ superposition state.

The range of quantum state yield, for $v=3$ is seen to be controllable over a range of 5 to 20%.

IV.   THEORY OF BIMOLECULAR LASER CATALYSIS

Bimolecular laser catalysis requires, almost by definition, a non-linear theory of the interaction with the radiation field. This follows because one has only a very short time, namely the collision time by which to influence the outcome of the chemical event. This is in sharp

contrast to the unimolecular case for which the molecule to be dissoci-
ated is available at all times.

The free-bound scheme which was developed by us[16],[33] and described above, is no exception, although the molecular matrix elements
are effectively much larger than those associated with free-free transitions.

In what follows, we consider the interaction of the $|\psi_{el}^1\rangle|n,\omega\rangle$ and
$|\psi_{el}^2\rangle|n-1,\omega\rangle$ electronic-field states, where 1 denotes the ground (scattering), and 2 the excited (bound) electronic state. In order to ac-
count for the interference between the "natural" non-radiative process
and the field-induced reaction, we draw upon the theory of scattering by
two potentials. Accordingly, we partition the Hamiltonian into two
parts,

$$H = H_1 + H_2 ,\tag{20a}$$

where,

$$H_1 = H_m ,\tag{20b}$$

and

$$H_2 = H_{rad} + H_{int} .\tag{20c}$$

The non-radiative part of the transition matrix, $\underline{\underline{T}}_1$, can be singled out

by partitioning the full transition matrix as[35],

$$\underline{\underline{T}} = \underline{\underline{T}}_1 + \underline{\underline{T}}_2,\tag{21}$$

$\underline{\underline{T}}_2$ contains the effects of the radiation field and the coupling between
the radiative and non-radiative processes.

This partitioning may be recast in a matrix-product form for the S-matrix[35],[36],

$$\underline{\underline{S}} = \underline{\underline{S}}_1^{1/2}\cdot\underline{\underline{S}}_2\cdot\underline{\underline{S}}_1^{1/2}\tag{22a}$$

where,

$$\underline{\underline{S}}_1 = \underline{\underline{I}} - 2\pi i\underline{\underline{T}}_1 ,\tag{22b}$$

$$\underline{\underline{S}}_2 = \underline{\underline{I}} - 2\pi i\underline{\underline{\Gamma}}_2\tag{22c}$$

with,

$$\underline{\underline{\Gamma}}_2 \equiv \underline{\underline{S}}_1^{-1/2}\cdot \underline{\underline{T}}_2\cdot\underline{\underline{S}}_1^{-1/2} .\tag{22d}$$

We now turn our attention to the non-perturbative evaluation of S_2, the radiative S-matrix. It is most conveniently done by first calculating the radiative reactance matrix - $\underline{\underline{R}}_2$, whose relation to $\underline{\underline{S}}_2$ is given by the well known expression,

$$\underline{\underline{S}}_2 = (\underline{\underline{I}} - i\pi\underline{\underline{R}}_2) \cdot (\underline{\underline{I}} + i\pi\underline{\underline{R}}_2)^{-1}. \tag{23}$$

The computation of $\underline{\underline{R}}_2$ proceeds by first partitioning $|E,v,\gamma\rangle$ - the standing wave solutions of the total (radiative + non-radiative) Schroedinger equation,

$$(E - H) \ |E,v,\gamma\rangle = 0, \tag{24}$$

into its two electronic components. Due to the nature of the interaction Hamiltonian, only $(\Psi_{el}^2 \ |E,v,\gamma\rangle$ must be considered explicitly. It is then possible to write $R_{2,v'v}^{\beta\gamma}$ as[36],[35],

$$R_{2,v'v}^{\beta\gamma} = -\langle E,v',\beta,1| \ \vec{\mu}_{1,2} \cdot \vec{\varepsilon} \ (\Psi_{el}^2 \ |E,v,\gamma\rangle. \tag{25}$$

where, $\vec{\mu}_{1,2} = (\Psi_{el}^1 | \ \vec{\mu} \ |\Psi_{el}^2)$. $\tag{26}$

In Ref. 36, a partitioning technique[35] was applied to the standing-waves case, using real optical potentials. Following that work, we define two orthogonal projectors Q and P, where,

$$P = |\Psi_{el}^1)|n,\omega\rangle\langle n,\omega|(\Psi_{el}^1 |, \tag{27a}$$

and

$$Q = |\Psi_{el}^2)|n-1,\omega\rangle\langle n-1,\omega|(\Psi_{el}^2 |. \tag{27b}$$

Obviously QP = 0 and since we only consider two electronic states, Q + P = I.

For the ω values considered, the states spanned by Q are closed. Hence the Q-projection of the total wavefunction - $|E,v,\gamma\rangle$, can be related to $P|E,v,\gamma,1\rangle$ - the the solution of the non-radiative motion on the ground state, in the following way,

$$Q|E,v,\gamma\rangle = -(E - Q\Xi Q)^{-1} Q \ \vec{\mu}\cdot\vec{\varepsilon} \ P|E,v,\gamma,1\rangle, \tag{28a}$$

where the standing-waves optical potential - $Q\Xi Q$ - is given as,

$$Q\Xi Q = QHQ + Q \ \vec{\mu}\cdot\vec{\varepsilon} \ P \ P_\upsilon(E - PHP)^{-1} \ P \ \vec{\mu}\cdot\vec{\varepsilon} \ Q \ , \tag{28b}$$

with P_υ signifying the Cauchy Principal Value.

Substituting Eqs. (28) into Eq. (26), we obtain that,

$$R_{2,v'v}^{\beta\gamma} = \langle E,v',\beta,1|P \ \vec{\mu}\cdot\vec{\varepsilon} \ Q(E - Q\Xi Q)^{-1}Q \ \vec{\mu}\cdot\vec{\varepsilon} \ P|E,v,\gamma,1\rangle, \tag{29}$$

where it was assumed that $\mu_{1,1} = \mu_{2,2} = 0$.

Because Q spans a closed manifold, one can compute $Q \Xi Q$ in a discrete representation. Writing Eq. (29) in matrix notation, we have that,

$$\underline{\underline{R}}_2 = \underline{\underline{\mu}}^\dagger(E) \cdot (E \underline{\underline{I}} - \underline{\underline{\Xi}})^{-1} \cdot \underline{\underline{\mu}}(E) , \qquad (30)$$

where † denotes a matrix-transpose, the $\approx$ underscore - a matrix having two pairs of indices. $\underline{\underline{\Xi}}$ is a matrix representation of the $Q \Xi Q$ optical potential. From Eq. (28b) we have that,

$$\underline{\underline{\Xi}} = \underline{\underline{h}} + P_\upsilon \int dE' \ \underline{\underline{\mu}}(E') \cdot \underline{\underline{\mu}}^\dagger(E')/(E - E') . \qquad (31)$$

Thus, the entire computation is based on knowing the $\underline{\underline{h}}$ and $\underline{\underline{\mu}}$ matrices. $\underline{\underline{h}}$ is given in terms of (two-dimensional) bound-bound integrals of the H_3 excited state Hamiltonian,

$$\{\underline{\underline{h}}\}_{nm,n'm'} \equiv \langle E_{n,m}|QHQ|E_{n',m'}\rangle. \qquad (32)$$

We use the SLTH[38] form for the $H + H_2$ ground-state potential and the empirical G1 surface of Maine et. al.[34] for the H_3 excited state potential. The continuum - $|E,v,\gamma,1\rangle$ and bound nuclear wavefunctions - $|E_{n,m}\rangle$ were approximated as products of <u>vibrationally-adiabatic</u> functions and, for the ground state, the uniform Airy scattering functions[33].

The $\underline{\underline{\mu}}(E)$ matrix is defined as,

$$\{\underline{\underline{\mu}}(E)\}_{nm,v\gamma} \equiv \langle E_{n,m}|\vec{\mu}_{2,1} \cdot \vec{\varepsilon}|E,v,\gamma,1\rangle . \qquad (33)$$

It involves the same transition-dipole bound-free integrals as in first order perturbation theory, used in the unimolecular case, except that the $\underline{\underline{\mu}}$ integrals <u>already</u> include the field amplitudes $\varepsilon(\omega)$, as required by the non-linear theory. Explicit formulae for both the bound-bound and bound-free integrals are given in Ref. 33.

$\underline{\underline{R}}_2$, obtained via Eqs. (37-40), is a real symmetric matrix having poles on the real energy axis. These poles become resonances, i.e., they acquire widths, only for the $\underline{\underline{S}}$ matrix, (obtained via Eqs. (22) and (23)).

As shown below, both the initial spread of collision energies and the laser band-widths are usually much larger than the absorption line widths. Under these circumstances the absorption line-strengths are simply proportional to the <u>area</u> under each line,

$$k_{abs}(E,v,\gamma|v) = \int d\omega \ |T_{2,vv}^{\gamma\gamma}(E,\omega)|^2 \qquad (34a)$$

where v denotes the center of the line, and $\underline{\underline{T}}_2$ is computed via Eqs.

(22), (23). Likewise, the laser-assisted reaction rates are proportional to the area under the _reactive_ lines, i.e., the frequency dependence of the (natural + radiative) reaction probabilities,

$$k_{reac}(E,v,\beta \leftarrow \gamma | v) = \int d\omega \; |T_{vv}^{\beta \gamma}(E,\omega)|^2 \qquad (34b)$$

In the three dimensional world, summation over the impact-parameter-dependent $|T_{vv}^{\beta \gamma}(E,\omega)|^2$ should also be performed.

In the weak-field limit, the lineshapes are strictly Lorentzian and the integrated transition rates are directly related[27] to the linewidths, given as,

$$k_{weak}(E,v,\gamma | v) = 2\pi |\langle E_{n,m}| \vec{\mu}_{2,1} \cdot \vec{\varepsilon} |E^+,v,\gamma,1\rangle|^2. \qquad (35)$$

$k_{weak}(E,v,\gamma | v)$ are essentially the μ matrix-elements squared, except that the _outgoing_ scattering solutions - $|E^+,v,\gamma,1\rangle$, replace $|E,v,\gamma,1\rangle$ - the standing waves solutions.

V. LASER CATALYSIS - A COMPUTATIONAL EXAMPLE

When considering higher field strengths, we must abandon the perturbative treatment and make use of the more exact version of our theory. In general, the integration over ω has to be done numerically, especially when the line-widths become comparable to the laser band-width, (which would necessitate using different field-strengths for different ω values), or when neighboring resonances begin to overlap. For the present system this occurs at very high field strengths (TW/cm^2). As shown elsewhere[33] and below, at intermediate field strengths, the reactive lineshapes are essentially of the "Fano-type"[39]. The ω integration can again be performed analytically.

Our calculated reactive lineshapes ($|T_{vv}^{\beta \gamma}(E,\omega)|^2$), at three laser intensities, are shown in Fig. 3.

At a first glance laser catalysis seems to be achievable at all field intensities, since the reaction probability reaches the value of 1 at some frequency, regardless of the laser intensity. However, as discussed above, the effect really depends on the _integrated_ lineshape, namely the line-strength. This magnitude is crucially dependent on the field strength, since the line broadens (while still reaching 1 at some point), as the field intensity grows. Roughly speaking the catalytic effect is therefore determined by the line-widths.

The increase in reaction rate (and line-width) is linear over a wide range of intensities. This is shown in Fig. 4, for a number of lines. The linearity in field intensity is due to the crucial role of the _excitation_ process in promoting the radiative reaction. Once the system is in the excited state it can react, irrespective of the rate of its down-transitions. Even if there were no external fields to de-excite the sys-

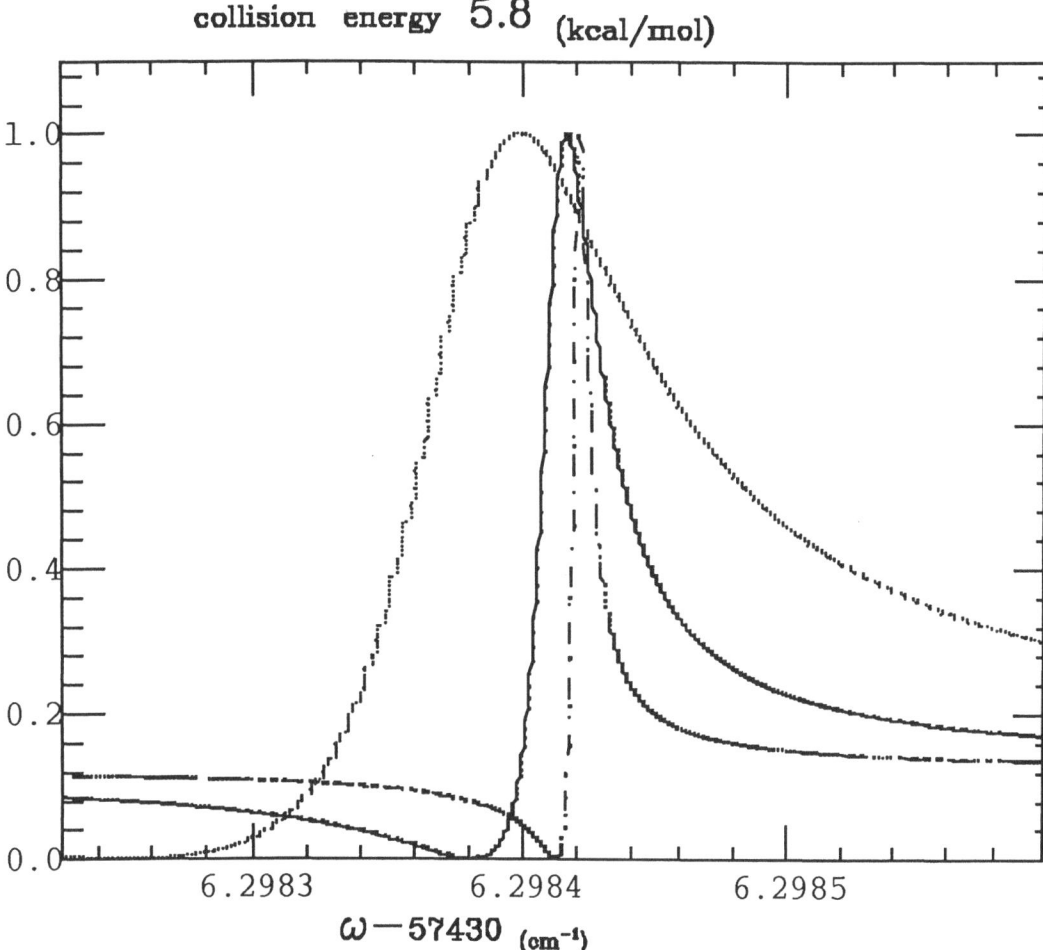

Figure 3: Calculated reactive lineshapes ($|T_{VV}^{\beta\gamma}(E,\omega)|^2$), at three laser intensities.
($-\cdot-\cdot$ 21MW/cm^2)
($\underline{\quad\quad}$ 83 MW/cm^2)
($\cdots$ 338 MW/cm^2)

tem to the ground state, the system, once excited, would still form products, (with a 50% probability - if all nuclei are identical), even if all coherence were lost.

As shown in Fig. 3 the main distinguishing feature of a <u>coherent</u> laser catalysis is that the lines are non-symmetric. The asymmetry is due to an interference between the (direct) non-radiative reactive route and the (complex) radiative one. Due to this interference the reaction is hindered on the red side of the line and enhanced on the blue side of the line.

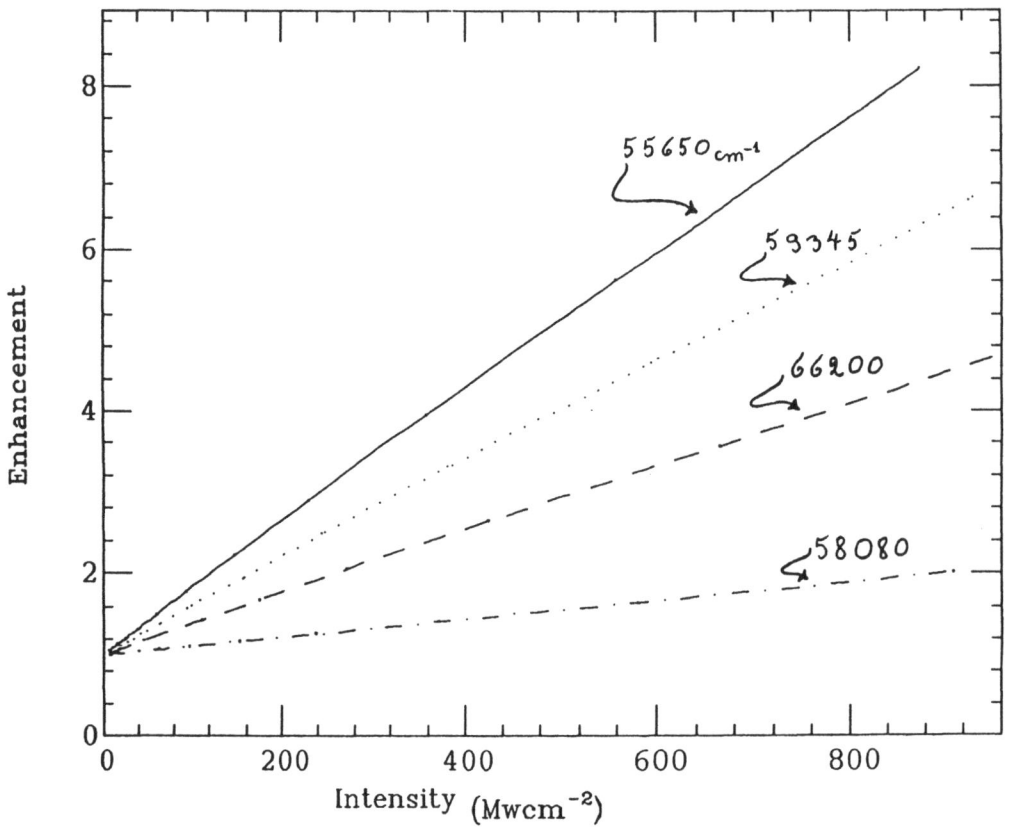

Figure 4: Enhancement factor of room-temperature H + H_2 reaction rates
as function of field intensity for four different lines.
(—— 55650 cm^{-1})
(-·-· 58080 cm^{-1})
(··· 59345 cm^{-1})
(--- 66200 cm^{-1})

IV. DISCUSSION

In this paper we have reviewed two ways of using laser to alter the natural flow of reactive events. One way, of relevance to unimolecular reactions, is to use a coherent two colour excitation to control the branching ratio in photodissociation reactions. The process is found to be very effective in altering electronic branching ratio in direct photodissociation such as that of CH_3I. The other way, pertinent to bimolecular reactions, ia to affect a reaction barrier crossing by a transient excitation to a bound excited state. We find that catalysis is, in principle, possible and that the effect scales linearly, over a wide range, with intensity.

Although the two methods seem quite different they both depend on coherence in one form or another. In the unimolecular case, control is achieved by the coherent interference between <u>two</u> optical pathways. Each pathway involves the absorption of a photon. The coherence, created between a pair of bound states, is transferred to a more useful form, that between a degenerate pair of continuum states.

In the bimolecular case, the laser-induced enhancement and hindrance of the reaction are also a result of interferences, this time between one path which is optically controlled and one path which is not. The optically controlled route involves a virtual two-step process, the absorption and emission of a photon. The non-radiative path is "natural" tunneling.

There are some important differences: The bimolecular effect increases linearly with laser power, hence relies, in practical applications, on the availability of powerful light sources. Control in the unimolecular case works well, its limits best understood, in the weak field limit. On the other hand the unimolecular effect completely relies on coherence. If the latter is not maintained the interference terms of Eq. (14) drop out and with it the entire effect. In the bimolecular case the asymmetry of the reactive lineshape is lost in the absence of coherence but not the entire effect. Once excited to the bound state, a process independent on coherence, (see Eq. (8)), the system has an even (50% in the symmetric cases) chance of reaction. To be sure, reaction probability will not reach 100% at some point, but the effect will still be substantial for reactions with high ground state barriers, such as "Woodward-Hoffman forbidden" reactions.

References

1. a) A. Ben Shaul, Y. Haas, K.l. Kompa and R.D. Levine, "Lasers and Chemical Change" (Springer-Verlag, Berlin, 1981).
b) N. Bloembergen and E. Yablonovitch, Physics Today, 31, 23 (1978).
c) R.L. Woodin and A. Kaldor, Adv. Chem. Phys. 47(b), 3 (1981).
2. M. Shapiro and P. Brumer, J. Chem. Phys., 84, 540 (1986)

3. P.Brumer and M. Shapiro, Chem. Phys. Lett., **126**, 541 (1986);
M. Shapiro and P. Brumer, J. Chem. Phys. **84**, 4103 (1986);
P. Brumer and M. Shapiro, Faraday Disc. Chem. Soc. **82**, xxx (1986).
4. N.M. Kroll and K.M. Watson Phys. Rev. A **8**, 804 (1973);
N.M. Kroll and K.M. Watson Phys. Rev. A **13**, 1018 (1976).
5. A.M.F. Lau Phys. Rev. A **13**, 139 (1976); ibid A **18**, 172 (1978); ibid A **25**, 363 (1981).
6. a) J.M. Yuan, T.F. George and F.J. McLafferty, Chem. Phys. Lett., **40**, 163 (1976).
b) J.M. Yuan, J.R. Laing and T.F. George, J. Chem. Phys., **66**, 1107 (1977).
c) T.F. George, J. Yuan, I.H. Zimmermann, Faraday Disc. Chem. Soc. **62**, 246 (1977);
d) P.L. deVries and T.F. George, Faraday Disc. Chem. Soc., **67**, 129 (1979).
e) T.F. George J. Phys. Chem. **86**, 10 (1982), and references therein.
7. J.I. Gerstein and M.H.Mittleman, J. Phys. B **9**, 383 (1976)
8. V.S. Dubov, L.I.Gudzenko, L.V. Gurvich and S.I. Iakovlenlo, Chem. Phys. Lett., **45**, 351 (1977).
9. H.J. Foth, J.C. Polanyi, H.H. Telle, J. Phys. Chem. **86**, 5027 (1982);
10. a) A.E. Orel and W.H. Miller J. Chem. Phys., **70**, 4393 (1979).
b) A.E. Orel and W.H. Miller, J. Chem. Phys., **73**, 241 (1980).
11. K.C. Kulander and A.E. Orel, J. Chem. Phys. **74**, 6529 (1981); ibid **75**, 675 (1981).
12. J.C. Light and A. Altenberger-Siczek, J. Chem. Phys. **70**, 4108 (1979).
13. M. Hutchinson and T.F. George, Mol. Phys., **46**, 81 (1982).
14. I. Last, M. Baer, I.H. Zimmerman and T.F. George, Chem. Phys. Lett., **101**, 163 (1983); M. Baer, I. Last, and Y. Shima, Chem. Phys. Lett., **110**, 163 (1984); I. Last and M. Baer, J. Chem. Phys., **82**, 4954 (1985).
15. T. Ho, C. Laughlin and S. Chu, Phys. Rev. A, **34**, 122 (1985).
16. M. Shapiro and Y. Zeiri, J. Chem. Phys. **85**, 6449 (1986).
17. S.E. Harris and J.C. White, IEEE J. Quantum Electron, **QE-13**, 972 (1977).
18. P.H. Cahuzac and P.E. Toscheck, Phys. Rev. Lett., **40**, 1087 (1978).
19. P. Hering, P.R. Brooks, R.F. Curl, R.S. Judson and R.S. Lowe, Phys. Rev. Lett., **44**, 687 (1980).
20. a) J.C. Polanyi, Faraday Disc. Chem. Soc., **67**, 129 (1979);
b) P. Arrowsmith, F.E. Bartoszek, S.H.P. Bly, T. Carrington, P.E. Chaters and J.C. Polanyi, J. Chem. Phys., **73**, 5896 (1980).
21. P. Polak=Dingles, J.-F. Delpech and J. Weiner, Phys. Rev. Lett., **44**, 1663 (1980); P. Polak-Dingles, R. Bonanno, J. Keller and J. Weiner, Phys. Rev. A **25**, 2539 (1982).
22. B.E. Wilcomb and R. Burnham, J. Chem. Phys., **74**, 6784 (1981).
23. H.P. Grieneisen, K. Hohla and K.L. Kompa, Opt. Commun. **37**, 97 (1981); H.P. Grieneisen, Hu Xue-jing and K.L. Kompa, Chem. Phys. Lett., **82**, 421 (1981).
24. P.R. Brooks, R.F. Curl and T.G. Maguire, Ber. Bunsenges. Phys. Chem. **86**, 401 (1982);

25. a) G. Inoue, J.K. Ku and D.W. Setser, J. Chem. Phys., **76**, 735 (1982);
b) J.K. Ku, D.W. Setser and D. Oba, Chem. Phys. Lett., **109**, 429 (1984).
26. D. Kleiber, A.M. Lyyra, K.M. Sando, S.P. Heneghan and W.C. Stwalley, Phys. Rev. Lett., **54**, 2003 (1985).
27. M. Shapiro and R. Bersohn, Ann. Rev. Phys. Chem. **33**, 409 (1982).
28. G.G. Balint-Kurti and M. Shapiro, Adv. Chem. Phys. **60**, 403 (1985); P. Brumer and M. Shapiro, Adv. Chem. Phys. **60**, 371 (1985).
29. a) R.K. Sparks K. Shobatake, L.R. Carlson and Y.T. Lee, J. Chem. Phys., **75**, 3838 (1981).
b) H.W. Hermann and S.R. Leone J. Chem. Phys., **76**, 4759 (1982).
c) G.N.A. van Veen, T. Baller, A.E. deVries and N.J.A. van Veen Chem. Phys., **87**, 405 (1984).
30. a) M. Shapiro and R. Bersohn J. Chem. Phys., **73**, 3810 (1980).
b) S.Y. Lee and E.J. Heller, J. Chem. Phys., **76**, 3035 (1982).
c) V. Engel and R. Schinke, Mol. Phys. **51**, 189 (1984).
31. M. Shapiro, J. Phys. Chem., **90**, 3644 (1986).
32. M. Shapiro and P. Brumer, in "Methods of Laser Spectroscopy" Y. Prior, A. Ben Reuven and M. Rosenbluth, editors, (Plenum Publishing Co. 1986), p. 239
33. T. Seideman and M. Shapiro, "Laser Catalysis and Transition State Spectra of the H + H$_2$ Exchange Reaction" submitted for publication.
34. H.R. Mayne, R.A. Poirier, J.C. Polanyi, J. Chem. Phys. **80**, 4025 (1984).
35. R.D. Levine, "Quantum Mechanics of Molecular Rate Processes" (Clarendon Press, Oxford, 1969).
36. H. Shyldkrot and M. Shapiro, J. Chem. Phys., **79**, 5927 (1983).
38. a) B. Liu, J. Chem. Phys., **58**, 1925 (1973). b) D.G. Truhlar and C.J. Horowitz, J. Chem. Phys., **68**, 2566 (1978); **71**, 1514(E) (1979).
39. U. Fano, Phys. Rev. **124**, 1866 (1961).

STUDIES ON THE STRUCTURE OF POROUS SILICAS: THE FRACTAL DILEMMA

J. M. Drake*, J. Klafter* and P. Levitz†
*Exxon Research and Engineering Company
Route 22 East, Annandale, NJ 08801, USA
†C.N.R.S.-C.R.S.O.C.I, 45071-Orleans Cedex 2,
France

ABSTRACT.   We report a series of characterization studies on a family of silica gels. Such silicas have been recently described as having self-similar (fractal) surfaces. Here we raise the possibility that modeling through regular geometrical shapes accounts well for the experimental observations.

INTRODUCTION

Due to the complexity of the structure of porous solids which are both physically and chemically disordered, there is a need to assume simple models in order to interpret experimental data. Such models have to capture the basic geometrical and chemical character-istics of the studied system[1-4].

When modeling the geometrical disorder, we face the dilemma of choosing between the classical approach towards spatial restrictions, which approximates the complex structures through regular geometrical shapes[1,2] (cylindrical, spherical, conical, bottle-like pores, etc.), and the more recent concept of fractals which introduces the idea of self similarity[5]. In the classical approach one usually starts from a primary picture of a local pore. A spatial distribu-tion of these primary elements is then added. An approximate rela-tionship between such models and particle packing is possible. Different materials have different local structures and pore networks[1,2]. In the fractal, self similar, approach one assumes the same roughness and same structural features for a given surface on all scales[6]. No local details enter the model. Two parameters are necessary in order to describe structural and dynamical proper-ties of fractals: the fractal dimension d, which is related to the density of sites on the structure and the spectral dimension $\tilde{d}$, which appears in connection with dynamical properties on fractals and which describes the dynamical connectivity of the system[7,8]. Being interested in porous systems one should distinguish between surface fractals and mass fractals (the pore network spatial organization)[9].

J. Jortner et al. (eds.), Large Finite Systems, 379–386.

The fractal concept appeals to one's intuition of rough surfaces and
its possible applicability to such cases has caught the attention of
several groups. The physical idea is that the general morphology of
the surface irregularities is independent of the magnification at
which the surface is observed. Increasing the magnification reveals
an increase in the number of irregularities that are morphologically
similar to that observed at the larger scale. This intuitive picture
has already been given by Adamson[10] and has been recently advocated
by Avnir et al.[6] Generally, fractal structures have been shown to
exist in nature[11]. Examples are: cluster-cluster aggregates,
silica particles and rocks. However, the applicability of the
concept must be tested in each case and confronted with the experimental data. Real fractal systems are characterized by two cutoffs:
a lower cutoff which corresponds to the elementary building block of
the structure and a upper cutoff beyond which the system is disordered but homogeneous.

There is a large number of methods to characterize porous materials[1,2]. We chose to study porous silica gels using small angle
x-ray and neutron scattering[3] and direct energy transfer (DET)
measurements[12]. In the following sections we show that regular
geometrical shapes as a description for local pore morphology account
well for the experimental observations. This does not rule out the
existence of fractal behavior in other systems.

SMALL ANGLE X-RAY AND NEUTRON SCATTERING

We have applied both small angle neutron and X-ray scattering
(SANS and SAXS) to characterize the morphological organization of a
family of silica samples which we then studied by direct energy
transfer. Applying SAXS and SANS in the study the pore structure of
silicas is important especially because of the large length scale
which can be probed by these methods, typically between 5 and 2000Å.
We are able therefore to characterize the interior pore surface and
the mass distribution of the silicas using a single technique. When
one compares scattering experiments to other approaches to characterize internal pore surfaces a number of obvious advantages emerge:
1) when compared to molecular tiling[6] scattering probes much larger
length scales, 2) scattering probes the pore boundary independent of
the chemical heterogeniety of the surface, 3) scattering can provide
a measure of the pore surface to volume ratio and 4) using an appropriate model the mean pore size and pore size distribution can be
obtained from the cord distribution used to fit the scattering data
over the whole q range.

The scattering functions s(q) for the different silicas are
shown in Fig. 1. Each scattering curve has basically the same shape
composed of two power law regions separated by a plateau at intermediate q. What is most striking is that these scattering curves can
be scaled both in q and intensity to give one curve of the form,

$$F(qr_0) = \frac{S(q)}{S_0(r_0)},$$ (1)

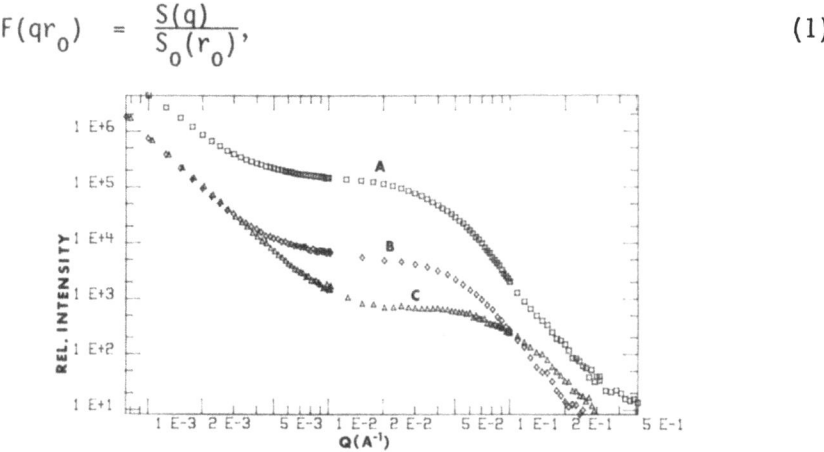

Figure 1.

where $F(qr_0)$ is a general scaling function and $S_0(r_0)$ and r_0 are appropriate scaling constants for each silica. The scaled scattering data is presented in Fig. 2. The parameter r_0 can be shown to be related to the mean pore size (Rp) obtained from N_2 desorption measurement[3]. Based on these results we conclude that the morphologies of these silicas, on length scales of the mean pore size are very similar and scale with Rp, the mean pore size. In fact, on the scale of Rp, $F(qr_0)$ is well approximated by Debye's model of scattering from a random interface between two otherwise uniform media[12]

$$F(qr_0) \sim \left(\frac{A}{1+q^2 r_0^2}\right)^2.$$ (2)

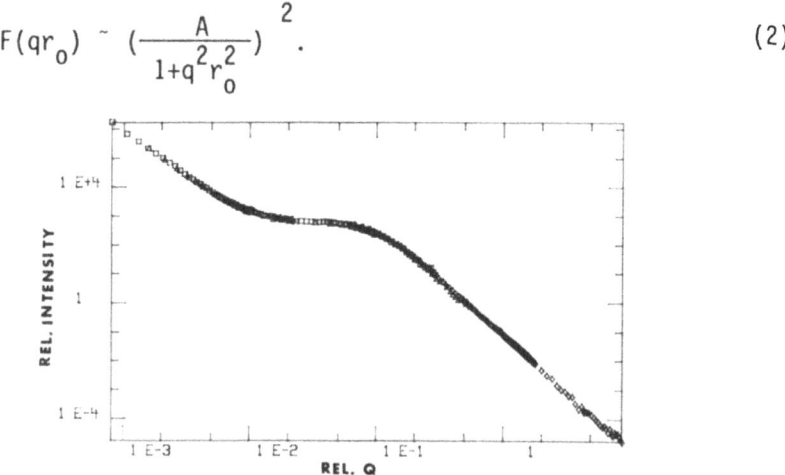

Figure 2.

In the limit of large q, at the scale of the pore surface the scattering is porod-like[3], namely power law dependent according to q^{-4}. This means flat surfaces at the corresponding length scale.

Our conclusions, based on the scattering data, are that these silicas are morphologically similar to each other on the scale of Rp. Their morphological structure scales with Rp and the pore surface is smooth at lengths greater than 10Å.

DIRECT ENERGY TRANSFER (DET)

It has been suggested that energy transfer methods can also be used to characterize complex structures such as porous mater- ials[13,14]. We note that the use of energy transfer probes for studying structural and dynamical properties of other disordered materials has become widespread in recent years. Here we apply DET measurements to the same family of porous silicas studied by SANS and SAXS as described in the previous section.

The basic idea behind the DET measurement is to tag the surface with a random distribution of donor molecules (rhodamine 6G in our case) and acceptor molecules (malachite green) at concentrations low enough to allow only for a one step transfer of the initially excited donors to the acceptors. Excitation transfer among the donors or among the acceptors is excluded in the concentration range and time scale of the experiment. One then follows the fluorescence of the donors[15]. Within a generalized Forster model for DET the survival probability of the donor has been calculated and it has the form[13,14]:

$$\Phi(t) = \exp[-\frac{t}{\tau_D} - p \int d\underset{\sim}{r} \, \rho(r)\{1-\exp[-tw(r)]\}] \qquad (3)$$

where p is the density of acceptors (p<<1), τ_D is the fluorescence lifetime of the isolated donor and $\rho(r)$ is the density of sites on the structure. Eq. (3) relates the time evolution of the excited donor to the spatial arrangement of the molecules involved in the energy transfer. For randomly distributed molecules we obtain morphological information about the underlying porous surfaces. The interaction donor-acceptor is assumed to be dipolar[16,17]:

$$W(r) = \frac{3}{2} K^2(\frac{R_0}{r})^6 \cdot \frac{1}{\tau_D} \qquad (4)$$

Here K is the anisotropy factors and R_0 is a critical radius which is determined from the spectral overlap of donor fluorescence and the absorption of the acceptor on the silica surface. R_0 provides a rough estimate for the length scale which can be probed by DET[16]. Usually R_0 does not exceed 100Å.

The survival probability of the donor depends on $\rho(r)$[13,14]. For $\rho(r)$ = constant we recover Forster type decays which depend on dimensionality[17]. In the case of fractal structures, where $\rho(r) = \rho_0 r^{d-\bar{d}}$ we obtain[12]:

$$\Phi(t) = \exp[-At^{\bar{d}/6} - t/\tau_D],\tag{5}$$

where A is time independent. Eq.(5) connects the dynamical behavior, $\Phi(t)$, to the geometry of the structure, d.

Eq.(3) applies also in cases of DET in regular geometrical shapes[14] (spheres, cylinders, etc.) which mimic simple pores. Here again $\rho(r)$ accounts for the corresponding geometrical restrictions and depends on the details of the local pore morphology. This is reflected in the DET decay expressions. The deviations from the Forster model in these pore model systems originate from crossovers due to the finiteness and confinement of the systems[14]. Still the overall decay may fit an expression for the form in Eq.(5) but with d being now an "effective dimension" (or apparent dimension)[18]:

$$\Phi(t) = \exp[-t/\tau_D - A c \Gamma(1-\bar{d}/6)(\frac{t}{\tau_D})^{\bar{d}/6}].\tag{6}$$

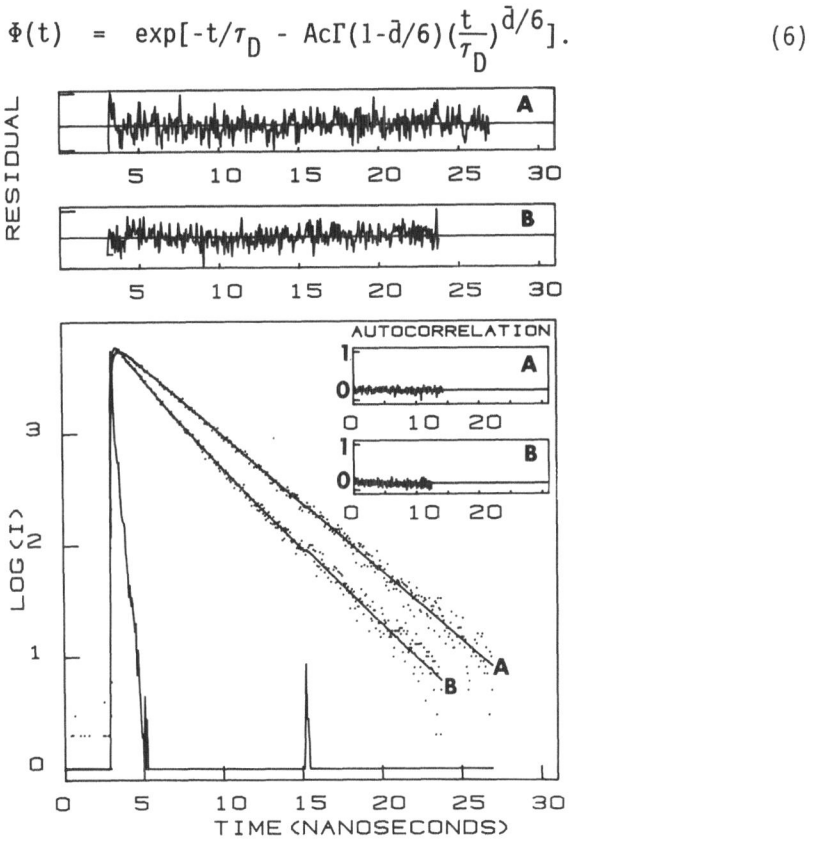

Figure 3.

Ac is related to the surface concentration of the acceptors[15]. The latter has been determined from the adsorption isotherm of the acceptors (malachite green) on each silica[15]. Fig. 3 is an example of the donor decay on a silica due to DET to acceptors.

DET provides a spectroscopic method to elucidate the special organization of adsorbed molecules in relationship to local pore geometries. DET is capable of sensing the location of acceptors relative to donors up to a time t_{max} ($\approx$ 2 nsec in our studies) which is determined by the experimental limitations of the detections system[15]. Using Eq.(4) a length scale R_{max} is determined;

$$R_{max} = R_0 \left(\frac{t_{max}}{\tau_D} \cdot \frac{2}{3K^2}\right)^{1/6} \tag{7}$$

which provides a more realistic estimate to lengths probed by DET. In the previous section on SAXS and SANS we have shown that on the scale of the mean pore size (Rp) the silica gels we are interested in have morphological features which scale with Rp. A basic picture, which is consistent with the scattering results, is to consider the internal surface of these materials as smooth (>10Å) random inter-faces between two media of uniform density[3]. The local curvature (Rp) is probably a remnant of spheres which were the primary building blocks from which the pore morphology was derived. This picture makes the classical pore description a plausible model. We adopt this approach for our DET data analysis. Both the donor and acceptor have a finite sizes which determine the lower length scale (R_{min} = 10Å) which can be probed by DET. The value of R_{max} estab-lishes the upper scale sensed. The relationship between Rp (from desorption and scattering) and R_{max} (from DET) determines what features of the morphology can be probed by the DET process[15], R_{max} < 100Å:

(a)   When 2 Rp/R_{max} >> 1, the DET probes a length scale less than Rp. The DET process therefore senses only a portion of the local spherical surface with a radius curvature similar to Rp. Here d is nearly 2. Deviations from value d=2 may occur because of crossovers to higher dimensional configuration at longer times.

(b)   For 2Rp/R_{max} << 1, the probing length scale is well above Rp. We are then sensitive to pore network. In other words, the local surface appears as a space filling interface with an effect d near 3. A transition to d=2 is possible at higher acceptor concentrations and short times. The experimental limitations of working at high surface concentrations make the observation of this crossover difficult.

(c)   In the limit of Rp/R$_{max}$=1, the problem is less clear. The DET
      process probes on average the local vicinity of the primary
      building blocks. Here, we expect a more pronounced crossover
      effect. The curious reader is referred to ref. 15 for more
      details.

Different interpretation has been recently given by Avnir et
al.[19] in terms of the fractal picture. Here we show, however, that
for our series of silica gels the scattering data when corroborated
with DET raises some doubts about the uniqueness of the fractal
explanation. Moreover, we have shown that the classical pore models
may approximate the reality quite well. Our experimental observations are well understood in terms of the classical approach.

In summary, we have shown that the DET process probes the pore
morphology of porous silicas within certain limitations which are
determined by τ_D, R$_0$ and R$_{max}$. The values of the effective dimension
used to fit the survival probability of the donor are associated with
the relationship between R$_0$ and Rp. The observations are consistent
with our SAXS and SANS results which show that on the scale of R$_p$ and
below the local surface of the building block is flat.

REFERENCES

(1)   R. K. Iler, The Chemistry of Silica. (John Wiley & Sons, New
      York, 1979).
(2)   K. K. Unger, Porous Silica (Elsevier, Amsterdam, 1978).
(3)   J. M. Drake, P. Levitz and S. Sinha in Better Ceramics Through
      Chemistry, ed. C. J. Brinkler (North-Holland, New York, 1986),
      Vol. 2
(4)   W. D. Dozier, J. M. Drake and J. Klafter, Phys. Rev. Lett. 56,
      197 (1986).
(5)   B. B. Mandelbrot, The Fractal Geometry of Nature (W. H.
      Freeman, San Francisco, 1982).
(6)   D. Avnir, D. Farin and P. Pfeifer, Nature 308, 261 (1984).
(7)   S. Alexander and R. Orbach, J. Phys. Lett. 43, L625 (1982).
(8)   A. Blumen, J. Klafter and G. Zumofen, in Optical Spectroscopy
      of Glasses, ed. I. Zschokke (D. Reidel, 1986).
(9)   S. Alexander, in Transport and Relaxation in Random Materials,
      ed. J. Klafter, R. J. Rubin and M. F. Shlesinger (World-
      Scientific, Singapore, 1986).
(10) A. W. Adamson, Physical Chemistry of Surfaces (John Wiley, New
      York, 1982).
(11) See Fractals in Physics, ed. L. Petronero and E. Tosatti
      (North-Holland, Amsterdam, 1986).
(12) P. Debye and A. M. Bueche, J. Appl. Phys. 20, 518 (1949).
(13) J. Klafter and A. Blumen, J. of Chem. Phys. 80, 875 (1984).
(14) J. Klafter and A. Blumen, J. of Lumin. 34, 77 (1985).
(15) P. Levitz and J. M. Drake, Phys. Rev. Lett. 58, 686 (1987).
(16) T. Forster, Z. Naturforsch. Teil A4, 321 (1949).

(17)   A. Blumen, Nuovo Cimento **63**, 50 (1981).
(18)   C. I. Yang, M. A. El-Sayed and S. L. Suib, J. Phys. Chem. (in press).
(19)   D. Rojanski, D. Huppert, H. D. Bale, Xie Dacai, P. W. Schmidt, D. Farin, A. Seri-Levy and D. Avnir, Phys. Rev. Lett. **56**, 2505 1986).

DOPPLER-FREE SPECTROSCOPY, AN OPPORTUNITY FOR STATE SPECIFIC INVESTIGATIONS IN LARGE MOLECULES

E. Riedle
Institut für Physikalische und Theoretische Chemie
Technische Universität München
Lichtenbergstr. 4
D-8046 Garching, West Germany

ABSTRACT.   Conventional high resolution spectroscopy in the gas phase allows the selection of single vibronic states of large molecules. The extreme rotational cooling in a supersonic jet can be of great help for this selection. However, many investigations point to the important role of rotation-vibration coupling for the dynamic behaviour of large molecules. Sub-Doppler experiments like Doppler-free two-photon spectroscopy permit the necessary resolution of single rovibronic transitions. It is demonstrated for the example of benzene, that experiments can now be performed, which completely resolve the rotational structure of the vibronic bands and permit the measurement of the line width and the decay behaviour of single rovibronic states. These investigations lead to a detailed understanding of the influence of rotation on the spectroscopy and the dynamics of the benzene molecule.

1. INTRODUCTION

Spectroscopy is the prime source of information on the energy levels, the structure and the dynamic behaviour of molecules. Especially high resolution optical spectroscopy has contributed a large deal to the present understanding of molecular systems. For large molecules the obtainable resolution in conventional studies is principally limited by the Doppler-broadening in the gas phase. This prevails in most cases the observation of single rovibronic transitions. The additional problem of hot band congestion and extensive rotational inhomogeneous broadening was partially overcome by the use of supersonic jets and their low rotational and vibrational temperatures. However, still the vast majority of all dynamic investigations of large molecules is only able to study single vibronic levels (SVL) instead of single rovibronic levels (SRVL).

387

J. Jortner et al. (eds.), Large Finite Systems, 387–398.

In recent years different Doppler-free techniques were developed to increase the resolution beyond the limit set by the Doppler-broadening and successfully tested for atomic systems and small molecules /1/. The application of these techniques to the study of large molecules is nontrivial due to the large number of thermally populated levels and the complexity of their spectra. However, recently a number of investigations have shown, that techniques like Doppler-free two-photon spectroscopy /2/, spectroscopy in highly collimated molecular beams /3/, intermodulated fluorescence /4/ and saturation spectroscopy /5/ render rotationally resolved spectra even for these molecules. In addition to the rotationally resolved spectra also line width measurements /6/ and decay time measurements /7,8/ of single rovibronic states were obtained and even the measurement of emission spectra of these states seems now possible. The experimental arrangements needed for these investigations and typical results, that show the importance of such ultra high resolution experiments, will be discussed for the example of Doppler-free two-photon spectroscopy of benzene.

2.   CW DOPPLER-FREE TWO-PHOTON SPECTROSCOPY

2.1. The experimental set up

The recording of Doppler-free two-photon spectra is based on the simultaneous absorption of two photons from conterpropagating light beams of an extremely narrow band tunable laser /9/. Contrary to the Doppler-broadened absorption of two photons from one laser beam, the Doppler-shift of the two photons balances itself excactly for this situation and a Doppler-free spectrum can be observed. Since the absorption probability in the two-photon process is very weak and depends quadratically on the laser intensity, a dramatic increase in signal is observed if the experiment is not performed by simple backreflection of the laser beam but rather within the standing wave field of an external resonator /6,10,11/. The experimental set up derived from this consideration and used in our experiments is shown in fig. 1 and has been described in detail /6,12/.

The high spectral resolution (1 MHz) is provided by a cw ring laser sytem (Coherent CR 699/21) pumped by the violet lines of a Kr^+ laser (CR 3000K UV) and providing typically 300 mW laser power around 5000 Å with Coumarin 102 dye. The benzene cell is placed within the concentric external resonator (mirrors of 70 % and 99 % reflectivity), whose length is locked to the laser frequency. Experiments are

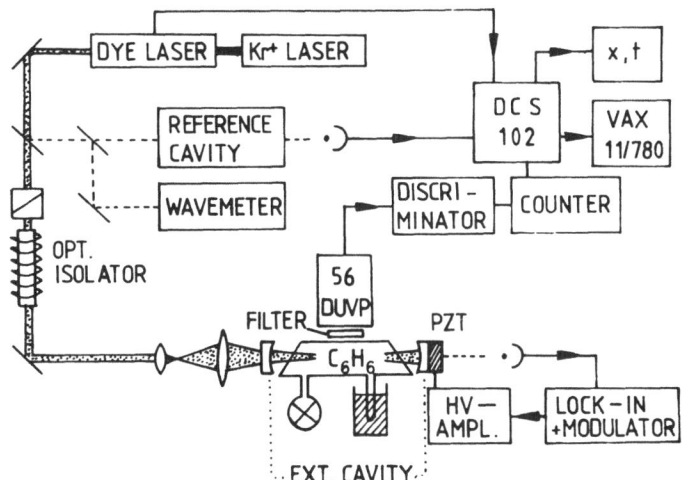

Figure 1. Experimental set up used for the recording of Doppler-free two-photon spectra of benzene /6,12/.

typically performed with benzene pressures below 1 torr to achieve nearly collisionless conditions. The fluorescence emitted by the excited molecules is monitored with a single-photon-counting system, i.e. fluorescence excitation spectra are recorded. For the calibration of the spectra the transmission of a highly stable 150 MHz Fabry-Perot-Etalon is recorded simultaneously with the molecular spectrum. The free spectral range of this etalon has been measured to 1 part in 10^6.

2.2. The rotationally resolved spectrum

As a typical result the first 6 cm^{-1} of the Doppler-free spectrum of the Q-branch of the 14^1_0-band of C_6H_6 at 39656.90 cm^{-1} is shown in fig. 2 /13,14/. The spectrum consists of 6 individual laser scans and the frequency scale has been linearized according to the transmission pattern of the 150 MHz FPI /15/. The observed lines correspond to well resolved single rotational lines. The Doppler-width of 1.7 GHz would not allow the resolution of the individual rovibronic transitions since typically 10 lines are located within the Doppler-width. All the lines can be assigned within the model of a semirigid symmetric top. A fit to the observed line positions renders the rotational constants of the excited state with an accuracy of 10^{-7} cm^{-1} and the quartic centrifugal distortion constants with an accuracy of 10^{-10} cm^{-1} /15/. A similar accuracy has even been reached for the asymmetric top molecule benzene-d_1 /16/.

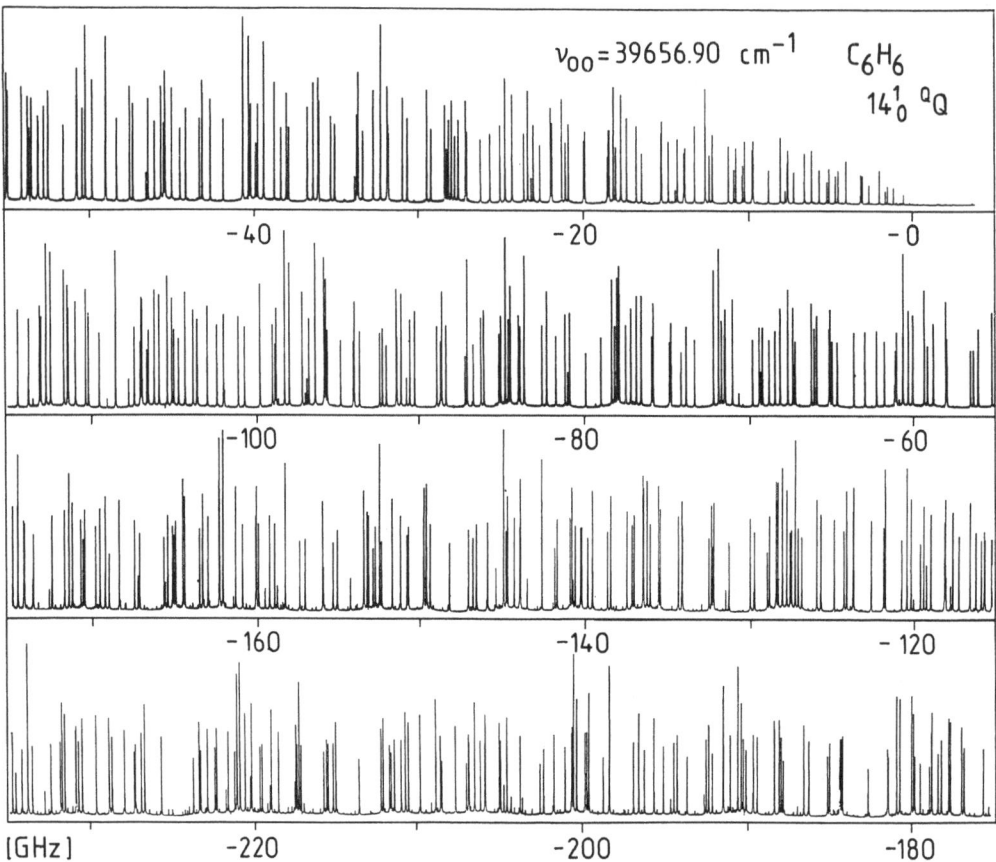

Figure 2. Part of the Doppler-free room temperature spectrum of the Q-branch of the 14^1_0-band of C_6H_6 /13,14/. Every line corresponds to an individual rovibronic transition. All the lines have been assigned.

The remaining deviations (residuals) between the calculated line positions and the observed ones are about 10 MHz. This extreme agreement is demonstrated in fig. 3 for the 14^1_0-band of 1,3,5-$C_6H_3D_3$ /15/. It is again seen, that Doppler-free two-photon spectroscopy allows the complete resolution of the electronic spectrum and the spectrum is extremely well reproduced with the simple model of a semi-rigid symmetric top.

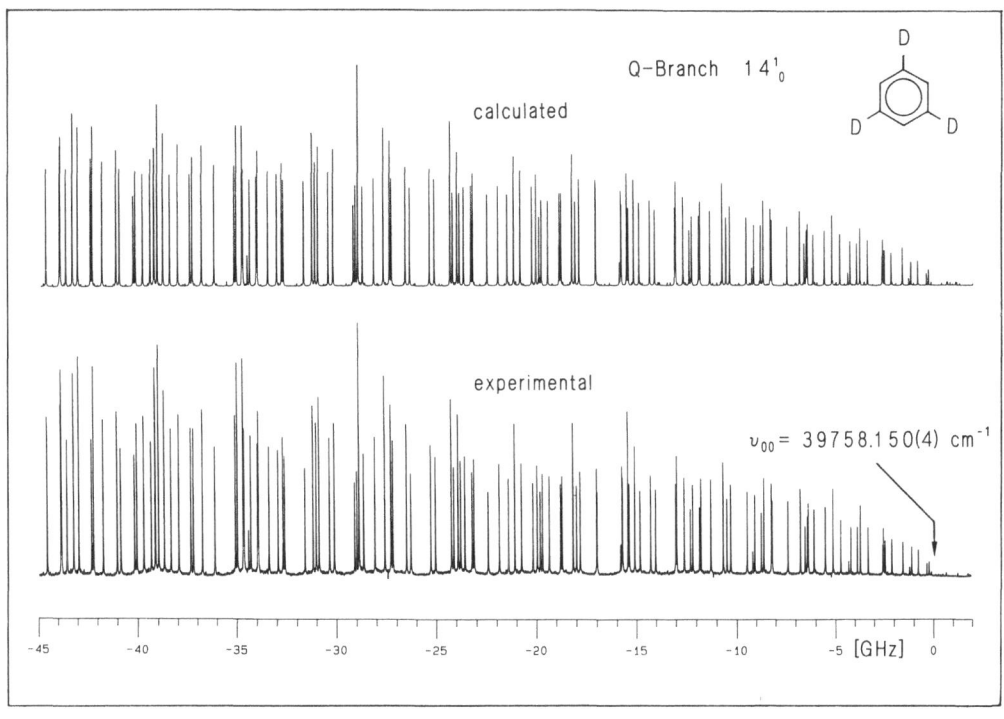

Figure 3. Part of the Doppler-free spectrum of the Q-branch
of the 14^1_0-band of $1,3,5-C_6H_3D_3$. The experimental spectrum
(lower trace) is compared to the calculated spectrum (upper
trace) of a semirigid symmetric top /15/.

2.3. Perturbations in the spectrum

In some parts of the spectrum, at higher rotational energy,
differences between the experimental spectrum and the calculated one are observed. Single lines of the calculated
spectrum are not observed in the experimental one, instead
two weaker lines are observed at slightly shifted positions
/17/. If these pairs of lines are labeled with the rotational quantum numbers of the missing line, the dependence
of the deviations on the quantum number J of total angular
momentum can be plotted for each value of K, the quantum
number of the projection of $\vec{J}$ on the figure axis of the
molecule. The result is shown in fig. 4 for the Q-branch
of the 14^1_0-Band of C_6H_6 /13,14/.

The typical J-dependence of the deviations found is that
of an avoided crossing. A careful analysis shows /17/,
that the observed perturbations are caused by the coupling
of light rotational states of the 14^1 vibronic state to

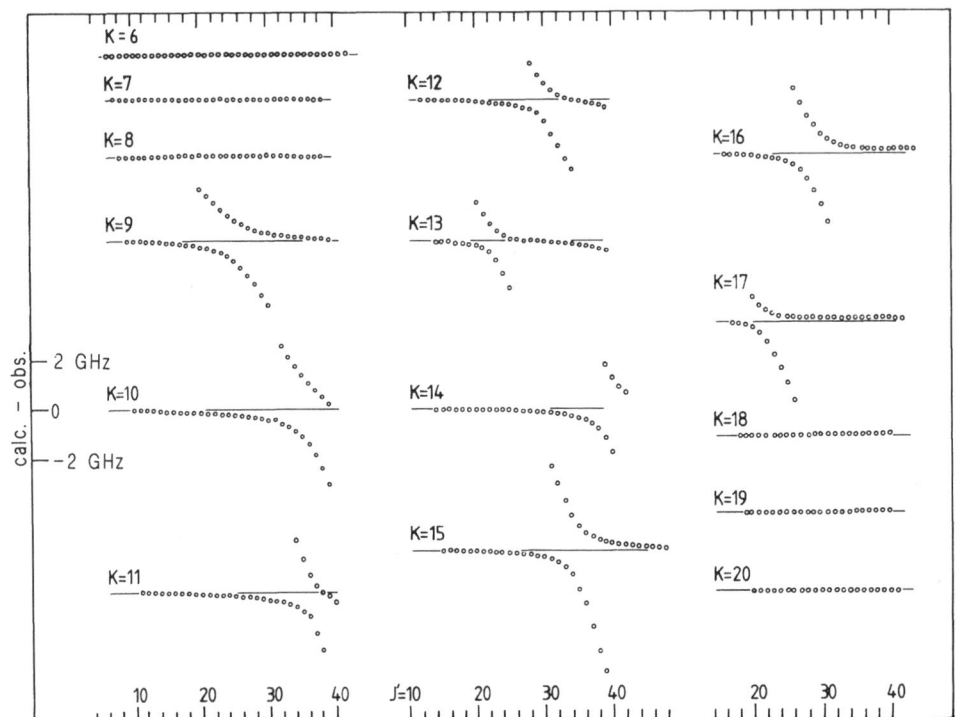

Figure 4. Residuals (calculated - observed) of the frequencies of rotational lines in the spectrum of the 14^1_0-band of C_6H_6 as a function of the final state quantum number J' for several values of K' /13,14/.

dark rovibronic states in the electronic S_1 state. As a result two eigenstates with mixed vibronic character result and both can be seen in the spectrum. These are the two lines observed in the experimental spectrum. From the positions of the two lines the coupling matrix element can be calculated for each pair /18/. The coupling shows a strong dependence on J and K. This leads to the conclusion that the observed coupling must be caused by perpendicular Coriolis coupling rather than by anharmonic or parallel Coriolis coupling /18/. This rotational dependence of the vibronic nature of the eigenstates can only be determined from rotationally resolved spectra.

2.4. Homogeneous line width measurements

The broadening of the transitions of large molecules due to fast nonradiative decay can normally only be inferred from the diffuseness of the vibronic bands /19/. Only for extremely strong broadening beyond the residual rotational contour in a supersonic jet and the favourable condition of widely spaced vibronic bands is a measurement of line broadening even for the GHz resolution of Doppler-limited spectroscopy possible /20/. However, such an experiment will not reveal the rotational dependence of the broadening. On the other hand, the extremely high resolution ($\Delta\nu$ below 10 MHz) of our cw set up allows the measurement of homogeneous line widths of single rovibronic transitions. This has been shown for the $14^1_0 12^2_0$-band of C_6H_6 at 3412 cm-1 vibrational excess energy /6/. In this band only the K = 0 lines are observed in our rotationally resolved fluorescence excitation spectra /21/, since all K ≠ 0 states decay very rapidly and no fluorescence is observed from these states. The remaining K = 0 lines are also broadened and the broadening is found to increase from 1.3 MHz for the J = 0 state to 46.1 MHz for the J = 14 state /6/. This strong rotational dependence of the homogeneous line width shows the importance of rotationally resolved observations for the determination of the decay behaviour of excited states of large molecules.

2.5. Emission spectra

The vibronic character of excited molecular states can be determined from SVL emission spectra. However, the analysis of line positions shows, that the vibronic character depends strongly on the rotational state of the molecule (see above). Therefore the SRVL emission spectrum is needed for the identification of the coupled background states. To obtain such an emission spectrum, a single rovibronic state has to be pumped in a Doppler-free experiment and the emission has to be spectrally resolved. This requires, that the dye laser frequency remains in resonance with the molecular transition frequency for a long time, typically to a precision of 1 MHz/h. In addition enough fluorescence intensity has to be produced to allow spectral resolution. These problems have recently been solved in our laboratory. First emission spectra of single rovibronic states will be presented in a forthcoming publication /22/.

3.   PULSED DOPPLER-FREE TWO-PHOTON SPECTROSCOPY

3.1.   The experimental set up

For the investigation of the decay of individual levels
pulsed excitation of the molecules has to be used. The
experimental set up is shown in Fig. 5 /7,13/. Extremely
narrow band pulsed laser light is produced by pulsed amplification of the cw light. With three stages of excimer
laser pumped amplifiers we can generate light pulses of
500 KW peak power and nearly Fourier transform limited
bandwidth. With an additional parasitic cavity around the
second amplifier the pulse length can be varied between
2.5 ns and 10 ns and the frequency width accordingly between
50 MHz and 180 MHz. After passing through the sample cell
the laser beam is reflected back into itself to allow the
Doppler-free absorption. The two beams are counterclockwise
circularly polarized to suppress the Doppler-broadened
background /23/. The resulting UV-fluorescence signal is
either integrated for the recording of spectra or its time
behaviour is recorded with a transient digitizer. It is
worth mentioning that in Doppler-free two-photon absorption
all molecules regardless of their velocity contribute to
the observed signal. Molecules are only excited to one
single level if the laser frequency is set to a resolved
rotational line in the spectrum regardless of the number
of lines within the Doppler-width. This allows the observation of the decay of an individual level /7,8/.

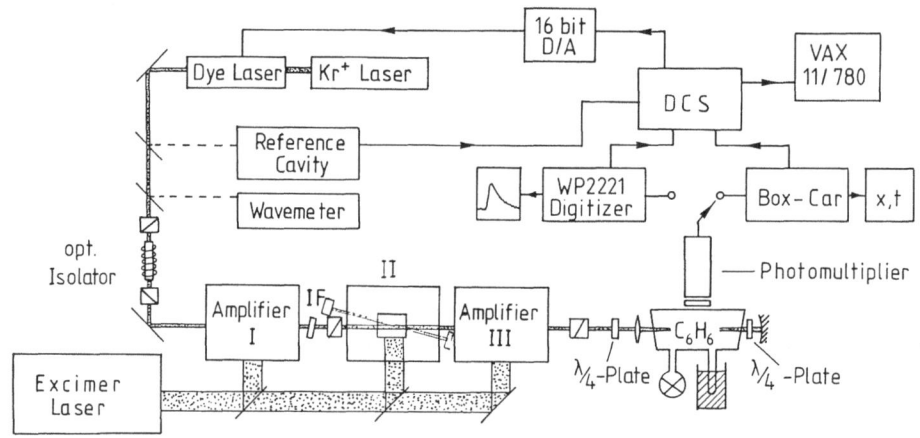

Figure 5.   Experimental set up for decay time measurements
of individual rotational levels of S_1 benzene. Fluorescence
decay is observed after pulsed Doppler-free two-photon excitation with the amplified light of the cw laser (taken
from ref. 7,13).

3.2. The resolution achieved with pulsed excitation

The resolution achieved with the pulsed set up is limited
by the length of the light pulse to about 100 MHz. This is
by one order of magnitude lower than the resolution achieved
with the cw set up. However, still most of the rotational
lines in the two-photon spectra of the Q-branch can be
resolved /23/, as can be seen from the upper part of fig.
6 /7/. The top spectrum shows part of the Doppler-free
spectrum of the 14^1_0-band of C_6H_6 as recorded with the cw
set up and the lower spectrum the same spectral range as
recorded with the pulsed set up. Clearly the resolution of
single rovibronic lines can be seen.

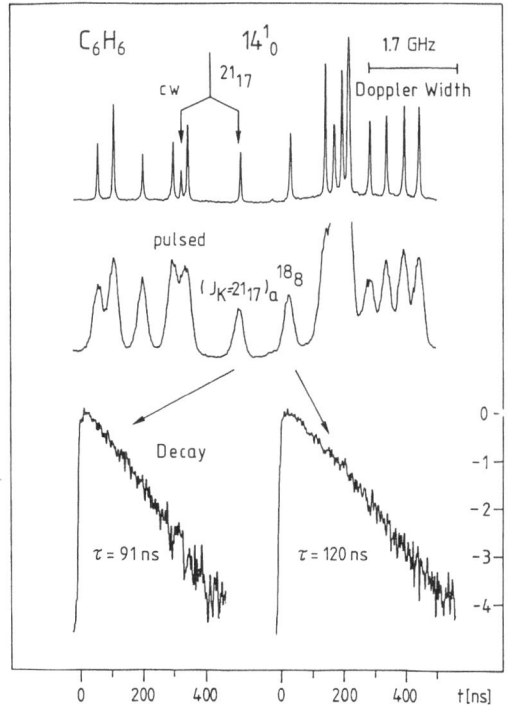

Figure 6. Part of the Q-branch of the 14^1_0-band of C_6H_6
under the high resolution of the cw set up (top) and the
lower resolution of the pulsed set up (middle trace). At
the bottom the fluorecence decay curves measured for the
two neighbouring lines $(J'_{K'}=21_{17})_a$ and $J'_{K'}=18_8$ are shown
on a half-logarithmic sacale. Both lines are well resolved,
even with the pulsed set up. $J'_{K'}=(21_{17})_a$ is one component
of the perturbed zero order state $J'_{K'}=21_{17}$, whereas
$J'_{K'}=18_8$ is an unperturbed state (taken from ref. 7).

3.3. Decay time measurements

The resolution of single rovibronic lines with the pulsed
set up allows the measurement of the decay behaviour of
single rovibronic states. These measurements have been
performed for a number of states in the 14^1 manifold /7/.
Two examples are shown in the lower part of fig. 6. All
decay curves measured were found single exponential. For
states found unperturbed in the spectral analysis a constant
decay time was found independent of the rotation of the
molecule. This is in agreement with theoretical predictions
for a inter-system-crossing (ISC) decay in the statistical
limit /24/.

The two decay curves shown in fig. 6, however, show a differing decay behaviour of the unperturbed $J'_{K'}=18_8$ and the
perturbed $(J'_{K'}=21_{17})_a$ state. The faster decay results from
the coupling of the light 14^1 rotational state to a dark
state, that decays much faster. Through the above mentioned
mixing of the states, the resulting eigenstate has a decay
time intermediate between the dark and the light state.
Since the spectral analysis allows us to determine the
degree of mixing of the states, we can determine the decay
time of the unperturbed dark state from the measured decay
time of the eigenstate. It is found to be much shorter than
the decay time of the 14^1 state. In this way does the measurement of the decay of perturbed states allow the determination of the decay behaviour of states normally not
observable (i.e. dark states).

4.   SUMMARY AND CONCLUSION

Sub-Doppler techniques can be used successfully for the
complete rotational resolution of the electronic spectrum
of large molecules. In this article this has been demonstrated for the example of Doppler-free two-photon spectroscopy
of benzene. The rotational resolution is important for the
investigation of dynamic processes, as rotation-vibration
coupling (seen as perturbations in the spectrum) influences
the decay behaviour of large molecules significantly. The
extremely high resolution of the cw set up allows the measurement of the homogeneous line width of single rovibronic
lines. Doppler-free excitation with extremely narrow band
pulsed laser light permits the measurement of the decay
time of single rovibronic states.

The increased selectivity in sub-Doppler spectroscopy will
lead to a more detailed understanding of the dynamical
processes in large molecules. In view of the presented
examples of strong rotational effects on the decay behaviour

of benzene, one should be very careful in the interpretation of experimental results with limited spectral resolution. The experimental set up needed for sub-Doppler investigations is nontrivial, but the existing problems can be overcome by careful design of the experiment. The wealth of additional information available from experiments with completely defined molecular states will certainly justify the invested effort.

5. ACKNOWLEDGEMENT

The author wishes to thank Prof. Neusser for careful reading of the manuscript and helpful comments. Financial support by the Deutsche Forschungsgemeinschaft is gratefully acknowledged.

REFERENCES

1.   For a review see, W. Demtröder, "Laser Spectroscopy", Springer, Berlin, 1981

2.   E. Riedle, H. J. Neusser, and E. W. Schlag, J. Chem. Phys. **75**, 4231 (1981)

3.   B. J. van der Meer, H. T. Jonkman, J. Kommandeur, W. L. Meerts, and W. A. Majewski, Chem. Phys. Letters **92**, 565 (1982)

4.   K. H. Fung and D. A. Ramsay, J. Phys. Chem. **88**, 395 (1984)

5.   A. Kiermeier, K. Dietrich, E. Riedle, and H. J. Neusser, J. Chem. Phys. **85**, 6983 (1986)

6.   E. Riedle and H. J. Neusser, J. Chem. Phys. **80**, 4686 (1984)

7.   U. Schubert, E. Riedle, and H. J. Neusser, J. Chem. Phys. **84**, 5326 (1986)

8.   U. Schubert, E. Riedle, H. J. Neusser, and E. W. Schlag, J. Chem. Phys. **84**, 6182 (1986)

9.   L. S. Vasilenko, V. P. Chebotayev, and A. V. Shishaev, JETP Letters **12**, 113 (1970)

10.   S. A. Lee, J. Helmcke, and J. L. Hall, in "Laser Spectroscopy IV", Springer Series in Optical Sciences, edited by H. Walther and K. W. Rothe (Springer, Berlin, 1979), Vol. 21, p. 130

11.   F. Biraben and L. Julien, Opt. Commun. **53**, 319 (1985)

12.   E. Riedle, H. Stepp, and H. J. Neusser, in "Laser Spectroscopy VI", Springer Series in Optical Sciences, edited by H. P. Weber and W. Lüthy (Springer, Berlin, 1983), Vol. 40, p. 144

13.   E. Riedle and H. J. Neusser, in "Stochasticity and Intramolecular Redistribution of Energy", NATO ASI Series, Series C, Vol. 200, edited by R. Lefebvre and S. Mukamel (D. Reidel, Dordrecht, 1987), pp. 203

14.   H. J. Neusser and E. Riedle, Comments At. Mol. Phys. **19**, 331 (1987)

15.   H. Sieber, Diplomarbeit, TU München, 1986

16.   A. E. Bruno, E. Riedle, and H. J. Neusser, Chem. Phys. Letters **126**, 558 (1986)

17.   E. Riedle, H. Stepp, and H. J. Neusser, Chem. Phys. Letters **110**, 452 (1984)

18.   E. Riedle and H. J. Neusser, in preparation

19.   J. H. Callomon, J. E. Parkin, and R. Lopez-Delgado, Chem. Pys. Letters **13**, 125 (1972)

20.   K. Aron, C. Otis, R. E. Demaray, and P. Johnson, J. Chem. Phys. **73**, 4167 (1980)

21.   E. Riedle, H. J. Neusser, and E. W. Schlag, J. Phys. Chem. **86**, 4847 (1982)

22.   F. Giesemann, U. Schubert, E. Riedle, and H. J. Neusser, to be published

23.   E. Riedle, R. Moder, and H. J. Neusser, Opt. Commun. **43**, 388 (1982)

24.   F. A. Novak and S. A. Rice, J. Chem. Phys. **73**, 858 (1980); W. E. Henke, H. L. Selzle, T. R. Hays, E. W. Schlag, and S. H. Lin, J. Chem. Phys. **76**, 1335 (1982)

INDEX OF SUBJECT